D1065329

Mobile Satellite Communication Networks

Mobile Satellite Communication Networks

Ray E. Sheriff and Y. Fun Hu
Both of
University of Bradford, UK

JOHN WILEY & SONS, LTD

e-mail (for orders and customer service enquiries): cs-books@wiley.co.uk
Visit our Home Page on http://www.wiley.co.uk or http://www.wiley.com

Other Wiley Editorial Offices

John Wiley & Sons, Inc., 605 Third Avenue,
New York, NY 10158-0012, USA

WILEY-VCH Verlag GmbH
Pappelallee 3, D-69469 Weinheim, Germany

John Wiley & Sons Australia Ltd, 33 Park Road, Milton,
Queensland 4064, Australia

John Wiley & Sons (Canada) Ltd, 22 Worcester Road
Rexdale, Ontario, M9W 1L1, Canada

John Wiley & Sons (Asia) Pte Ltd, 2 Clementi Loop #02-01,
Jin Xing Distripark, Singapore 129809

British Library Cataloguing in Publication Data

A catalogue record for this book is available from the British Library

ISBN 0471 72047 X

Typeset in Times by Deerpark Publishing Services Ltd, Shannon, Ireland.
Printed and bound in Great Britain by T. J. International Ltd, Padstow, Cornwall.

This book is printed on acid-free paper responsibly manufactured from sustainable forestry, in which at least two trees are planted for each one used for paper production.

Contents

Preface

The last decade proved to be hugely successful for the mobile communications industry, characterised by continued and rapid growth in demand, spurred on by new technological advances and innovative marketing techniques. Of course, when we refer to mobile communications, we tend to implicitly refer to cellular systems, such as GSM. The plight of the mobile-satellite industry over the last decade, although eventful, has, at times, been more akin to an out of control roller coaster ride. From an innovative start, the industry is now in the process of re-assessing, if not re-inventing itself. While niche satellite markets have continued to grow steadily over the last 10 years, the significant market penetration derived from personal mobile services via satellite that was anticipated at the start of the decade has failed to materialise. With this in mind, it may seem like a strange time to bring out a book on mobile-satellite communication networks.

Certainly, if we were to have produced a book merely describing the technology behind mobile-satellite networks, then in many respects, we would have failed to address the real issues that affect mobile-satellite systems, namely the influence of terrestrial mobile communications and an assessment of the market, which ultimately decides whether a system is viable or not. With this in mind, we have put together a book which aims to highlight these key issues, while at the same time covering the fundamentals of the subject. We believe this combination provides a unique approach to the subject that is relevant to those involved in the mobile-satellite industry.

With this approach in mind, the first chapter reflects on the status of the mobile communications industry. It is now accepted that mobile-satellite communications will largely play complementary roles to their terrestrial counterparts. Consequently, it is important to have an understanding of the different types of mobile communications systems that are presently available. This chapter covers the development of cellular and cordless communications from their initial beginnings to the present day.

This is followed in Chapter 2 by a review of the current state of the mobile-satellite industry. Usually, when we discuss mobile communications, invariably we will associate this with voice communications. However, satellites are also used for other innovative services such as store-and-forward data relaying and vehicular fleet management. This chapter considers all aspects of mobile-satellite communications from the established geostationary systems to the latest "little" and "big" low Earth orbit systems.

The introduction of non-geostationary satellite systems often requires the design of complex multi-satellite constellations. Chapter 3 presents the theory behind the design of such networks. This is followed in Chapter 4 by a discussion of the properties of the communication channel, from both a mobile and fixed perspective. The mobile channel in particular offers a hostile environment in which to design a reliable communications link. This chapter reviews the present status of understanding with regard to the characterisation of the channel, indicating the methods of prediction.

The characteristics of the radio interface are considered in Chapter 5. Here, the transmission chain is analysed presenting the link budget method of analysis. The chapter includes a description of applicable modulation and coding techniques and multiple access schemes.

In Chapter 6, the network procedures associated with a mobile-satellite network are presented. In order to facilitate a smooth integration with terrestrial mobile networks, it is important that as many of the procedures between the two systems are as similar as possible. This chapter focuses on two key areas of mobility management, namely location management and handover management, as well as resource management techniques. In Chapter 7, the requirements for integration with fixed and mobile networks are presented, highlighting the requirements for integration with the GSM network.

Chapter 8 presents an analysis of the market potential for mobile-satellite communications. The methodology for deriving the market is presented, followed by a series of market predictions. Finally, in Chapter 9, we attempt to predict how the mobile-satellite market will develop over the coming decade.

Certainly, it has been an interesting time to produce a book of this type. The next 10 years promise to be as innovative, if not more so, than the last, with the introduction of mobile multimedia services and a greater influence of the Internet on mobile service evolution. It will be interesting to see how the mobile-satellite communications industry adapts to meet these new markets over the next few years.

Acknowledgement

The texts extracted from the ITU material have been reproduced with the prior authorisation of the union as copyright holder.

The sole responsibility for selecting extracts for reproduction lies with the beneficiaries of this authorisation alone and can in no way be attributed to the ITU.

The complete volumes of the ITU material, from which the texts reproduced are extracted, can be obtained from:

International Telecommunication Union
Sales and Marketing Service
Place des Nations – CH-1211 Geneva 20 (Switzerland)
Telephone: +41-22-730-6141 (English)/+41-22-730-6142 (French)/
+41-22-730-6143 (Spanish)
Telex: 421 000 uit ch/Fax: +41-22-730-5194
X.400 : S = sales; P = itu; A = 400net; C = ch
E-mail: sales@itu.int/http://www.itu.int/publications

Acknowledgments

Figures

Tables

1

Mobile Communication System Evolution

1.1 Historical Perspective

The mobile phone has proved to be one of the most outstanding technological and commercial successes of the last decade. Since its introduction in the 1980s, the phone's place in the market place has rapidly progressed from a minority, specialised item to virtually an essential commodity for both business and leisure use. Over the last two decades, advances in mobile technology, combined with the significant reduction in operating costs and the development of new applications and services, have ensured a buoyant market. By mid-2000, there were over 220 million mobile subscribers in Europe and over 580 million mobile subscribers world-wide. In the UK, every other person owns a mobile phone; while in Finland the number of mobile phones per capita now exceeds that of households with fixed phone lines.

As with most technological innovations, the mobile phone's marketability is not based on overnight success but rather a systematic, evolutionary development involving multi-national co-operation at both technical and political levels. In fact, the concept of a mobile phone is not new. As early as 1947, the cellular concept was discussed within Bell Laboratories [YOU-79]. However, it was not until the 1970s that technology had developed sufficiently to allow the commercial implementation of such a system to be investigated.

The evolution of mobile communications can be categorised into generations of development. Presently, we are on the verge of the third-generation (3G) of mobile systems. Broadly speaking, first-generation (1G) systems are those that paved the way and are generally categorised as being national networks that are based on analogue technology. Such networks were introduced into service in the 1980s. These networks were designed to provide voice communications to the mobile user.

Second-generation (2G) systems are categorised by digital technology. They are supported by international roaming agreements, allowing the possibility to operate a mobile phone across national boundaries. With the introduction of 2G systems, in addition to digital voice telephony, a new range of low data rate digital services became available, including mobile fax, voice mail and short message service (SMS) [PEE-00]. Also at this stage in the evolution, new types of systems began to emerge which catered for particular market needs; not only cellular mobile, but also cordless, public mobile radio, satellite and wireless-local area network (W-LAN) solutions. 2G systems are synonymous with the globalisation of mobile systems, and in this respect the importance of standardisation is clear. For example, GSM, which was standardised in Europe by the European Telecommunications Standards Institute (ETSI), is now recognised as a global standard, with its adoption in most countries of the world. The final evolutionary phase of 2G networks, in recognition of the importance of the Internet and as a stepping stone towards the introduction of 3G technology, introduced packet-oriented services, providing the first opportunity to introduce mobile-multimedia services.

Within the next few years, it is expected that mobile users will wish to access broadband multimedia services, such as those provided by fixed networks. This demand for broader bandwidth services is driven by the need to provide services and applications comparable with those presently available to personal computers (PCs). The phenomenal growth in the Internet, with over 500 million users predicted by 2005, perfectly illustrates the need for access to broadband services and applications. These types of services are beyond the capability of present 2G systems, which offer voice and low data rate services. The convergence of mobile and Internet protocol (IP) based technologies is now the major driving force behind the development of 3G systems. The 3G mobile communications systems will be capable of delivering services and applications at data rates of up to and beyond 2 Mbit/s.

The standardisation of 3G systems comes under the overall responsibility of the International Telecommunication Union (ITU). Globally, this will be known as international mobile telecommunications 2000 (IMT-2000) and will consist of a family systems providing cellular, cordless, W-LAN and satellite services. In Europe, the 3G system will be known as the Universal Mobile Telecommunications System (UMTS). Although voice is still likely to be the dominant application in the first few years of 3G networks, there will also be the possibility to operate mobile-multimedia applications, such as video-telephony, file transfer protocol (ftp) file access, Web browsing and so on. As 3G technology evolves, new broader bandwidth applications will enter the market to such an extent that the transmission of data will provide the greatest volume of traffic.

Research is now addressing the requirements of fourth-generation (4G) mobile networks. Mobile data rates beyond 2 Mbit/s, and possibly up to 155 Mbit/s in some environments, will further extend the services and applications that could be delivered. Improvements in quality of service (QoS), bandwidth efficiency and the move to an all IP-based, packet-oriented environment can be envisaged, based on the emerging standards of Mobile IP, under development by the Internet Engineering Task Force (IETF) [PER-98, SOL-98]. 4G mobile networks are likely to be introduced sometime after 2005, possibly as late as 2010.

Although this book is primarily focused on mobile-satellite networks, initially in order to appreciate the context in which satellite technologies have developed, and the likely applications for such technologies, it is important to have an understanding of where we are at present in terms of mobile technology. This chapter aims to provide a flavour of the underlying technological developments that have driven the mobile communication industry to the brink of the establishment of the mobile information society.

1.2 Cellular Systems

1.2.1 Basic Concepts

Cellular networks operate by dividing the service coverage area into zones or cells, each of which has its own set of resources or channels, which can be accessed by users of the network.

Usually cellular coverage is represented by a hexagonal cell structure to demonstrate the concept, however, in practice the shape of cells is determined by the local topography. Sophisticated planning tools are used extensively by terrestrial cellular operators to assist with the planning of their cellular networks.

The shape and boundary of a cell is determined by its base station (BS), which provides the radio coverage. A BS communicates with mobile users through signalling and traffic channels (TCH). Signals transmitted in the direction from the BS to the mobile are termed the *forward link* or *downlink*, and conversely, the *reverse link* or *uplink* is in the direction of mobile to BS. Signalling channels are used to perform administrative and management functions such as setting up a call, while TCHs are used to convey the information content of a call. The allocation of channels to a cell is therefore divided between the TCHs, which form the majority, and signalling channels. These are allocated for both forward and reverse directions.

In order to increase the capacity of a network, there are three possibilities, either:

1. a greater number of channels are made available;
2. more spectrally efficient modulation and multiple access techniques are employed; or
3. the same channels are re-used, separated by a distance which would not cause an unacceptable level of co-channel interference.

Cellular networks, which are limited in terms of available bandwidth, operate using the principal of frequency re-use. This implies that the same pool of frequencies is re-used in cells that are sufficiently separated so as not to cause harmful co-channel interference. For a hexagonal cell structure, it is possible to cluster cells so that no two adjacent cells are using the same frequency. This is only achievable for certain cell-cluster sizes, which can be determined from the relationship

$$N = i^2 + ij + j^2 \tag{1.1}$$

where $i, j = 0, 1, 2, 3$, etc.

A seven-cell frequency re-use pattern is shown in Figure 1.1. The total bandwidth available to the network is divided between cells in a cluster, which can then be used to determine the number of calls that can be supported in each cell. By reducing the number of cells per cluster, the system capacity can be increased, since more channels can be available per cell. However, a reduction in the cluster size will also result in a reduction in the frequency re-use distance, hence the system may become more prone to co-channel interference.

The frequency re-use distance can be determined for a given cell cluster size from the equation

$$\frac{D}{R} = \sqrt{3N} \tag{1.2}$$

where D is the mean re-use distance, R is the cell radius and N is the cluster size.

In a terrestrial mobile radio environment, the strength of the received carrier power, at a distance R from the transmitter is related by the following expression:

$$C \propto \frac{1}{R^{\gamma}} \tag{1.3}$$

where γ is a constant related to the terrain environment, usually assumed to be equal to 4.

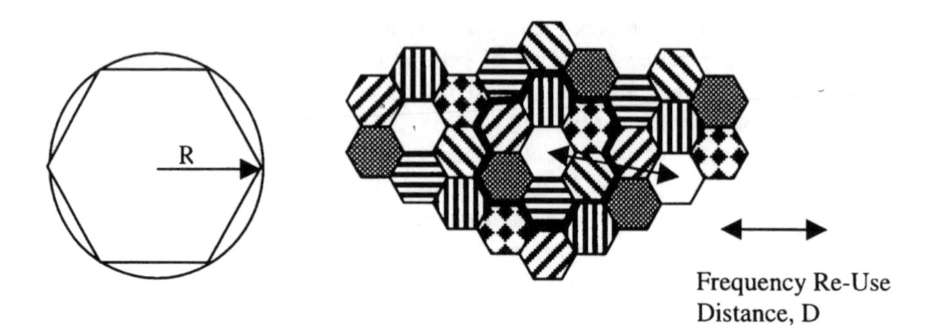

Figure 1.1 Seven-cell frequency re-use pattern.

For a seven-cell re-use configuration, the ratio of the carrier-to-interference experienced by a mobile from the six cells located at a minimum re-use distance of D from the mobile, that is on the first tier of the cell cluster re-use pattern, is given by

$$\frac{C}{I} = \frac{\sum D^{\gamma}}{R^{\gamma}} = \frac{D^{\gamma}}{6R^{\gamma}} = \frac{q^{\gamma}}{6} \tag{1.4}$$

From the above, q, termed the *co-channel interference reduction factor*, is given by [LEE-89]

$$q = \frac{D}{R} \tag{1.5}$$

Note: the above assumes that equal power is radiated by all cells and that the interference received from cells operating using the same frequency in the second tier of the cell cluster, can be neglected. Thus, for $\gamma = 4$, a seven-cell cluster pattern can provide a C/I ratio of 18 dB. In order to minimise the effect of co-channel interference, power control techniques are employed at the mobile terminal and the BS to ensure that power levels are maintained at the minimum level needed to maintain the target QoS.

How the mobile user gains access to the available channels within a cell is governed by the multiple access technique used by the network. Analogue cellular networks employ frequency division multiple access (FDMA), whereas digital networks employ either time division multiple access (TDMA) or code division multiple access (CDMA). For FDMA, a seven cell re-use pattern is generally employed, whereas for CDMA a single-cell frequency re-use pattern is achievable. Further discussions on the advantages and drawbacks of each technique, in the context of satellite communications, can be found in Chapter 5.

In a terrestrial mobile environment, reception cannot rely on line-of-sight communications and is largely dependent upon the reception of signal reflections from the surrounding environment. (Note: This is the opposite of the mobile-satellite case, which is reliant on line-of-sight operation, and is discussed in detail in Chapter 4.) The resultant scattering and multipath components arrive at the receiver with random phase. The propagation channel can be characterised by a combination of a slow-fading, long-term component and a fast-fading, short-term component. As a consequence of the local terrain, the change in a mobile's position relative to that of a transmitting BS will result in periodic nulls in the received signal strength. This is due to the fact that the vector summation of the multipath and scattering components at the receiver results in a signal envelope of the form of a standing wave pattern, which has signal nulls at half-wave intervals. For a signal transmitting at 900 MHz, which is typical for cellular applications, a half-wavelength distance corresponds to approximately 17 cm. This phenomenon is known as slow-fading and is characterised by a log-normal probability density function.

As the mobile's velocity, v, increases, the variation in the received signal envelope becomes much more pronounced and the effect of the Döppler shift on the received multipath signal components also has an influence on the received signal, where Döppler shift, f_d, is given by

$$f_d = \frac{v}{\lambda}\cos(\alpha) \qquad \text{Hz} \tag{1.6}$$

where α is the angle of arrival of the incident wave.

This phenomenon is termed fast-fading and is characterised by a Rayleigh probability density function. Such variations in received signal strength can be as much as 30 dB below or 10 dB above the root mean square signal level, although such extremes occur infrequently.

In rural areas, where the density of users is relatively low, large cells of about 25 km radius can be employed to provide service coverage. This was indeed the scenario when mobile communications were first introduced into service. In order to sustain the mobile to BS link over such a distance requires the use of a vehicular-type mobile terminal, where available transmit power is not so constrained in comparison with hand-held devices. With an increase in user-density, the cell size needs to reduce in order to enable a greater frequency re-use and hence to increase the capacity of the network. Urban cells are typically of 1 km radius. This reduction in cell size will also correspond to a reduction in BS and mobile terminal transmit power requirements. This is particularly important in the latter case, since it paves the way for the introduction of hand-held terminals.

When a mobile moves from one cell to another during the course of an on-going call, a handover (also termed handoff) of the call between BSs must be performed in order to ensure that the call continues without interruption. Otherwise the call will be dropped and the mobile user would need to re-initiate the call set-up sequence. Handover between BSs involves monitoring of the signal strength between the mobile to BS link. Once the signal strength reduces below a given threshold, the network initiates a procedure to reserve a channel through another BS, which can provide a channel of sufficient signal strength (Figure 1.2).

A number of BSs are clustered together via a fixed-network connection to a mobile switching centre (MSC), which provides the switching functionality between BSs during handover and can also provide connection to the fixed or core network (CN) to allow the routing of calls. The clustering of BSs around a MSC is used to define a Location Area, which can be used to determine the latest known location of a mobile user. This is achieved by associating Home and Visitor Location Areas to a mobile. Each mobile is registered with a single home location register (HLR) upon joining the network. Once a mobile roams outside of its Home Location Area into a new designated Location Area, it temporarily registers with the network as a visitor, where its details are stored in a visitor location register (VLR) associated with the MSC. Each MSC in the network has an associated VLR and HLR. The mobile's location is relayed back to its HLR, a database containing various information on the mobile terminal, some of which is then forwarded to the VLR. The network also comprises of other databases

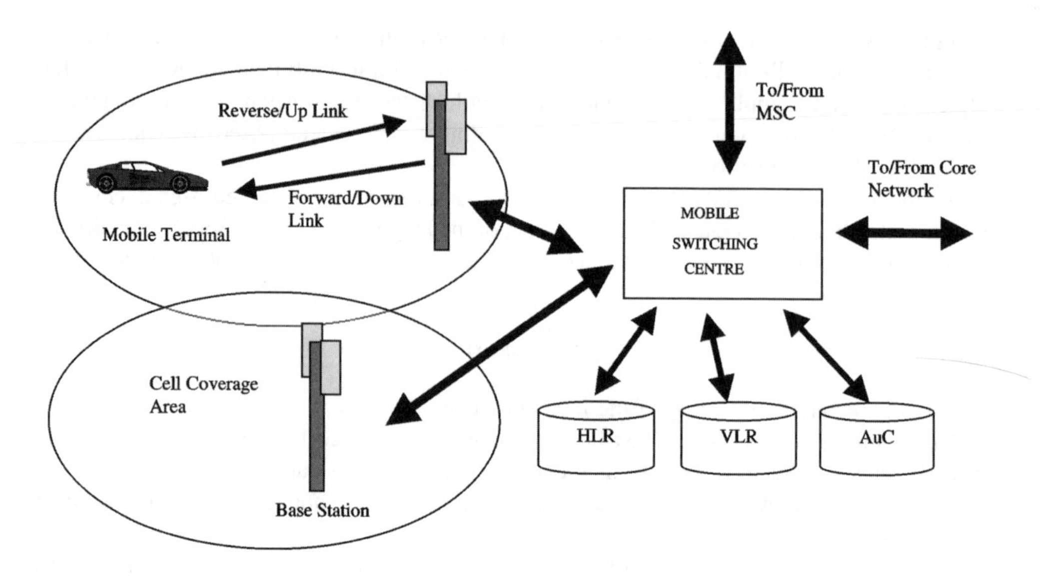

Figure 1.2 Basic cellular network architecture.

that can be used to verify that the mobile has access to the network, such as the Authentication Centre (AuC), for example. These procedures are described later in the chapter for the GSM system.

1.2.2 First-Generation (1G) Systems

1.2.2.1 Introduction

In the future mobile information society, where mobile-multimedia delivery will be the major technological driving force, analogue cellular technology has little, if any significance. Indeed, in many countries across Europe, mobile operators are now switching off their analogue services in favour of digital technology. However, analogue technologies still play an important role in many countries around the world, by being able to provide established and reliable mobile voice telephony at a competitive price. This section considers three of the major analogue systems that can still be found with significant customer databases throughout the world.

1.2.2.2 Nordic Mobile Telephone (NMT) System

On 1 October, 1981, the Nordic NMT450 became the first European cellular mobile communication system to be introduced into service [MAC-93]. This system was initially developed to provide mobile communication facilities to the rural and less-populated regions of the Scandinavian countries Denmark, Norway, Finland and Sweden. NMT450 was essentially developed for in-car and portable telephones. By adopting common standards and operating frequencies, roaming between Scandinavian countries was possible. Importantly, the introduction of this new technology provided network operators and suppliers with an early market lead, one that has been sustained right up to the present day.

As is synonymous of 1G systems, NMT450 is an analogue system. It operates in the 450 MHz band, specifically 453–457.5 MHz (mobile to BS) and 463–467.5 MHz (BS to mobile). FDMA/FM is employed as the multiple access scheme/modulation method for audio signals, with a maximum frequency deviation of ±5 kHz. Frequency shift keying (FSK) is used to modulate control signals with a frequency deviation of ±3.5 kHz. NMT450 operates using a channel spacing of 25 kHz, enabling the support of 180 channels. Since its introduction, the NMT450 system has continued to evolve with the development of the NMT450i (where i stands for improvement) and NMT900 systems.

NMT900 was introduced into service in 1986, around about the same time as other Western European countries were starting to introduce their own city based mobile cellular-based solutions. NMT900 is designed for city use, catering for hand-held and portable terminals. It operates in the 900 MHz band with the ability to accommodate higher data rates and more channels.

The NMT system continues to hold a significant market share throughout the world and, significantly, the system continues to evolve, through a series of planned upgrades. In Europe, the NMT family has a particularly large market share in Eastern European countries, where mobile telephony is only now starting to become prevalent.

The next phase in the evolution of the NMT450 network, as initiated by the NMT MoU, is the digitisation of the standard. This is considered an important and necessary evolutionary phase, in light of competition from existing 2G and future generation mobile networks. This will be achieved through the down banding of the GSM network, and will be known as GSM400. The possibility to provide dual-band GSM phones in order to support global roaming is considered particularly attractive. Towards the end of 1999, Nokia and Ericsson combined to demonstrate the first call made on a dual-mode GSM400/1800 prototype mobile phone.

Since 1981, Nordic countries have continued to lead the way with now over 60% of the population in Finland and Norway having a mobile phone. The Scandinavian-based companies Nokia and Ericsson are world leaders in mobile phone technology and both are driving the phone's evolution forward.

1.2.2.3 Advanced Mobile Phone Service (AMPS)

Bell Labs in the US developed the AMPS communications system in the late 1970s [BEL-79]. The AMPS system was introduced into commercial service in 1983 by AT&T with a 3-month trial in Chicago. The system operates in the US in the 800 MHz band, specifically 824–849 MHz (mobile to BS) and 869–894 MHz (BS to mobile). These bands offer 832 channels, which are divided equally between two operators in each geographical area. Of these 832 channels, 42 channels carry only system information. The AMPS system provides a channel spacing of 30 kHz using FM modulation with a 12 kHz peak frequency deviation for voice signals.

Signalling between mobile and BS is at 10 kbit/s employing Manchester coding. The signals are modulated using FSK, with a frequency deviation of ±8 kHz. The AMPS system specifies six one-way logical channels for transmission of user and signalling information. The Reverse TCH and Forward TCH are dedicated to the transmission of user data on a one-to-one basis. Signalling information is carried to the BS on the channels reverse control channel (RECC) and reverse voice channel (RVC); and to the mobile using the channels forward control channel (FOCC) and forward voice channel (FVC).

The forward and reverse control channels are used exclusively for network control information and can be referred to as Common Control Channels. To safeguard control channels from the effect of the mobile channel, information is protected using concatenated pairs of block codes. To further protect information, an inner code employs multiple repetition of each BCH (Bose–Chadhuri–Hocquenghem) code word at least five times, and 11 times for the FVC.

In order to identify the BS assigned to a call, AMPS employs a supervisory audio tone (SAT), which can be one of three frequencies (5970, 6000 and 6030 Hz). At call set-up, a mobile terminal is informed of the SAT at the BS to which it communicates. During a call, the mobile terminal continuously monitors the SAT injected by the BS. The BS also monitors the same SAT injected by the mobile terminal. Should the received SAT be incorrect at either the mobile terminal or the BS, the signal is muted, since this would imply reception of a source of interference.

Like NMT450, the AMPS standard has continued to evolve and remains one of the most widely used systems in the world. Although market penetration did not reach Europe, at least in its unmodified form, it remains a dominant standard in the Americas and Asia.

Narrowband-AMPS Motorola developed the narrowband-AMPS (N-AMPS) system in order to increase the available capacity offered by the network. This was achieved by dividing the available 30 kHz AMPS channel into three. N-AMPS employs frequency modulation with a maximum deviation of 5 kHz from the carrier. From the outset, mobile phones were developed for dual-mode operation allowing operation with the AMPS 30 kHz channel.

Due to the narrower bandwidth, there is a slight degradation in speech quality when compared to AMPS. In order to optimise reception, N-AMPS employs a radio resource management technique called *Mobile Reported Interference*. This procedure involves the mobile terminal monitoring the received signal strength of a forward narrow TCH and the BER on the control signals of the associated control channel. A BS sends the mobile a decision threshold on the reserve associated control channel, below which handover can be initiated.

Signalling control channels are transmitted using a continuous 100 bit/s Manchester coded in-band sub-audible signal. In addition to signalling messages, alphanumeric messages can also be transmitted to the mobile.

N-AMPS was standardised in 1992 under IS-88, IS-89 and IS-90. In 1993, IS-88 was combined with the AMPS standard IS-553 to form a single common analogue standard.

1.2.2.4 Total Access Communications System (TACS)

By the mid-1980s, most of Western Europe had mobile cellular capability, although each country tended to adopt its own system. For example, the C-NETZ system was introduced in Germany and Austria, and RADIOCOM 2000 and NMT-F, the French version of NMT900 could be found in France. This variety of technology made it impossible for international commuters to use their phones on international networks, since every national operator had its own standard. In the UK, Racal Vodafone and Cellnet, competing operators providing technically compatible systems, introduced the TACS into service in January 1985. TACS was based on the American AMPS standard with modifications to the operating frequencies and channel spacing. TACS offers a capacity of 600 channels in the bands 890–905 MHz (mobile to BS) and 935–950 MHz (BS to mobile), the available bandwidth being divided equally between the two operators. Twenty-one of these channels are dedicated for control channels per operator. The system was developed with the aim of serving highly populated urban areas as well as rural areas. This necessitated the use of a small cell size in urban areas of 1 km. In TACS, the cell size ranges from 1 to 10 km. TACS provides a channel spacing of 25 kHz using FM modulation with a 9.5 kHz peak deviation for voice signals. In highly densely populated regions, the number of available channels is increased to up to 640 (320 channels per operator) by extending the available spectrum to below the conference of European Posts and Telegraphs (CEPT) cellular band. This is known as extended TACS (ETACS). Here, the operating frequency bands are 917–933 MHz in the mobile to BS direction and 872–888 MHz in the BS to mobile.

Fifteen years after TACS was first introduced into the UK, the combined Vodafone and Cellnet customer base amounted to just under half a million subscribers out of a total of 31 million. The future of analogue technology in developed markets is clearly limited, particularly with the re-farming of the spectrum for the 3G services. Nevertheless, analogue systems such as TACS have been responsible for developing the mobile culture and in this respect, their contribution to the evolution of the mobile society remains significant.

Within Europe, TACS networks can also be found in Austria, Azerbaijan, Ireland, Italy, Malta and Spain. A variant of TACS, known as J-TACS, operates in Japan.

1.2.3 Second-Generation (2G) Systems

1.2.3.1 Global System for Mobile Communications (GSM)

Development Following a proposal by Nordic Telecom and Netherlands PTT, the Group Spécial Mobil (GSM) study group was formed in 1982 by the CEPT. The aim of this study group was to define a pan-European public land mobile system.

By the middle of the 1980s, the mobile industry's attention had focused on the need to implement more spectrally efficient 2G digital type services, offering a number of significant advantages including greater immunity to interference, increased security and the possibility of providing a wider range of services. Unlike the evolution of the North American AMPS, which will be discussed shortly, the implementation of GSM took a more revolutionary approach to its design and implementation.

In 1987, 13 operators and administrators signed the GSM memorandum of understanding (MoU) agreement and the original French name was changed to the more descriptive Global System for Mobile communications (GSM), although the acronym remained the same. By 1999, 296 operators and administrators from 110 countries had signed the GSM MoU. Significantly, in 1987, following the evaluation of several candidate technologies through laboratory and field trial experiments, agreement was reached on the use of a regular pulse excitation-linear predictive coder (RPE-LPC) for speech coding and TDMA was selected as the multiple access method.

In 1989 responsibility for the GSM specification was transferred to the ETSI and a year later Phase 1 GSM specifications were published. Commercial GSM services began in Europe two years later in mid-1991. In addition to voice services, the SMS was created as part of the GSM Phase 1 standard. This provides the facility to send and receive text messages from mobile phones. Messages can be up to 160 characters in length and can be used to alert the user of an incoming e-mail message, for example. It is a store-and-forward service, with all messages passing through an SMS centre. The SMS has proved to be hugely popular in Europe, with the transmission of in excess of 1 billion messages per month as of April 1999.

In 1997, Phase 2 specifications came on-line, allowing the transmission of fax and data services.

At the end of 1998, ETSI completed its standardisation of GSM Phase 2+ services high speed circuit switched data (HSCSD) and general packet radio service (GPRS). These two new services are aimed very much at exploiting the potential markets in the mobile data sector, recognising the influence of the Internet on mobile technologies. HSCSD and GPRS will be discussed shortly.

Responsibility for the maintenance and future development of the GSM standards is now under the control of the 3G partnership project (3GPP).

Radio Interface The ITU allocated the bands 890–915 MHz for the uplink (mobile to BS) and 935–960 MHz for the downlink (BS to mobile) for mobile networks. As has already been seen, analogue mobile services were already using most of the available spectrum, however, the upper 10 MHz in each band was initially reserved for the introduction of GSM operation, with coexistence in the UK with TACS in the 935–950 and 890–905 bands.

The modulation method adopted by GSM is Gaussian-filtered minimum shift keying (GMSK) with a BT (3 dB bandwidth × bit period) value of 0.3 at a gross data rate of 270 kbit/s. This enables a compromise between complexity of the transmitter (which is important when trying to maintain a low-cost terminal), increased spectral efficiency and limited spurious emissions (which is necessary to limit adjacent channel interference).

GSM specifies five categories of terminal class, as shown in Table 1.1. The power level can be adjusted up or down in steps of 2 dB to a minimum of 13 dBm. The power control is achieved by the mobile station (MS) measuring the signal strength or quality of the mobile link, which is then passed to the base transceiver station (BTS). The BTS, in turn determines if and when the power level should be adjusted. BTSs are categorised, in a similar manner, into eight classes ranging from 2.5 to 320 W in 3-dB steps. In order to limit co-channel interference, both the mobile and the BTS operate at the minimum power level required to maintain signal quality.

Table 1.1 GSM terminal classes

Class	Peak transmit power (W)	Peak transmit power (dBm)
1	20	43
2	8	39
3	5	37
4	2	33
5	0.8	29

GSM's multiple access scheme is based on a TDMA/FDMA approach, combined with optional slow frequency hopping, which can be used to counteract multipath fading and co-channel interference [HOD-90]. Each band is divided into 124 carrier frequencies using FDMA, and separated by 200 kHz. Each carrier frequency is divided in time, using a TDMA scheme, into eight time-slots for full-rate operation (or 16 for half-rate). GSM supports both full-rate and half-rate TCHs, referred to as TCH/F and TCH/H, respectively. The full-rate channel supports a gross data rate of 22.8 kbit/s and allows data to be transmitted at 12, 6 or 3 kbit/s. The half-rate channel, which occupies half a TDMA slot, supports a gross data rate of 11.4 kbit/s. Data can be transmitted at 6 or 3.6 kbit/s. The full-rate TDMA frame structure is shown in Figure 1.3.

The GSM full-rate speech coder has an output rate of 13 kbit/s. Speech is handled in blocks of 20 ms duration, hence the output of the speech coder will produce streams of 260 bits. Each block of 260 bits is then subject to error correction. GSM divides speech bits into two classes: "Class 1" bits have a strong influence on perceived signal quality and are subject to error correction; "Class 2" bits are left unprotected. Of the 260 bits in a 20-ms frame, 182 bits are termed "Class 1". The Class 1 bits are further divided into 50 Class 1a bits, which are most sensitive to bit errors, and Class 1b bits, which account for the other 132 bits. Three cyclic redundancy code bits are added to the Class 1a bits, which are then added to the Class 1b bits, before adding four tail bits resulting in 189 bits in total. These bits are then subject to half-rate convolutional coding. This results in 378 bits being output from the encoder, which are then added to the 78 unprotected bits, resulting in a total of 456 bits in a 20-ms frame, equivalent to a coded rate of 22.8 kbit/s (Figure 1.4).

The output of the error correction device is fed into the channel coder, which performs interleaving of the bits. Interleaving and associated de-interleaving at the receiver are used to disperse the effect of bursty errors introduced by the mobile transmission environment. (This technique is also employed in mobile-satellite communications (see Chapter 5).) The coder

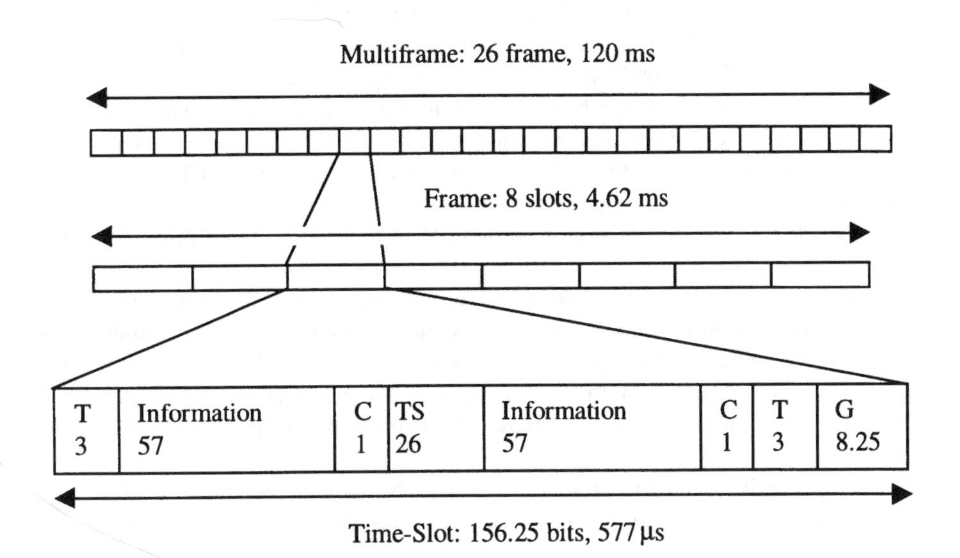

Figure 1.3 GSM TDMA 26-frame structure.

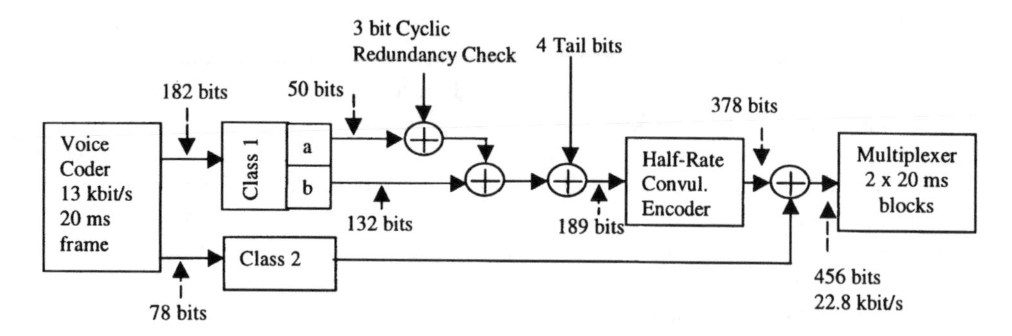

Figure 1.4 GSM full-rate speech coder.

takes two 20-ms time frames, equivalent to 912 bits, and arranges them into eight blocks of 114 bits. Each block of 114 bits is then placed in a time-slot for transmission.

Two 57-bit fields are allocated for the transmission of information within each GSM time-slot. In addition to information content, each time-slot comprises three tail-bits (all logical zeros) located at the beginning and the end of each time-slot. These bits are used to provide a buffer between time-slots; two control bits, which follow each 57 bit of user information, that are used to distinguish between voice and data transmissions; and a 26-bit training sequence, located in the middle of the slot. The training sequence is used for identification purposes and to perform channel equalisation. A guard-time of 30.5 μs, corresponding to 8.25 bits, is then added to the time-slot prior to transmission [GOO-91]. Each time-slot lasts for 0.577 ms, during which time 156.25 bits are transmitted, resulting in a gross bit rate of 270.833 kbit/s.

A group of eight time-slots is called a TDMA frame, which is of 4.615 ms duration. Each time-slot is used to communicate with an individual mobile station, hence each TDMA frame can support eight users at a time. TDMA frames are grouped into what is known as multi-frames, consisting of either 26 or 51 TDMA frames. In the 26-frame format, 24 frames are allocated to TCHs (TCH/F), with each TCH occupying one of the eight timeslots per frame, and one frame (frame-12) for the eight associated slow associated control channels (SACCH). The other frame (frame-25) remains spare unless half-rate operation is employed, in which case it is occupied by the eight other SACCHs associated with the extra TCHs. An SACCH is used for control and supervisory signals associated with a TCH. In addition, a fast associated control channel (FACCH) steals slots from a TCH in order to transmit power control and handover signalling messages.

GSM employs a number of logical control channels to manage its network. These channels are grouped under three categories: broadcast control channel (BCCH); common control channel (CCCH); and dedicated control channel (DCCH). The GSM logical control channels are summarised in Table 1.2. With the exception of the SACCH and FACCH, which are transmitted on the 26-frame structure, these channels are transmitted using the 51-frame structure.

Fifty-one of the 26-frame and 26 of the 51-frame formats are combined to form a GSM superframe, and the TDMA hierarchy is complete when 2048 superframes are combined to form a hyperframe [WAL-99].

Table 1.2 GSM logical control channels

Group	Channel	Description
DCCH	Slow associated control channel (SACCH)	Control and supervisory signals associated with the TCH
DCCH	Fast associated control channel (FACCH)	Steals timeslots from the traffic allocation and used for control requirements, such as handover and power control
DCCH	Stand-alone dedicated control channel (SDCCH)	Used for registration, location updating, and authentication and call set-up
BCCH	Frequency correction channel (FCCH) – downlink	Allows mobiles to achieve initial frequency synchronisation with the local BTS
BCCH	Synchronisation channel (SCH)	Allows mobiles to achieve initial time synchronisation with the local BTS
BCCH	Broadcast control channel (BCCH) – downlink only	Provides the MS with information such as BS identity, frequency allocation and frequency hopping sequences
CCCH	Paging channel (PCH) – downlink only	Alerts the mobile that the network requires to signal it
CCCH	Random access channel (RACH) – uplink only	A slotted-ALOHA channel used by the MS to request access to the network
CCCH	Access grant channel (AGCH) – downlink only	Used to allocate a stand-alone DCCH to a mobile for signalling, following a request on the RACCH

Network Architecture A simplified form of the GSM network architecture is shown in Figure 1.5.

Mobile Station (MS) A subscriber uses an MS to access services provided by the network. The MS consists of two entities, the mobile equipment (ME) and subscriber identity module (SIM). The ME performs the functions required to support the radio

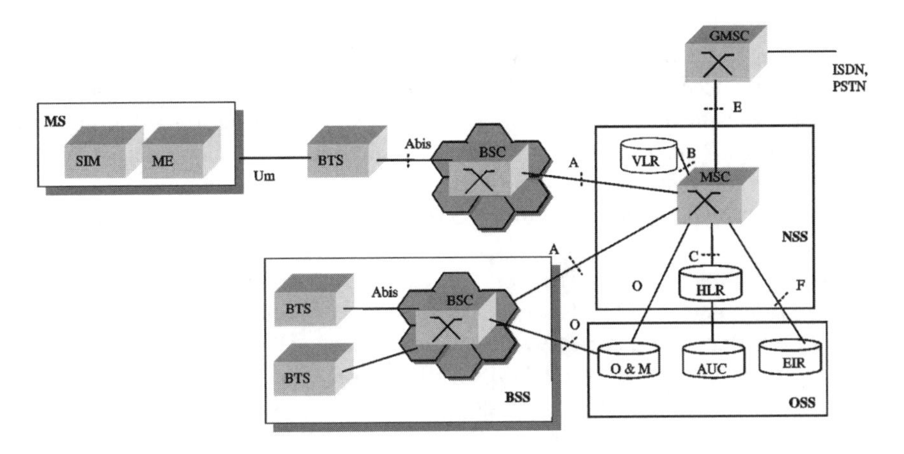

Figure 1.5 GSM simplified network architecture.

channel between the MS and a BTS. These functions include modulation, coding, and so on. It also provides the application interface of the MS to enable the user to access services. A SIM card provides the ability to personalise a mobile phone. This is a smart card that needs to be inserted into the mobile phone before it can become operational. The SIM card contains the user's international mobile subscriber identity (IMSI), as well as other user specific data including an authentication key. Similarly, a terminal is identified by its international ME identity (IMEI). The IMSI and IMEI provide the capability for personal and terminal mobility, respectively. The radio interface between the MT and the BTS is termed the Um-interface and is one of the two mandatory interfaces in the GSM network.

Base Station System (BSS) The BTS forms part of the base station system (BSS), along with the base station controller (BSC). The BTS provides the radio coverage per cell, while the BSC performs the necessary control functions, which include channel allocation and local switching to achieve handover when a mobile moves from one BTS to another under the control of the same BSC. A BTS is connected to a BSC via an Abis-interface.

Network Management and Switching Subsystem (NMSS) The NMSS provides the connection between the mobile user and other users. Central to the operation of the NMSS is the MSC. A BSS is connected to an MSC via an A-interface, the other mandatory GSM interface. The coverage area of an MSC is determined by the cellular coverage provided by the BTSs that are connected to it. The functions of an MSC include the routing of calls to the appropriate BSS, performing handover between BSSs and inter-working with other fixed networks. A special type of MSC is the gateway MSC (GMSC), which provides connection to fixed telephone networks and *vice versa*. A GMSC is connected to an MSC via an E-interface. Central to the operation of the MSC are the two databases: HLR, connected via the C-interface; and VLR, connected via the B-interface.

The HLR database contains management information for an MS. Each MS has an associated HLR. The information contained within an HLR includes the present location of an MS, and its IMSI, which is used by the authentication centre (AuC) to authorise a subscriber's access to the network. A service profile is stored on the HLR for each MS, as is the last known VLR and any subscriber restrictions.

Each MSC has an associated VLR. Whenever an MS is switched on in a new location area or roams into a new location area covered by an MSC, it must register with its VLR. At this stage, the visiting network assigns a MS roaming number (MSRN) and a temporary mobile subscriber identity (TIMSI) to the MS. The location of the MS, usually in terms of the VLR's signalling address, is then conveyed to the HLR. The VLR, by using subscriber information provided by the HLR, can perform the necessary routing, verification and authentication procedures for an MS that would normally be performed by the HLR.

An MSC also provides connection to the SMS centre (SMSC), which is responsible for storing and forwarding messages.

Operation Subsystem (OSS) The OSS provides the functions for the operation and management of the network. The Network Operation and Maintenance Centre performs all the necessary functionalities necessary to monitor and manage the network. It is connected to all of the major network elements (BTS, MSC, HLR, VLR) via an O-interface using an X.25 connection. The equipment interface register (EIR) is used by the network to identify any equipment that may be using the network illegally. The MSC-EIR connection is specified by the F-interface. The AuC also forms part of the OSS.

Mobility related and other GSM signalling in the CN is performed by the mobile application part (MAP), developed specifically for GSM. Importantly, the GSM MAP will be used to provide one of the CNs for IMT-2000.

1.2.3.2 Digital Cellular System 1800 (DCS1800)

The first evolution of GSM came about with the introduction of DCS1800, which is aimed primarily at the mass-market pedestrian user located in urban, densely populated regions. DCS1800 was introduced under the personal communications network (PCN) concept, also known as personal communication services (PCS) in the US. In 1989, the UK Government's Department of Trade and Industry outlined its intention to issue licenses for personal communication networks in the 1700–2300 MHz band [DTI-89]. It was recognised that the new service, which would be aimed primarily at the pedestrian user, would be an adaptation of the GSM standard. Subsequently, ETSI produced the Phase 1 DCS1800 specification in January 1991, which detailed the generic differences between DCS1800 and GSM. This was followed by a Phase 2 specification detailing a common framework for PCN and GSM. DCS1800 operates using largely the same specification as GSM, making use of the same network architecture but, as its name implies, it operates in the 1800 MHz band. Here, parallels can be drawn with the evolution of the NMT450 to the NMT900 system from earlier discussions.

The bands that have been allocated for DCS1800 operation are 1710–1785 MHz for the mobile to BS link and 1805–1880 MHz for the BS to mobile link. Taking into account the 200-kHz guard-band, 374 carriers can be supported. Apart from the difference in the operating frequency, the only other major difference is in the transmit power specification of the mobile station. Two power classes were defined at 250 mW and 1 W peak power, respectively. As DCS1800 is intended primarily for urban environments, the cells are much smaller than those of GSM; hence the transmit power requirements are reduced.

The UK was the first country to introduce PCN into operation, through two network operators: Mercury's One2One, which was introduced into service in September 1993; and Hutchinson's Orange, which was introduced into service in April 1994.

Dual-mode 900/1800 MHz terminals are now available on the market, as are triple-mode 900/1800/1900 MHz terminals, allowing roaming into North America (in the US, the 1900 MHz band is used for PCS). Towards the end of 1999, Orange and One2One accounted for a third of the UK digital cellular market at just under 5 million subscribers, with virtually an equal market share between the two of them [MCI-99].

1.2.3.3 Digital Advanced Mobile Phone Services (D-AMPS)

D-AMPS was specified by the Telecommunications Industry Association (TIA) Interim Standard-54 (IS-54) in 1990, as an intermediate solution while a fully digital standard was specified. IS-54 retained the analogue AMPS control channels and gave operators the opportunity to provide digital cellular services while the complete digital standard was being developed. A fully digital standard was specified as IS-136, which uses the same digital radio channels as IS-54, and also includes digital signalling channels. Mobility related and other signalling in the CN is specified by the IS-41 standard. Importantly, as with the GSM MAP, IS-41 will be used to provide one of the CNs for IMT-2000.

Due to regulatory constraints, D-AMPS operates alongside AMPS in the same frequency bands in the US. As with the European digital system, GSM, the multiple access technique is based on TDMA, however, a reduced hierarchical approach to the frame structure is implemented, resulting in a simpler implementation. As with AMPS, the carrier spacing is 30 kHz. D-AMPS employs Pi/4-shifted DPSK as the modulation method. Carriers are transmitted at 48.6 kbit/s and root raised cosine filtering with a roll-off factor of 0.35 is employed at the transmitter and receiver.

A TDMA frame consists of six time-slots, each of 6.67 ms duration; a frame is of 40 ms duration. Each slot contains 324 bits including 260 bits of user information. In the mobile to BS direction, these information bits are divided into three packets of 16, 122 and 122 bits, respectively. In the opposite direction, data are divided equally into two 130-bit packets. In addition to user information, the mobile to BS time-slot contains [GOO-91]:

- Six guard bits, six ramp time bits and 28 bits of synchronisation information containing a known bit pattern, one for each time-slot in a frame.
- A 12-bit digital verification colour code, to assist with the management of time-slot assignments.
- And 12 bits of system control information, transmitted on the slow access control channel (SACCH).

In the BS to mobile direction, the capacity allocated to the guard and ramp time bits is reserved for future use. The structure of the slots is shown in Figure 1.6.

D-AMPS adopts vector sum excited linear prediction (VSELP) as the speech coding technique. The output from the coder produces a source rate of 7.95 kb/s, which is then subject to coding. Output bits are processed in 20-ms bursts, or in other words blocks of 159 bits. Of these 159 bits, 77 are termed class 1 bits, which are considered to have the greater influence on speech quality, using a similar approach to GSM. These 77 bits are subject to an error detecting code, in the form of the addition of seven parity checks and five tail bits, and half-rate convolutional coding. The resultant 178 bits are then multiplexed with the other 82

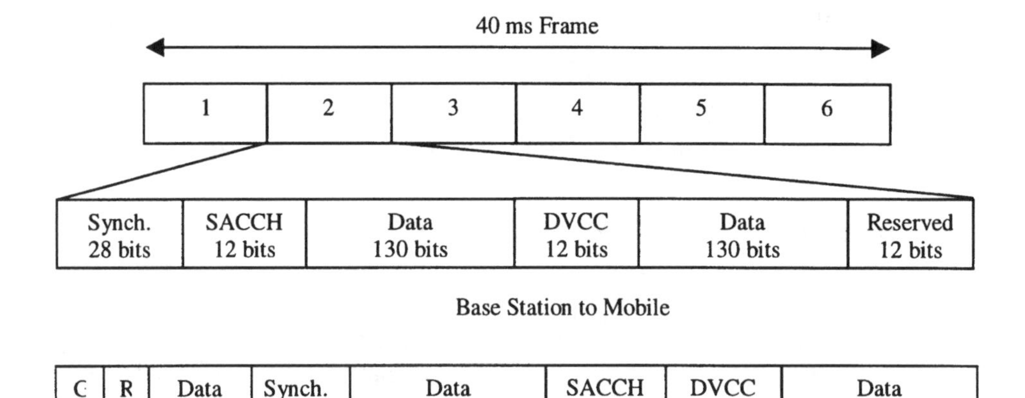

Figure 1.6 D-AMPS time-slot structure.

bits to produce 260 bits. These blocks of 260 bits are then interleaved in order to protect the information from bursty errors caused by the mobile channel, and then placed in the assigned slots in the TDMA frame.

D-AMPS has been developed as a dual-mode system, providing in effect backward compatibility with its analogue counterpart. As far as control signals are concerned, D-AMPS uses the SACCH and the FACCH to communicate with the network. Control messages have a length of 65 bits. For the FACCH, where immediate control information needs to be transmitted, 1/4-rate convolutional coding is applied, resulting in 260 bits, which are then inserted into a single time-slot. For the SACCH, half-rate convolutional coding is applied and the resultant bits are dispersed over 12 time-slots, resulting in a delay of 240 ms.

1.2.3.4 cmdaOne

In 1991, Qualcomm announced the development of a CDMA based mobile system. This was subsequently specified as Interim Standard-95A (IS-95A) and later became known under the commercial name cmdaOne, as established by the CDMA Development Group (CDG). Interestingly, one of Qualcomm's main products is OMNITRACS, a geostationary satellite based fleet management system incorporating two-way mobile communications and position reporting services. Qualcomm is also a partner in the GLOBALSTAR project, a non-geostationary satellite system, which will be discussed in the following chapter. GLOBALSTAR incorporates a modified version of the cmdaOne radio access technique. In 1995, Hutchinson became the first operator to launch cmdaOne in Hong Kong.

Like D-AMPS, cmdaOne is a dual-mode system, which operates alongside the AMPS service in the same band. Digital transmissions are used by default and the mobile automatically switches to analogue mode when no digital coverage is available. Mobility related and other signalling in the CN is specified by the IS-41 standard. The mobile transmits at 45 MHz below the BS transmit frequency and channels are separated by 30 kHz. Specifically, the operational frequency bands are 869–894 MHz (BS to mobile) and 824–849 MHz (mobile to BS). cmdaOne also operates in the PCS band (1930–1980 MHz uplink, 1850–1910 MHz downlink) and dual-band phones are available. It can be seen that in the PCS band, the mobile transmits at 80 MHz below the BS, moreover, channels are separated by 50 kHz. The system operates using a bandwidth of 1.23 MHz, this being equivalent to 41 different 30-kHz AMPS channels. One of the advantages of CDMA is the increase in available capacity when compared to other cellular technologies. In this respect a more than ten-fold capacity increase when compared to AMPS, and a three-fold increase when compared to a TDMA system is possible.

Unlike other cellular systems, the Qualcomm CDMA system operates using different transmission techniques on the forward and reverse directions.

In the IS-95A forward direction, each BS has access to 64 CDMA channels. These channels are categorised as either common control channels, which are broadcast to all mobile terminals, or broadcast channels, which are dedicated to a particular terminal. Each channel is derived from one row of a 64×64 Walsh Hadamard matrix, with the row numbers transcending from 0 to 63. One of the common control channels, Walsh function 0, which comprises all zeros, is used to transmit the pilot signal. Another of the common control channels, Channel

32, acts as a synchronisation (SYNC) channel, providing mobile terminals with important system information such as the BS identifier, the time-offset specific to the BS introduced in the radio modulator and system time. BSs are synchronised using the Global Positioning System (GPS) satellite system [GET-93]. The SYNC channel always transmits at a data rate of 1200 bit/s.

The remaining 62 channels are available for traffic over a broadcast channel, with up to the first seven of these being available for paging, using a common control channel. The paging channel operates at either 4.8 or 9.6 kbit/s and is used to alert the mobile of an incoming call.

Variable rate voice coding produces data rates in the range 1.2–9.6 kbit/s for Rate Set 1, or 1.8–14.4 kbit/s for Rate Set 2, depending on speech activity. Rate Set 2 is optional on the forward TCH and is only available in the PCS band, which also supports Rate Set 1. Data are grouped into 20-ms frames. This is then encoded using a half-rate convolutional encoder with a constraint length of 9 for Rate Set 1. Rate Set 2 differs only in that it employs a 3/4-rate convolutional encoder of constraint length nine. In order to ensure a constant rate of 19.2 kbit/s, repetition of lower rate code bits is performed prior to interleaving and spreading by a pseudo-random sequence derived from the long code, which repeats itself after $2^{42}-1$ chips, and a long code mask, which is unique to each terminal and contains the mobile's electronic serial number. This 19.2-kbit/s signal is then multiplexed with power control bits, which are applied at 800 bit/s. The Walsh code assigned to the user's TCH is then used to spread the signal. Quadrature phase shift keying (QPSK) is used to modulate the carrier, introducing a time-offset associated with the BS to the in-phase and quadrature components. This delay is introduced since BSs with different time offsets appear as background noise to the mobile, hence reducing interference levels. A quadrature pair of pseudo-noise (PN) sequences is then used to spread the signal at 1.2288 Mchip/s. Baseband filtering is then applied to ensure that the components of the modulation remain within the channel before modulating the in-phase and quadrature signals onto the CDMA channel. BSs vary radiated power in proportion to data rate in a frame. So, for example, a 1.2-kbit/s signal would be transmitted at an eighth of the power of a 9.6-kbit/s signal. This ensures that all bits are transmitted with the same energy [GAR-96] (Figure 1.7).

The reverse direction has two associated channels: the access channel and the TCH. The access channel is used by the mobile to request a TCH, respond to a paging message, or for location updates. This operates at 4.8 kbit/s. An access channel is associated with a particular paging channel, and from the above discussion on the forward link, it can be seen that up to seven access channels can exist.

As before, the TCH supports Rate Set 1 and may support Rate Set 2. In the reverse direction, variable rate voice coding produces data rates in the range 1.2–9.6 kbit/s for Rate Set 1, or 1.8–14.4 kbit/s for Rate Set 2, depending on speech activity. So, for example, a data rate of 1.2 kbit/s corresponds to the transmission of no speech. Data are organised into 20-ms frames. For Rate Set 1, this is then encoded using a 1/3-rate convolutional encoder with a constraint length of 9. As in the forward direction, in order to ensure a constant rate of 28.8 kbit/s, repetition of lower rate code bits is performed. Rate Set 2 employs a half-rate convolutional encoder of constraint length 9, again employing repetition to ensure a constant rate of 28.8 kbit/s. The encoded bits are then interleaved prior to Orthogonal Walsh modulation, by which a block of six code symbols is used to generate a sequence

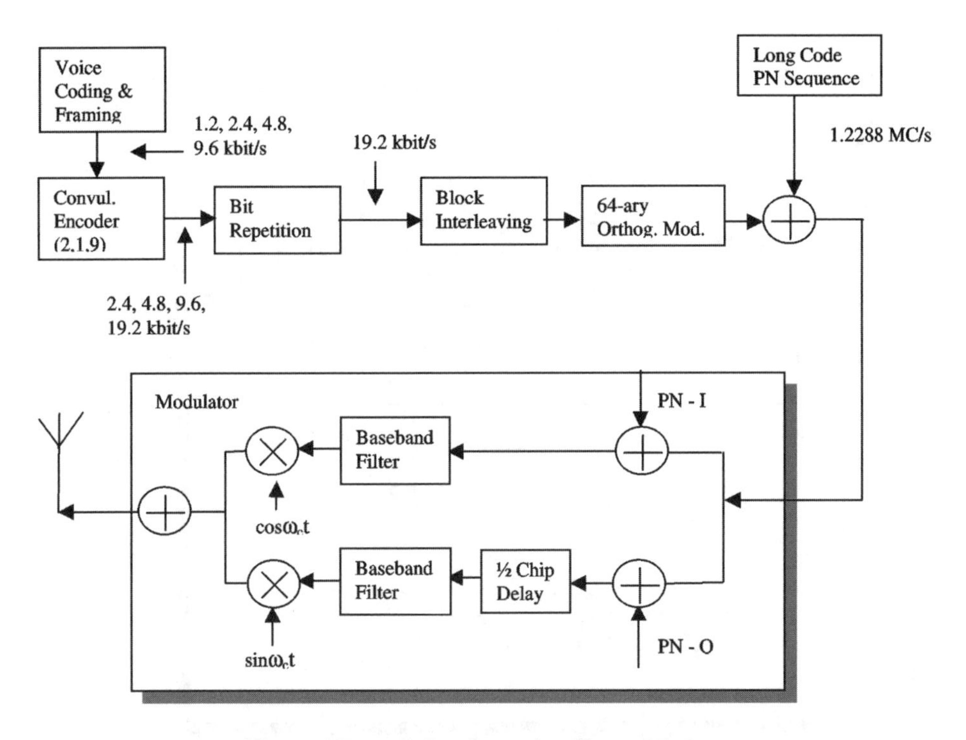

Figure 1.7 cmdaOne forward traffic modulation.

of 64 bits, corresponding to a row of a 64 × 64 Walsh Hadamard matrix. Each mobile transmits a different Walsh code, enabling the BS to identify the transmitter. A data burst randomiser is then applied with a duty cycle inversely proportional to the data rate. This results in the random distribution of energy over each frame, independent of other potential interfering terminals operating in the same cell. This is different from the technique described above for the BSs. The long code, which repeats itself after $2^{42}-1$ chips, and long code mask, which is unique to each terminal and contains the mobile's electronic serial number, are then used to spread the signal at 1.2288 Mchip/s. Offset-QPSK is used to modulate the carrier. This involves delaying the quadrature channel by half a chip. The same quadrature pair of PN sequences, as used by the BS, is then used to spread the signal at 1.2288 Mchip/s with a periodicity of $2^{15}-1$ chips. Baseband filtering is then applied to ensure that the components of the modulation remain within the channel before modulating the in-phase and quadrature signals onto the CDMA channel (Figure 1.8).

In September 1998, Leap Wireless International Inc., an independent Qualcomm spin-off company was formed, its purpose being to deploy CDMA based networks. As well as North America, cmdaOne is deployed throughout the world including the major markets of Australia, China, South Korea and Japan; and as of 1999, there were 43 wireless local loop (WLL) systems in 22 countries using cmdaOne technology. The co-existence of cmdaOne and GSM networks in Australia and China, the latter a potentially huge market, has led to the possibility of the development of a dual GSM/cmdaOne handset.

cmdaOne is now only second to GSM in the world market stakes. This is despite the fact that no significant market penetration has been achieved in Europe. The proponents of

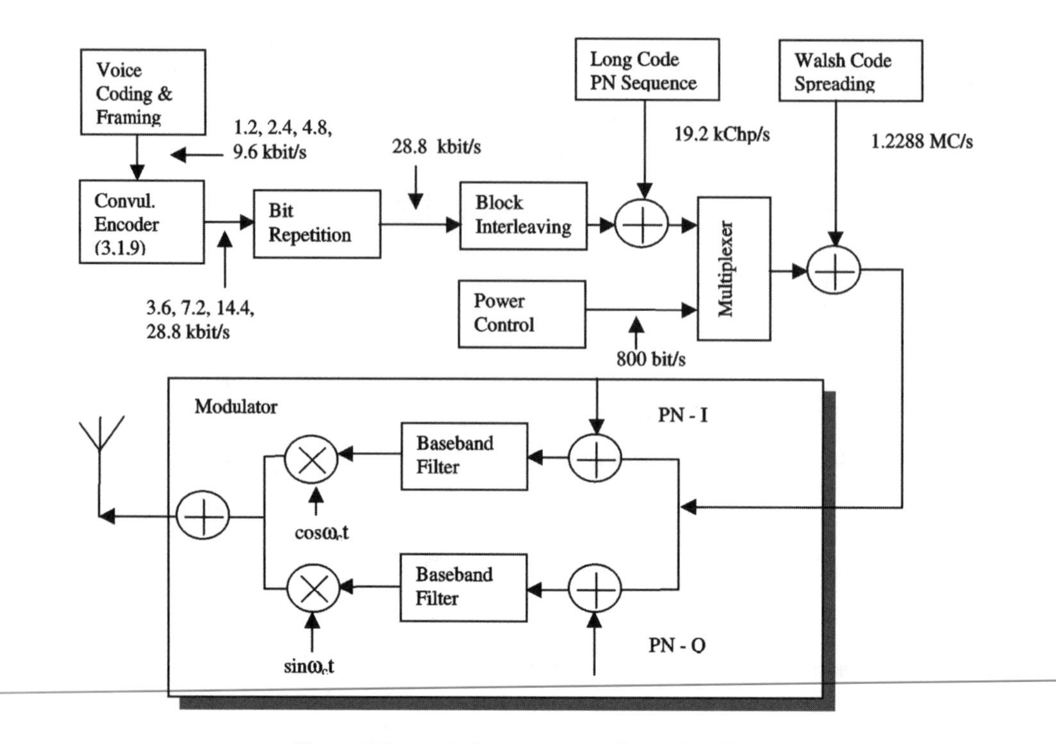

Figure 1.8 cmdaOne reverse traffic modulation.

cmdaOne see it as an important stepping stone towards the development of 3G networks. Indeed, the evolution of cmdaOne, known as multicarrier CDMA, was selected as one of the family of radio interfaces for IMT-2000.

1.2.3.5 Personal Digital Cellular System (PDC)

The air interface of the PDC, formerly known as the Japanese digital cellular system (JDC), was specified in April 1991, following 2 years of investigation. PDC operates in two frequency bands: 800 MHz with 130 MHz duplex operation; and 1.5 GHz with 48 MHz duplex operation. The bands are divided into 25-kHz channels. Nippon Telegraph and Telephone (NTT) introduced PDC into service in 1993 at the lower of the two aforementioned frequency bands, before adding the higher band operation in 1994.

The multiple access scheme employed is three- or six-channel TDMA/frequency division duplex (FDD), depending on whether a full or half-rate codec is employed. A TDMA frame is of 20 ms duration, which is divided into six slots. The information bit rate is at 11.2 kbit/s at full-rate or 5.6 kbit/s at half-rate. The modulation method used is Pi/4-differential (D)QPSK, which is transmitted at a bit rate of 42 kbit/s.

1.2.4 Evolved Second-Generation (2G) Systems

1.2.4.1 Overview

So-called evolved 2G networks are aimed at exploiting the demand for mobile data services, with the emphasis being on the availability of higher data rates in comparison to 2G services. The majority of these service offerings also point the way towards packet-oriented delivery.

1.2.4.2 High Speed Circuit Switched Data (HSCSD)

In recognition of the needs of the market, the HSCSD service is aimed very much at the mobile user who needs to access or to transmit large data files while on the move. Transmission of large files, of the order of megabytes, is clearly unattractive when using the GSM rate of 9.6 kbit/s. However, by making available up to eight GSM full-rate TCHs to a single user, HSCSD can achieve rates of at least 76.8 kbit/s and if overhead reduction techniques are employed, significantly higher data rates of 115 kbit/s and higher could be achievable. Clearly such data rates will only be viable through network operators with a large amount of spare capacity that are willing to address niche market applications.

The HSCSD makes use of the GSM network architecture and no modifications to the physical infrastructure of the network are required, only software upgrades. This is clearly an attractive opportunity for network operators, by recognising the potential to create new, potentially lucrative markets for a minimum investment. However, the long-term future of HSCSD is unclear as the future of data transmission appears to be based on packet-switched technology.

1.2.4.3 General Packet Radio Service (GPRS)

The introduction of the GPRS represents an important step in the evolution of the GSM network. Significantly, GPRS is a packet-switched system, unlike GSM and HSCSD, which are circuit-switched. The aim of GPRS is to provide Internet-type services to the mobile user, bringing closer the convergence of IP and mobility. Indeed, apart from addressing a significant market in its own right, clearly GPRS can be considered to be an important stepping stone between GSM and UMTS.

In packet-switched networks the user is continuously connected but may only be charged for the data that is transported over the network. This is quite different from circuit-switched networks, such as GSM, where a connection is established at call set-up and the user is billed for the duration of the call, irrespective of whether any information is transported. In this respect, packet-switched technology can be considered to be more spectrally efficient and economically attractive. Essentially, channel resources are made available to all users of the network. A user's information is divided into packets and transmitted when required. Other users are also able to access the channels when necessary.

A GPRS MS comprises mobile terminal (MT), which provides the mechanism for transmitting and receiving data and terminal equipment (TE), a PC-like device upon which applications run. Within a GSM/GPRS network, a GPRS MS has the capability to function in three operational modes, as shown in Table 1.3.

Table 1.3 GPRS MS operational modes

Class	Operation
A	Simultaneous packet and circuit-switched modes
B	Automatic selection of circuit or packet-switched mode
C	Packet mode only

GPRS offers non-real time services at data rates from 9 up to 171 kbit/s, with most networks offering about half the maximum data rate. The available bit rate is dependent upon the number of users of the network and the type of channel coding employed. GPRS specifies four types of coding (CS-1, CS-2, CS-3 and CS-4), the choice of which depends on the operating environment. The theoretical maximum data rate is achieved by employing CS-4, whereby no coding is employed, and only a single user is gaining access to channel resources. The data rates offered allow a full range of services to be accessed, from simple messaging to efficient file transfer and web browsing. Inter-working with TCP/IP and X.25 bearer services will be available. Once connected to the network, a call can be established in less than 1 s, which is significantly faster than GSM. GPRS, in addition to providing point-to-point services, has the capability to deliver point-to-multipoint calls, where the receiving parties can be either closed-group or users within a particular broadcast area.

GPRS makes use of the same radio interface as GSM. To achieve a variable data rate, a MS can be allocated from one to eight time-slots within a particular TDMA frame. Moreover, the time-slot allocation can be different in the forward and reverse directions. Hence, the MS can transmit and receive data at different rates, thus enabling asymmetric service delivery. This approach is the key to the efficient provision of Internet-type services. Channels are dynamically allocated to MSs on a per-demand basis from a common pool of resources available to the network. Upon completing a communication, the channels are released by the MS and returned to the pool, thus enabling multi-users to share the same channel resources. Packets are transmitted on unidirectional packet data TCHs (PDTCH), which are similar to the full-rate TCHs used by GSM for speech. These channels are temporarily allocated to a particular MS when data transfer occurs. TCHs are transmitted over a 52-multiframe physical channel, which is made up of two GSM 26 multiframes. A 52-multiframe comprises of 12 blocks of four TDMA frames plus two frames for the packet broadcast control channel (PBCCH) and two spare. Hence, a 52-multiframe lasts for 240 ms. A 51-multiframe is also defined for the exclusive transmission of the packet call control channel (PCCCH) and PBCCHs. In addition, GPRS provides a number of logical channels for control and signalling purposes, grouped under PBCCCH, packet dedicated control channel (PDCCH) and PCCCH. These are summarised in Table 1.4.

The network architecture is based on GSM with some additions. The architecture consists of the following components: Servicing GPRS (CHECK) support node; gateway GPRS support node; MSC/VLR; HLR; BSS with packet control unit (PCU) – the PCU is a new functional entity introduced by ETSI to support introduction of GPRS in GSM networks; GPRS MS; and a GPRS IP backbone. The GPRS network architecture is shown in Figure 1.9.

The servicing GPRS support node (SGSN) is used to perform location monitoring of individual mobile terminals, security functions and access control. An SGSN is connected

Table 1.4 GPRS logical control channels

Group	Channel	Description
PDCCH	Packet associated control channel (PACCH)	Provides signal information related to a specific MS. Can be used for paging for circuit-switched services, when MS is in packet-switched mode
PDCCH	Packet timing advance control – uplink direction (PTCCH/U)	Used by MS to transmit random access burst, from which BS estimates timing advance
PDCCH	Packet timing advance control – downlink direction (PTCCH/D)	Used by BS to transmit timing advance information to MSs
PBCCH	Packet broadcast control channel (PBCCH) – downlink only	Used by the BS to provide the MS with specific system information relating to packet data
PCCCH	Packet notification channel (PNC)	Used to send a point-to-multipoint multicast (PTM-M) to a group of MSs
PCCCH	Packet paging channel (PPCH) – downlink only	Alerts the mobile prior to packet transfer
PCCCH	Packet random access channel (PRACH) – uplink only	Used by the MS to initiate data transfer or signalling
PCCCH	Packet access grant channel (PAGCH) – downlink only	Used to assign resources to the MS prior to packet transfer

to a BSS by G.703/G.704 frame relay. This is termed the Gb-interface. When a session needs to be established, the SGSN establishes a "tunnel" between the mobile and a given packet data network using the packet data protocol (PDP). This protocol provides the routing information and QoS profile required to transfer protocol data units between the MS and the gateway GSN (GGSN). To provide closer co-ordination with GSM, a Gs-interface connecting the SGSN with an MSC is specified.

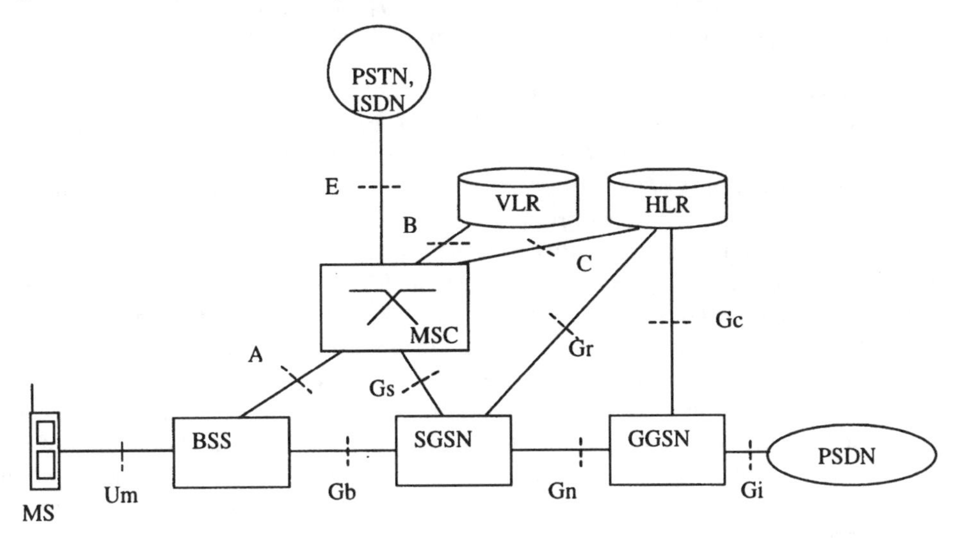

Figure 1.9 GPRS network architecture.

The GGSN performs the functions necessary to allow inter-working with external packet-switched networks. Each external packet-switched network is referenced by a unique access point name (APN). GGSNs are connected to SGSNs via an IP-based GPRS backbone network. This is termed a *Gn* connection. The GGSN uses a GPRS tunnelling protocol (GTP) to encapsulate IP packets received from external IP networks and to tunnel to the SGSN. The connection between a GGSN and an IP network is termed *Gi*. An optional Gc-interface between the GGSN and the HLR is also specified, again to facilitate closer co-operation with the GSM network.

In addition to the new hardware elements introduced by GPRS, some software upgrades are also necessary to the components of the GSM network elements, specifically the HLR with GPRS subscriber information and optionally the MSC/VLR for more efficient co-ordination of services, for example, between GSM and GPRS subscribers.

An MS must perform a GPRS *attach procedure* before it can gain access to the services provided by the network. This involves registering the MS with the local SGSN. Once authorisation to use the services of the network has been verified and the user profile has been obtained from the HLR, the SGSN supplies the MS with a packet temporary mobile subscriber identity (P-TMSI).

Once an attach procedure has been performed, in order to be able to send and receive packets, the MS must create a *PDP context*. This is used to describe the service characteristics and comprises the following: a PDP address, which is an IP address allocated by the PDN to which correspondence will take place; the PDP type (e.g. IPv6); the address of the GGSN serving the PDN; and the required QoS. The PDP context is stored at the MS, SGSN and GGSN and communication can then take place with the PDN. A user is not limited to a single PDP context at any particular time. PDP addresses can be statically or dynamically allocated to an MS. In the latter case, the GGSN is responsible for allocation/de-allocation, while in the former, the user's home network permanently assigns the PDP address.

A key feature of GPRS is its ability to support various categories of QoS, which, as noted above, can be set for each PDP context. A QoS profile is derived from four criteria, service precedence; delay; reliability; and throughput [BET-99]. Such a process allows the flexible negotiation of required QoS based on the status of the network at the time of the session.

IS-136 and GPRS Cellular digital packet data (CDPD) was developed to provide packet data services as a network overlay to the AMPS network [SAL-99]. Subsequently, this was adapted to also provide solutions for the IS-136 and IS-95 networks. CDPD provides mobility management features that are independent of those provided by IS-41. Consequently, a mobile terminal would need to perform procedures associated with both networks. This is unlike GPRS, which derives its mobility management features from the GSM MAP. As we will see shortly, in the future mobile environment, the CN will comprise of IS-41 and GSM MAP, which will be developed to allow network to network interworking. With this in mind, it has been decided that a more elegant solution to the requirements for the evolution of the IS-136 network towards a packet-oriented environment can be achieved by adapting the GPRS network to the IS-41 standard [FAC-99]. Moreover, the adoption of the GPRS network architecture offers two other major advantages: the world-wide availability of the GSM network facilitates global roaming; and the 3G standard UWC-136, which is an evolution of IS-136, makes use of the enhanced data rates for GSM evolution (EDGE) radio interface.

1.2.4.4 Enhanced Data Rates for GSM Evolution (EDGE)

The introduction of EDGE technology will further increase the network capacity and available data rates of both circuit-switched (HSCSD) and packet-switched (GPRS) GSM derivatives. This is achieved by the implementation of a new radio interface incorporating 8-phase shift keying (8-PSK), which will co-exist with the existing GSM modulation technique, GMSK. This new method of modulation enables available HSCSD and GPRS data rates to be extended by up to three-fold per channel. Alternatively, existing data rates can be achieved by using fewer time-slots, hence increasing network efficiency.

Importantly, no further modifications to the underlying GSM network are required. The only updates are through software upgrades to take into account the new radio interface technology. New multimedia-type terminals will also be introduced into the market in order to exploit the new service capabilities offered by the network.

The Phase 1 EDGE standard, specified by ETSI, considers both enhanced HSCSD (ECSD) and enhanced GPRS (EGPRS) services, with data rates of up to 38.4 kbit/s/time-slot and 60 kbit/s/time-slot, respectively. Higher data rates can be achieved by combining TCHs, so for example, a 64-kbit/s service could be achieved by combining two ECSD channels. Rates of over 400 kbit/s can be achieved for EGPRS.

Clearly at the rates offered, the boundary between EDGE and UMTS (to be discussed shortly) becomes less and less distinct. EDGE will be able to provide the vast majority of services envisaged by UMTS and an early market lead will be gained. EDGE and UMTS are envisaged to operate alongside each other, particularly in the early years of UMTS deployment. Significantly, EDGE is also seen as a means for GSM operators without a UMTS licence to supply personal multimedia services using existing GSM band allocations.

In fact, the division between GSM's EDGE and 3G technologies is even less distinct, as EDGE forms the basis of the UWC-136 solution that will provide the TDMA component of the IMT-2000 family of radio systems.

1.2.4.5 IS-95B

IS-95B, the packet-mode version of IS-95A, is capable of delivering data rates of up to 115.2 kbit/s. This is achieved by combining up to a maximum of eight TCHs. A TCH comprises a fundamental code channel (FCC) and between 0 and seven supplemental code channels (SCC). The transmission rate of the FCC can vary in line with the Rate Sets 1 and 2 discussed previously (see section on IS-95A), where as the SCC channels operate at the maximum bit rates associated with the particular Rate Set (9.6 or 14.4 kbit/s). The transmission of the signals is as described for IS-95A.

The IS-95B standard forms the basis for the 3G cdma2000 solution, which is one of the family of radio interfaces selected for IMT-2000.

1.2.4.6 I-Mode

In February 1999, NTT DoCoMo launched its proprietary commercial mobile Internet service, i-mode. Within two years, the service had attracted just under 20 million subscribers, well in excess of original market expectations. The i-mode network is essentially a PDC overlay, providing packet data rates at 9.6 kbit/s up to a maximum of 28.8 kbit/s. In this respect, the concept is similar to the relationship between GSM and GPRS. I-mode phones, which have large, 256-colour displays, are able to browse the web and send e-mail messages in a similar manner to that when using a PC, as well as play games, gain access to location based material, download karoke songs, perform commercial transactions and so on. To achieve this, i-mode equipment operates using a compact form of HTML, the language of the web. This enables i-mode website developers to easily implement content. This a different approach to that adopted by GSM, and its derivatives, which has defined its own language under the wireless application protocol (WAP).

Being a packet-oriented network, i-mode terminals are permanently connected to the network, but users are only charged for the packets received. The success of i-mode in Japan clearly demonstrates the potential for mobile IP type services and paves the way for the launch of 3G services in Japan.

1.3 Cordless Telephones

1.3.1 Background

In comparison with cellular phones, the market take-up of the cordless phone has been relatively modest, despite the significant potential markets. Like cellular technology, cordless telephone technology has been around since the mid-1980s, however, it has only been in the last few years that cordless technology has been able to make significant inroads into the market, with the WLL market being particularly buoyant.

The cordless market initially addressed three specific sectors: domestic; business; and telepoint.

In its simplest form, such as would be used in the home, a cordless network comprises of a single BS, and a low-power cordless terminal. The BS provides the connection to the local fixed network and performs the required protocol conversions between the fixed network (e.g. PSTN) and the cordless access network; the low-power terminal communicates with the BS over a radio link. More than one terminal device could be connected to the BS.

As more, distributed users enter the network, each with their own individual extension number, such as in a typical business environment, for example, more BSs are required in order to provide the necessary coverage. As with cellular systems, each BS will provide coverage over a specific area, providing several channels. To enable effective routing of in-coming calls, it is necessary to ensure that the local private business exchange (PBX) is aware of the BS associated with each cordless telephone. This necessitates the use of a local database containing the relation between BS and local extension number.

Telepoint takes the concept of the cordless environment to the outdoors. Such a system was introduced in the UK at the start of the 1990s, with fixed stations (FS) located at post offices, banks, etc. providing an operating radius of 300 m from the transmitter. The legislation at the time restricted the use of UK telepoint services to outgoing calls only. The telepoint service proved unpopular in the UK, no doubt partly due to the increasing popularity of cellular mobile phones, and was eventually withdrawn in 1993. Telepoint technology, however, has proved popular in France, Italy and Hong Kong.

In the following, three of the major cordless telephone systems are briefly described. Further information on cordless technology can be found in Refs. [ITU-97, TUT-90].

1.3.2 Cordless Telephone-2 (CT-2)

The British CT-2 system was the first digital cordless telephone system to be introduced into service. CT-2 is aimed at the home, office and telepoint environments and builds on previous analogue cordless telephone initiatives (CT-1, CT-1+).

CT-2 supports a common air interface (CAI), which was specified by ETSI in 1991. This allows the interworking of various manufacturers' equipment on the same network.

CT-2 operates using an FDMA/time division duplex (TDD) approach, in the band 864.1–868.1 MHz. This allows 40 100-kHz channels to be supported. The system operates using a form of dynamic channel allocation. Signals are modulated using frequency shift keying with Gaussian filtering. A frequency deviation of 14.4–25.2 kHz above the carrier equates to a "1" and equal shifts below equates to a "0".

The frame structure is shown in Figure 1.10. Information is transmitted at 72 kbit/s in frames of 144 bits, corresponding to a frame length of 2 ms. The first half of a frame is used for FS to portable station (PS) transmissions and the latter half is used in the opposite direction.

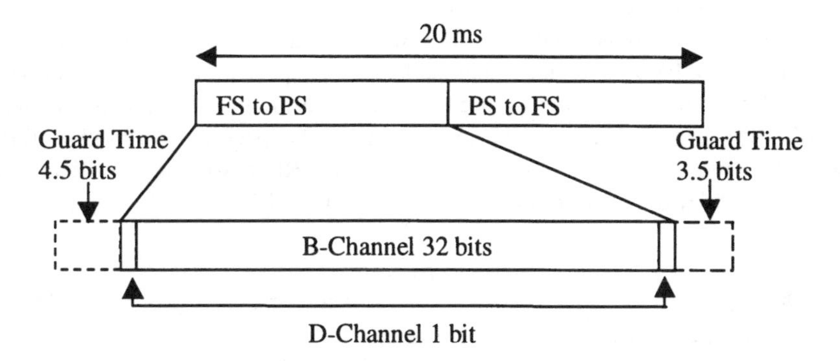

Figure 1.10 CT-2 frame structure.

In a frame, the B channel is used to convey user information, while a D channel is used for signalling information at either 1 or 2 kbit/s, this being specified by the manufacturer. The FS to PS half of the frame is preceded by a guard-time equivalent to 4½ bits, and the PS to FS preceded by a guard-time of 3½ bits. Selection of the lower of the D channel data rates allows a greater guard-time to be employed. Speech is coded at 32 kbit/s using adaptive differential pulse code modulation (ADPCM), as used in fixed terminal telephony. This frame format is known as Multiplex One (M1).

CT-2 employs three frame formats: M1, Multiplex Two (M2) and Multiplex Three (M3). The M3 frame is used by the PS to initiate call set-up. In this case, rather than using a TDD approach as in M1 and M2 frame formats, the PS transmits five consecutive frames of 144 bits, followed by a 4-ms pause in which it waits for a response from the FS.

The M2 frame structure is used when a FS is setting up a call with a PS. Here, the B channel is not required, since no user information is conveyed. In this case, the D channel operates at 16 kbit/s. A preamble signal, comprising 10 bits, alternating between 0 and 1, precedes the SYNC channel comprising 24 bits. When the PS synchronises to the FS it is able to transmit the SYNC word and forward its identification code in the D channel, by using the uplink part of the frame.

PSs operate with a maximum power of 10 mW, as do FSs. Errors can be detected and corrected for on the D channel but no error correction is employed on the B channel.

Further information can be found in Ref. [GAR-90].

1.3.3 Digital Enhanced Cordless Telecommunications (DECT)

The DECT system, standardised by ETSI in 1992 and introduced into service in 1993, is aimed at the residential, high capacity office and WLL environments. By the end of 1998, 25 million phones had been sold [MCI-99b]. The WLL market has become increasingly significant with 32% of WLL lines in 1998 using DECT technology. In Italy, the first DECT telepoint-type service was commercialised in 1998 under the name Fido, and this had attracted in the region of 100 000 subscribers by the end of the first year.

DECT employs frequency division/TDMA/TDD in its transmissions. The basic modulation method adopted by DECT is Gaussian frequency shift keying (GFSK) with a BT (3 dB bandwidth × bit period) value of 0.5 at a gross data rate of 1152 kbit/s. Speech is encoded using 32 kbit/s ADPCM. The modulation techniques Pi/2-DBPSK, Pi/4-DQPSK and Pi/8-D8PSK have since been added to the specification in recognition of the importance of data services and to increase the available data rate. The system operates in the 1880–1900 MHz[1] band utilising ten carriers with a carrier spacing of 1728 kHz. Terminals operate with a peak power of 250 mW and a mean RF power of 5 mW at 8 kbit/s. The DECT TDMA frame structure is shown in Figure 1.11.

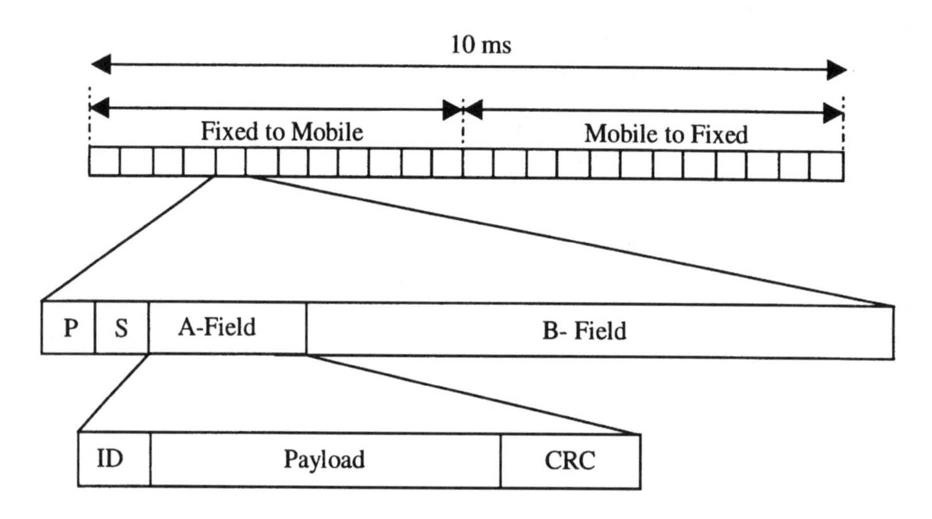

Figure 1.11 DECT frame structure.

A DECT frame is of 10 ms duration. In full-slot mode, each frame is divided into 24 slots; the first 12 being fixed to mobile transmissions and the latter 12 for the reverse operation.

[1] 1900–1920 MHz in China, 1910–1930 MHz in North America.

Each time-slot has a duration of 416.7 μs. A slot is divided into the following: initially a 16-bit preamble followed by a 16-bit synchronisation burst. The next 64 bits, forming what is known as the A-field, are used to transmit control information; this is followed by 320 bits of information, denoted by the B-field.

The A-field is divided into an 8-bit identifier, followed by a 40-bit payload and then a 16-bit cyclic redundancy check. In DECT, five logical channels are used, denoted by C, M, N, P and Q. The Q-channel is used to transmit system information to the portable units. It is inserted once every 16 frames, giving a bit rate of 400 bit/s. The M-channel is used to co-ordinate channel allocation and handover procedures and can take up to eight slots per frame. The C-channel is used for call management procedures and can occupy the same number of slots as the M-channel. The N-channel acts in a similar manner to the SAT in AMPS, and can occupy up to 15 frames in each multi-frame or all of the multi-frame if no other control channels are in operation. The P-channel is used for paging and can operate at up to 2.4 kbit/s.

The B-field can contain two types of user information. A normal telephone transmission is transported in an unprotected information channel, denoted as I_N. Other types of user information are transmitted in protected information channels, denoted as I_P. In the protected mode of operation, the I_P channel is divided into four 80-bit fields. Each field contains 64 bits of user information plus a 16-bit frame check sequence. Four parity check bits and a guard period equivalent to 60 bits then follow. As can be seen from these figures, the equivalent information rate is 32 kbit/s (320 bits/10 ms) and the signalling channel rate is equivalent to 6.4 kbit/s. Designed with data services in mind, reflecting the office environment scenario, multiple bit rates can be achieved by using more than one slot per frame. It is also possible to allocate a different number of slots in the forward and return links, resulting in different transmission rates in each direction.

DECT also provides radio frequency bit rates of 2304 and 3456 kbit/s. These rates are achieved by using the four-level modulation technique Pi/4-DQPSK and the eight-level modulation technique Pi/8-D8PSK, respectively. These higher order modulation techniques are applied only to the B-field. The use of multiple time-slots and/or carriers are used to provide higher user data rates, which have been standardised up to 552 kbit/s. Table 1.5 summarises the slot types employed by DECT [KLE-98].

Table 1.5 DECT slot types

Slot type	A-field (number of bits/slot)	Number of slots per frame			B field (number of bits/slot)
		2-level	4-level	8-level	
Short slot	64	0	0	0	
Half slot	64	80	160	24	2 × 24
Full slot	64	320	640	960	2 × 12
Double slot	64	800	1600	2400	2 × 6

The DECT network comprises only three unique elements: fixed network termination, portable radio termination, and interworking unit, as shown in Figure 1.12. The portable radio termination is equivalent to a mobile terminal's radio part, and the fixed network

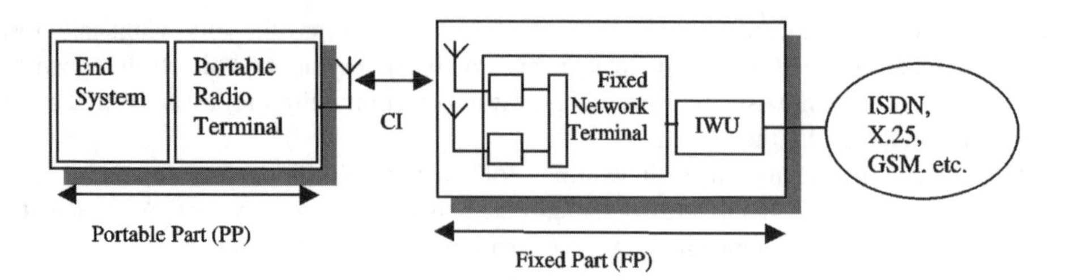

Figure 1.12 DECT network architecture.

termination is equivalent to a BS. DECT supports intracell and intercell handover. The former being the handover within a cell covered by a BS, the latter being the handover between cells covered by a different BS. Handover is seamless, implying that there is no disruption to a call during the event.

DECT is intended as an access network, allowing different networks such as ISDN, X.25 and GSM to operate through DECT terminals. This is achieved through the interworking unit, which converts the protocols associated with non-DECT networks to DECT protocols and *vice versa*.

The DECT radio interface has been selected as one of the five radio interfaces that will form the family of radio interfaces operating under the umbrella of the IMT-2000 3G system. The other radio interfaces are discussed in the following section.

1.3.4 Personal Handyphone System (PHS)

The PHS is the Japanese specified cordless telephone system. It provides connections in the home, office and outdoor environments. As with DECT, PHS employs frequency division/TDMA/TDD in its transmissions. PHS employs a common air interface standard. One TDMA frame has a duration of 5 ms, which is divided into eight time-slots. Hence, four duplex carriers are supported per carrier.

The modulation method adopted by PHS is Pi/4-DQPSK with a roll-off factor of 0.5. Bit-streams are transmitted at a gross data rate of 384 kbit/s. Speech is encoded using 32 kbit/s ADPCM.

The system operates in the 1895–1918 MHz band with a carrier spacing of 300 kHz. Terminals operate with a peak power of 80 mW and a mean RF power of 10 mW. The network comprises cell stations (CS), which provide connection to the fixed network and PS. Direct calls between PSs are supported by the system.

1.4 Third-Generation (3G) Systems

1.4.1 International Mobile Telecommunications-2000 (IMT-2000)

1.4.1.1 Background

The concept of IMT-2000 began life in 1985 as the Future Public Land Mobile Telecommunication Systems, leading to the commercially unfriendly and difficult to pronounce acronym FPLMTS. Task Group 8/1 was responsible within the ITU for defining FPLMTS.

From the outset, FPLMTS was considered to comprise of both terrestrial and satellite components, noting that the provision of such services to maritime and aeronautical users could be solely dependent on satellite delivery. One of the aims being to provide an environment in which global access to services of a quality comparable to the fixed network would be achievable. Personal terminals were intended to be small, pocket-sized and low-cost in order to facilitate mass market penetration; to a large extent this aim has already been achieved by existing manufacturers and network operators. Vehicular-mounted and fixed terminals were also envisaged. The provision of dual-mode satellite/terrestrial terminals, allowing inter-segment handover between the respective environments was anticipated. Services were also to be provided to fixed users.

A two-phase implementation of FPLMTS was considered, where in the first phase data rates of up to approximately 2 Mbit/s were to be made available in certain operating environments. The second phase would introduce new services and possibly higher data rates.

Along with the name change in the mid-1990s, the nature of IMT-2000 has evolved over the years, while still retaining many of the original, fundamental principles of the system. This evolution was necessary in order to take into account the variety and wide availability of mobile systems that are now on the market, as well as the huge investments that have been made by the wireless industry in establishing mobile networks around the world. An evolutionary approach to establishing IMT-2000, rather than adopting a revolutionary new system, was therefore established.

1.4.1.2 Radio Interfaces

Rather than provide a single radio solution, IMT-2000 comprises a family of radio interfaces encompassing existing 2G systems as well as the evolutionary 3G systems that are now being standardised. At the end of June 1998, 15 candidate proposals for the IMT-2000 radio interface were submitted to the ITU for technical evaluation by independent evaluation groups, located around the world. Of these 15 proposals, five were aimed at the satellite component. The satellite radio interface included two proposals from the European Space Agency (ESA). The terrestrial candidates included DECT for the indoor and pedestrian environments, ETSI's UMTS terrestrial radio access (UTRA), the Japanese W-CDMA, a TD-SCDMA proposal from China, plus TDMA UWC-136 and cdma2000 from the US. Following the evaluation of these proposals, the ITU TG 8/1 concluded that for the terrestrial component both CDMA and TDMA methods would be employed, as well as a combination of the two. The use of space diversity multiple access (SDMA), in combination with the aforementioned multiple access technologies, was also considered a possibility.

Both FDD and TDD are to be employed in order to improve spectral efficiency. IMT-2000 will therefore provide a choice of multiple access standards covered by a single, flexible standard, the eventual aim being to minimise the number of terrestrial radio interfaces while maximising the number of common characteristics through harmonisation of the respective standards. For example, China's TD-SCDMA proposal, which uses a single frequency band on a TDD basis for transmit and receive, underwent harmonisation with the TDD version of UTRA.

Eventually, for wideband-CDMA, a single global standard comprising of three modes of operation was defined, comprising direct-sequence (IMT-DS), multicarrier (IMT-MC) and TDD.

- Direct-sequence is based on W-CDMA UTRA FDD. This has been specified by 3GPP and operates in the IMT-2000 paired bands at a chip rate of 3.84 Mcps, spread over approximately 5 MHz. This mode of operation will be used in the UMTS macro- and micro-cellular environments. A W-CDMA frame has a period of 10 ms and is divided into 15 slots. A superframe comprises of 72 frames. Signals are QPSK modulated, and root-raised cosine filtering with a roll-off factor of 0.22 is employed at the transmitter and receiver. Detailed further information can be found in [HOL-00].
- Multicarrier, also known under the commercial name cmda2000, has been specified by 3GPP2 and can operate as an IS-95 spectrum overlay. Multicarrier operates on the downlink at a basic chip rate of 1.2288 Mcps, occupying 1.25 MHz of bandwidth. BSs provide RF channels in multiples of 1.25 MHz bandwidths, where the multiple will be either 1X or 3X, the former option being available in the first instance. Eventually multiplication factors of 6, 9 and 12 will be used to extend the achievable data rate. The 3X option, which will become available in the second phase of launch, enables the IMT-2000 higher data rates to be supported by combining three carriers within a 5-MHz band. This results in an achievable data rate of 1–2 Mbit/s. Terminals need not necessarily receive all three carriers but if this were the case, then the full data rate options made possible by the 3X mode would not be available in this instance. On the uplink, direct spread is used to minimise terminal complexity at a chip rate of 3.6864 Mcps. Muticarrier relies on two types of TCH to transport its data: the forward/reverse-fundamental channel (F-FCH, R-FCH) and the forward/reverse-supplemental channel (R-FCH, R-SCH). The fundamental channel is used to provide user data rates of up to 14.4 kbit/s. Data rates above this are provided over the supplemental channel, which offers rates from 9.6 kbit/s to 1 Mbit/s, this being dependent upon the selected radio configuration (RC) [SAR-00]. RC is used to specify the link parameters in terms of data rate, coding rate and modulation method. Turbo coding is applied to rates above 14.4 kbit/s, whereas the fundamental channel always utilises convolutional coding. In total, there are six RCs for the reverse link and nine in the forward direction. Data are transmitted over the fundamental and supplemental channels in frames of 20 ms duration. Further information on the other channels employed for broadcast and control messaging can be found in Ref. [HOL-00].
- UTRA TDD (based on a harmonised UTRA TDD/TD-SCDMA solution), as specified by 3GPP. This operates in unpaired spectral bands at a chip rate of 3.84 Mcps, spread over approximately 5 MHz. As its name suggests, the transmit and receive signals are separated in the time domain. This necessitates some form of synchronisation between BSs to co-ordinate transmissions, otherwise significant interference would occur. The physical layer characteristics are the same as the W-CDMA FDD format. A radio frame is divided into 15 slots, each of which accommodates a number of channels, separated using CDMA. Signals are QPSK modulated, and root raised cosine filtering with a roll-off factor of 0.22 is employed at the transmitter and receiver. The TDD format is expected to be used for the pico-cellular environment.

The UTRA solutions, i.e. those that will be used to provide the UMTS radio interface, will in Phase 1 deployment use the evolved GSM MAP CNs, with the possibility to allow inter-working with the IS-41 core. Network signalling to operate directly with IS-41 will be introduced in Phase 2 implementation. Similarly, multicarrier will make use of the evolved

IS-41 core in Phase 1, with the possibility to interwork with GSM MAP. Direct connection to the GSM MAP will be implemented in Phase 2.

UWC-136, as proposed by the Universal Wireless Communication (UWC) Consortium and Telecommunication Industry Association (TIA), represents a convergence between the TDMA-136, GSM and EDGE standards. UWC-136 will adopt the GPRS packet data network architecture, while enhancing the TDMA-136 radio interface to include GSM/EDGE compatibility and a high data rate indoor solution. In the ITU terminology, this radio interface is known as IMT-DS. The UWC-136 radio interface comprises three components: an enhancement of the existing TDMA-136 30-kHz channels (this is known as 136+); the addition of a 200-kHz GSM/EDGE compatible carrier for high speed mobility (known as 136HS Outdoor); and a 1.6-GHz component for low mobility, indoor applications (136HS indoor).

- The 136+ bearer employs two types of modulation: Pi/4-DQPSK and 8-PSK, with voice and data being operable using both schemes. As with TDMA-136, channels are spaced 30 kHz apart and use a frame length of 40 ms, divided into six time-slots.
- The 136HS outdoor mode supports GMSK and 8-PSK modulation at a symbol rate of 270.833 ksymbols/s. Channels are separated by 200 kHz. The frame length is 4.615 ms and is divided into eight time-slots. This is radio compatible with GSM/EDGE.
- The 136HS indoor mode supports binary-offset-quadrature amplitude modulation (B-O-QAM) and quarternary-offset-QAM (Q-O-QAM) at a symbol rate of 2.6 Msymbols/s. Carriers are spaced 1600 kHz apart. The frame length is 4.615 ms and is divided into between 16 and 64 time-slots.

The UWC-136 solution provides backward compatibility with the AMPS, IS-54, IS-136 and GSM networks.

As was noted earlier, DECT will provide the other radio interface (IMT-FT).

Due to the global nature of the satellite component, such a harmonisation was not considered appropriate in the first phase of the introduction of satellite IMT-2000 services. Closer harmonisation is anticipated with the 2G of satellite IMT-2000 system proposals. As noted earlier, initial IMT-2000 deployment will make use of the evolved 2G GSM MAP and IS-41 CNs and eventually an all IP-based network solution (Figure 1.13).

1.4.1.3 Spectrum

The initial aim was to have a common frequency allocation for all three regions of the world for 3G services. The spectrum was first allocated to FPLMTS at WARC 92, assigning the bands 1885–1980, 2010–2025 and 2110–2170 MHz on a global basis for the terrestrial component and 1980–2010 (uplink) and 2170–2200 MHz (downlink) for the satellite component. These bands were to be made available for FPLMTS from 2005. At WRC 95, the allocation for the satellite component was revised to bring forward the availability of the spectrum to 2000, apart from Region 2 (Americas and the Caribbean), which retained the 2005 start. Furthermore, the frequency allocation in Region 2 was modified to the 1990–2025/2160–2200 MHz bands. Moreover, in the US, the 1850–1990 MHz band was allocated to the PCS.

In Europe, terrestrial UMTS will occupy the 1900–1980 and 2110–2170 MHz bands and the unpaired 2010–2025 MHz band. As noted earlier, DECT occupies the 1880–1900 MHz band in Europe.

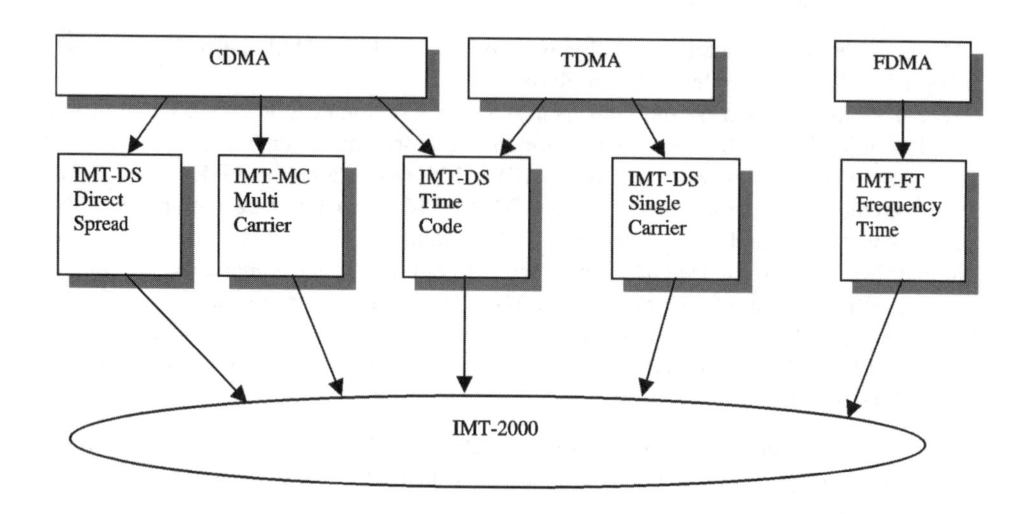

Figure 1.13 IMT-2000 family of radio systems.

The spectrum allocated to IMT-2000 services appeared on the agenda of WRC 2000. At this meeting, additional spectrum for the terrestrial component of IMT-2000 was allocated in the following bands: 806–960; 1710–1885; and 2500–2690 MHz. The allocation of the band below 1 GHz was aimed at facilitating the evolution of 1G and 2G networks, which already use this band (Figure 1.14).

The original bands for the satellite component were updated at the WRC 2000 meeting to also include an additional allocation in the 1525–1544; 1545–1559; 1610–1626.5; 1626.5–1645.5; 1646.5–1660.5; and 2483.5–2500 MHz bands. These bands were to be shared with the existing mobile-satellite services and other services, already assigned to these bands. Similarly the 2500–2520 and 2670–2690 MHz bands were identified for the satellite component of IMT-2000, with the possibility that the 2500–2520 and 2670–2690 MHz bands could be made available to the terrestrial component of IMT-2000 sometime in the future.

In addition to the terrestrial and satellite components of IMT-2000, spectral allocation was made for the first time to high altitude platform stations (HAPS), that could be used to provide terrestrial IMT-2000 services. An HAPS is defined by the ITU as a station located on an object at an altitude of 20–50 km and at a specified, nominal fixed point relative to the Earth. HAPS, when used as BSs with the terrestrial component of IMT-2000, may be used to provide terrestrial IMT-2000 services in the 1885–1980, 2010–2025 and 2110–2170 MHz bands in Regions 1 and 3; and 1885–1980 and 2110–2160 MHz bands in Region 2. This form of service delivery will be discussed further in Chapter 9.

The clamour for spectrum by competing candidate 3G operators has provided huge financial windfalls for European governments. Government agencies allocate spectrum on either a "beauty contest" basis usually with an associated fixed entry cost, where candidate operators are evaluated purely on technical merits, or on the basis of a competitive auction, which can undergo many rounds of bids (305 rounds in The Netherlands, for example). The breakdown between these two approaches across Europe has been slightly in favour of the auction route, although Scandinavian countries, with the exception of Denmark, plus Ireland, France, Spain

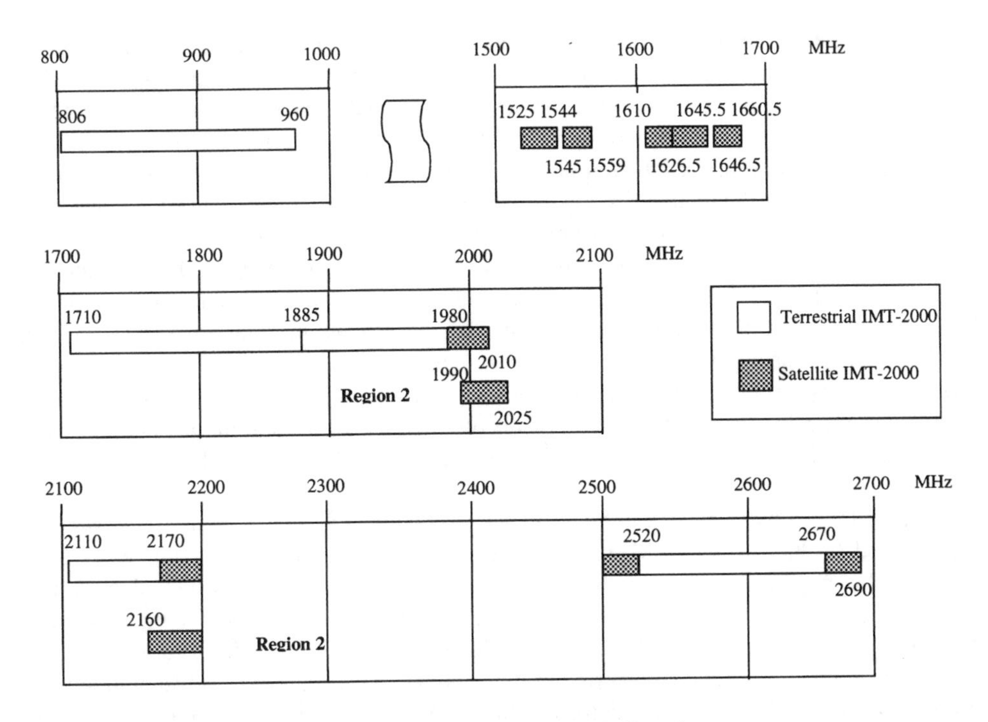

Figure 1.14 IMT-2000 spectral allocation.

and Portugal have all opted for the beauty contest approach. The auction approach has proved to be particularly lucrative. For example, in the UK, five UMTS licenses were auctioned off to the four existing GSM operators plus the new operator Hutchinson Telecom for a total of £23 billion. This figure is, of course, simply for the right to provide 3G services within a given band. Additional investment will also be associated with developing the costs of the supporting network infrastructure, for example. This is no small consideration, given the fact that the UK awarding body expects a minimum of 80% population coverage by the end of 2007.

1.4.2 Universal Mobile Telecommunications System (UMTS)

1.4.2.1 Objectives

UMTS is the European 3G system and will be one of the family of radio interfaces that will form the IMT-2000, operating in the frequency bands allocated to FPLMTS at WARC 92 and to IMT-2000 at WRC 2000. Like IMT-2000, UMTS will consist of both satellite and terrestrial components.

Research began on UMTS as early as 1988, as part of the EU's Research and Development in Advanced Communications in Europe (RACE) programme [DAS-95]. Indeed, the EU has continued to play a prominent role in support of research and development throughout the 1990s, notably through the Fourth Framework Advanced Communications Technologies and Services (ACTS) programme (1994–1998) and the Fifth Framework Information Society Technologies (IST) programme (1998–2002). Through these initiatives, a culture of international co-operation between manufacturers, operators, service providers, research establishments and universities has been successfully established. Certainly, the 4-year ACTS programme played a prominent role in the development and standardisation of UMTS technologies, particularly in the areas of radio interface and network technologies. Its successor, the IST programme, promises to continue with the success of previous research initiatives.

An important development in the establishment of UMTS emerged in 1996, when an association of telecommunications operators, manufacturers and regulators joined together to create the UMTS Forum. The Forum was established to promote and accelerate the development of UMTS through the definition of necessary policy actions and standards. To achieve this aim, the Forum defined four working groups: the Market Aspects Group; the Regulatory Aspects Group; the Spectrum Aspects Group; and the Terminal Aspects Group. As a result of these groups' activities several reports have been generated dealing with issues including the spectral requirements of UMTS, the potential market and licensing conditions [UMT-97, UMT-98a, UMT-99b, UMT-99c, UMT-00a, UMT-00b, UMT-00c, UMT-01].

Initially, UMTS was intended to be implemented by the end of the century, however, this was revised to 2002. Like GSM, originally UMTS was to be designed completely from scratch. This approach worked with GSM because at the time of its design, there was a need for a new, revolutionary system, i.e. the implementation of a continental digital service. However, by the time that UMTS is targeted for implementation, GSM type services will have been around for a significant period of time and will have achieved a substantial level of market penetration. Indeed, it is anticipated that GSM will continue to carry the vast majority of voice and low data rate traffic for the first few years after the introduction of UMTS. Indeed, UMTS will evolve from the GSM and ISDN networks. This GSM-UMTS migration path, referred to as G-UMTS, is the focus of the initial phase of UMTS implementation of 3GPP.

3GPP was formed in 1998 with the aim of providing globally applicable technical specifications for a 3G mobile system. These specifications are based on an evolved GSM CN and the UTRA [HUB-00].

In summary, UMTS must serve two functions:

1. to support all those services, facilities and applications currently provided by existing 2G systems and, as far as possible, of a QoS equivalent to that of the fixed network; and
2. to provide a new range of broadband multimedia type services. These services will be available at bit rates of between 64 and 2048 kbit/s, providing image transfer, remote database access, high definition fax, low resolution video, web access, etc. Both circuit-switched and packet-switched services will be supported by UMTS. Some services will necessitate the need for bandwidth-on-demand, that is the dynamic allocation of bandwidth to the end-user when needed.

1.4.2.2 Cell Types

UMTS is required to operate in a variety of transmission environments, each of which will have an impact on the type of services offered. For example, it is likely that in the first few years after operation the highest data rate services may only be found in the indoor environment. Typical environments include office, home, urban-vehicular and -pedestrian, rural-

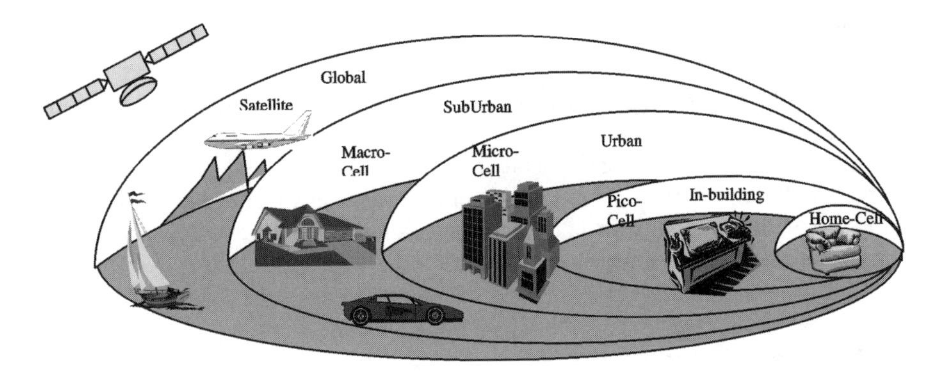

Figure 1.15 UMTS cell types.

vehicular and -pedestrian, satellite-fixed, satellite-rural, aeronautical and maritime. A hierarchical cell-structure is necessary in order to optimise the use of the available radio spectrum. With this in mind, five basic cell types have been identified to support possible UMTS transmission environments [COS-95], which are in agreement with those of IMT-2000 (Figure 1.15).

- *Satellite Cells*: Provide coverage to areas of sparse population where it is uneconomic to supply terrestrial coverage, or to areas where there is no terrestrial coverage (e.g. maritime and aeronautical users). Satellite cells will be formed by beam forming networks on-board the satellite. Such beams will have a radius in the region of 500–1000 km. These cells may appear stationary or moving with respect to a user on the ground, this being dependent on the type of satellite employed to provide coverage. Geostationary satellites will provide fixed cells, whilst the use of non-geostationary satellites result in the cell moving with respect to the user, necessitating the need to perform handover between satellite cells in order to ensure the continuation of an on-going call. Satellite cells would cover all outdoor UMTS operating environments, however, coverage in built-up, urban areas may only offer limited availability.
- *Macro Cells*: Serve areas that have a low population density. Typically rural areas, and will be of radius of up to 35 km. Distances could be larger than this if directional antennas are employed. They will also provide a back-up, "umbrella" type, coverage for the smaller UMTS type cells. They will additionally be used to provide coverage for high speed mobiles, such as trains.
- *Micro Cells*: Serve areas that have a high traffic density. Typically of cell radius less than 1 km, they will be implemented in built-up urban areas. Cells may be elongated, again by employing directional antennas, to provide optimum coverage to traffic areas, for example. Micro cells could also provide additional capacity for macro cells.
- *Pico Cells*: Serve areas largely indoor, providing a full set of high bit rate services. They will have a cell radius of less than 100 m, the maximum distance from the terminal to the indoor BS being determined by the transmit power (battery) and service bit rate requirements.
- *Home Cells*: Serve the residential sector, providing a full set of high bit rate services.

UMTS aims to provide services at the following bit rates via terrestrial access networks, depending on the operating environment:

- 144 kbit/s (with an eventual goal of achieving 384 kbit/s) in rural outdoor environments, at a maximum speed of 500 km/h;
- 384 kbit/s (with an eventual goal of achieving 512 kbit/s) with limited mobility in macro- and micro-cellular suburban outdoor environments, at a maximum speed of 120 km/h;
- 2 Mbit/s with low mobility in home and pico-cellular indoor and low-range outdoor environments at a maximum speed of 10 km/h [ETS-99].

1.4.2.3 Satellite-UMTS

The satellite component of IMT-2000 is anticipated to deliver services at rates of up to 144 kbit/s, although the maximum rate is only likely to be achievable in rural areas, to terminals of restricted mobility.

In Europe, the IMT-2000 satellite component will provide satellite-UMTS services. Work on the satellite component of UMTS began with the SAINT project, which was funded under the EC's RACE II programme. The SAINT project completed its 2-year research at the end of 1995, resulting in a number of recommendations particularly with regard to the radio interface, potential market and integration with the terrestrial network. European research activities on satellite-UMTS continued under the EC's Fourth Framework ACTS programme, through a number of collaborative projects including SINUS and INSURED [GUN-98]. The SINUS project, which used the results of the SAINT project as a technical baseline, resulted in the development of a satellite-UMTS test-bed, which allowed the performance of multimedia services to be investigated for various satellite constellations. The INSURED project used IRIDIUM satellites to experiment on inter-segment handover between satellite and terrestrial mobile networks. The European Space Agency also supported S-UMTS activities, focussing particularly on the satellite-UMTS radio interface.

Satellite-UMTS will deliver a range of services based on both circuit and packet-switched technology. Unlike satellite-PCN, which is discussed in the next chapter, voice will no longer be the dominant service, as data services become more and more prevalent. In this respect, satellite-UMTS mirrors the service evolution of its terrestrial counterpart. The implementation of satellite-UMTS will not be constrained to any particular type of orbit (that is geostationary, low earth, medium earth or elliptical). Indeed, it is likely that, given the broad range of services and applications that will need to be provided in the future, no particular orbit will be singularly suited for meeting all the QoS requirements of the user.

1.4.2.4 Architecture and Domains

The UMTS architecture is shown in Figure 1.16. It can be seen that the architecture consists of three basic components: user equipment (UE); UMTS terrestrial radio access network (UTRAN); and CN. The UE communicates with UTRAN using the Uu-interface. Communication between the UTRAN and the CN is achieved via the Iu-interface. The UTRAN architecture is shown in Figure 1.17.

The radio network subsystem (RNS) comprises a radio network controller (RNC) and one or more Node Bs. Communication between a Node B and an RNC is via an Iub-interface.

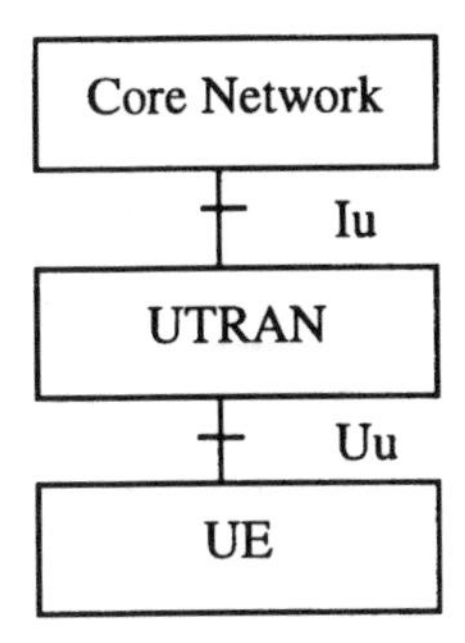

Figure 1.16 UMTS network architecture.

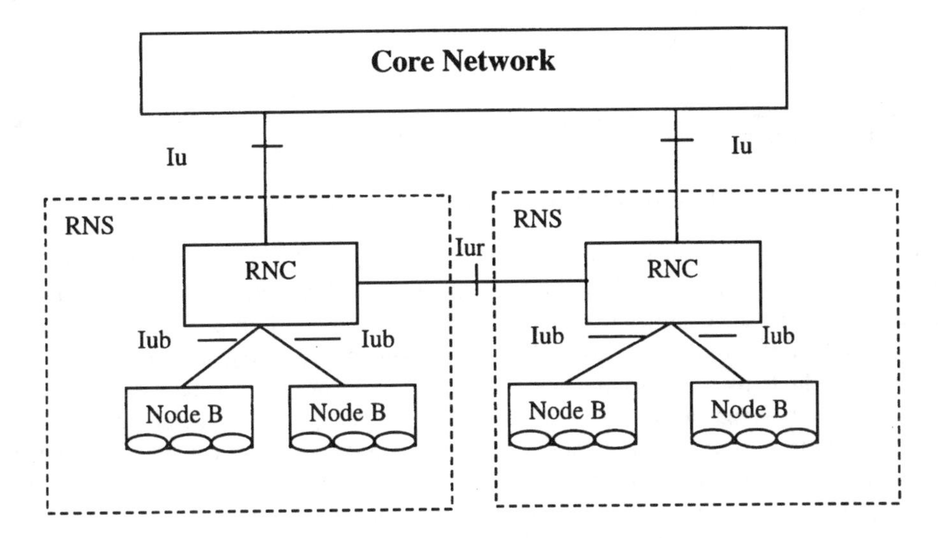

Figure 1.17 UTRAN architecture.

Communication between RNCs is achieved via an Iur-interface. In comparison to the GSM network architecture, the RNS can be considered to be the 3G equivalent of the BSC. In this respect, the RNS provides the call control functions, power control and handover switching, as well as the connection to the CN. A Node B is equivalent to a BTS, providing the radio coverage in support of 3G radio interfaces.

The division of the UMTS network into domains, or the grouping of physical entities is reported in Ref. [3GP-99] and shown in Figure 1.18.

The UE domain comprises of two domains, the ME domain and the user services identity module (USIM) domain. Communication between the USIM and the ME is via the Cu interface. The ME domain is further sub-divided into several entities, such as the mobile termination (MT), which performs radio transmission and terminal equipment (TE), which contains the UMTS applications. The USIM is associated with a particular user and is typically in the form of a smart card, containing information specific to a user. A USIM subscribes to a home network domain.

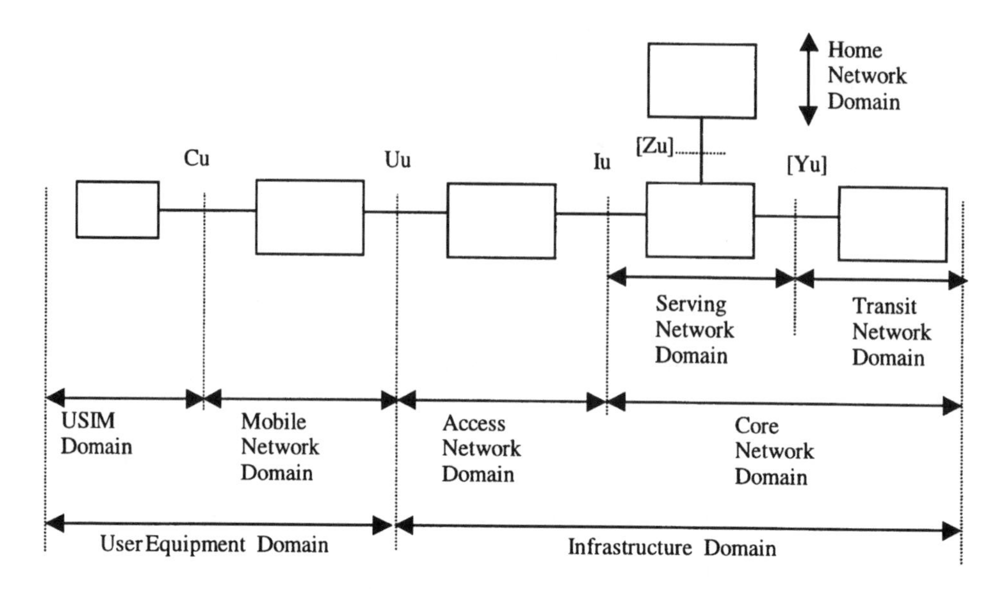

Figure 1.18 Grouping of UMTS physical entities.

The infrastructure domain is divided into an access network domain and a CN domain. The access network domain provides the user with the functionality to access the CN domain and consists of the physical entities required to manage the resources of the access network.

The CN domain is further sub-divided into the serving network domain, the home network domain and the transit network domain. The serving network domain provides the access network domain with a connection to the CN. It is responsible for routing calls and transporting user data from source to destination. The home network domain contains user specific data. The service network domain interacts with the home network domain when setting up user specific services. The transit network domain is the part of the CN between the service network domain and the remote party. This part of the infrastructure domain is only required when the calling and called parties are located on separate networks.

Standardisation of the UMTS access network is based on a GSM network evolutionary approach, also taking into account the importance of the Internet and the existence of the virtual home environment (VHE), enabling the user to roam seamlessly from one network to another.

1.5 Fourth-Generation (4G) Systems

With the completion of many aspects of the standardisation of 3G systems, attention has now focused on the definition and standardisation of 4G technologies. The influence of the Internet will have a significant bearing on 4G capabilities, as operators move towards an all IP-environment. In this scenario, the legacy of 2G technologies, in particular the CN and radio interface solutions will diminish, although perhaps not to the extent to which 1G influenced 3G. As technology continues to develop and evolve, the ability to deliver faster, broadband services at a premium QoS will be implicit requirements of next-generation technologies.

While 3G can rightly claim to have brought forward the convergence of mobile and Internet technologies, 4G will herald the convergence of fixed, broadcast and mobile technologies. The possibility of converging UMTS and digital video broadcasting (DVB) and digital audio broadcasting (DAB) is an area for further investigation. Such a solution would allow broadcast quality television to be beamed directly to the mobile user, for example. It is in such an environment that cellular, cordless, WLL and satellite technologies will combine to open up new possibilities for the telecommunications sector.

References

[3GP-99] 3rd Generation Partnership Project; Technical Specification Group Services and System Aspects *General UMTS Architecture*, 3G TS 23.101 3.0.1 (1999–2004).

[BEL-79] "Advanced Mobile Phone Service", *Bell System Technical Journal*, 58(1); 1–269.

[BET-99] C. Bettstetter, H.-J. Vögel, J. Eberspächer, "GSM Phase 2+ General Packet Radio Service GPRS: Architecture, Protocols and Air Interface", *IEEE Communications Surveys*, 2(3), Third Quarter 1999.

[COS-95] J. Cosmas, B. Evans, C. Evci, W. Herzig, H. Persson, J. Pettifor, P. Polese, R. Rheinschmitt, A. Samukic, "Overview of the Mobile Communications Programme of RACE II", *Electronics & Communication Engineering Journal*, 7(4), August 1995; 155–167.

[DAS-95] J.S. DaSilva, B.E. Fernandes, "Addressing the Needs of the European Community – The European Research Program for Advanced Mobile Systems", *IEEE Personal Communications*, 2(1), February 1995; 14–19.

[DTI-89] *Phones on the Move*, Department of Trade and Industry, UK Government, January 1989.

[ETS-99] ETSI Technical Report, *UMTS Baseline Document: Positions on UMTS agreed by SMG including SMG#26*, UMTS 30.01 V3.6.0 (1999–2002).

[FAC-99] S. Faccin, L. Hsu, R. Koodli, K. Le, R. Purnadi, "GPRS and IS-136 Integration for Flexible Network and Services Evolution", *IEEE Personal Communications*, 6(3), June 1999; 48–54.

[GAR-90] J.G. Gardiner, "Second Generation Cordless (CT-2) Telephony in the UK: Telepoint Services and the Common Air Interface", *Electronics & Communication Engineering Journal*, 2(2), April 1990; 71–78.

[GAR-96] V.K. Garg, J.E. Wilkes, *Wireless and Personal Communications Systems*, Prentice-Hall, Englewood Cliffs, NJ, 1996.

[GET-93] I.A. Getting, "The Global Positioning System", *IEEE Spectrum*, 30(12), December 1993; 36–38, 43–47.

[GOO-91] D.J. Goodman, "Second Generation Wireless Information Networks", *IEEE Transactions on Vehicular Technology*, 40(2), May 1991; 366–374.

[GUN-98] A. Guntsch, M. Ibnkahla, G. Losquadro, M. Mazzella, D. Roviras, A. Timm, "EU's R&D Activities on Third Generation Mobile Satellite Systems (S-UMTS)", *IEEE Communications Magazine*, 36(2), February 1998; 104–110.

[HOD-90] M.R.L. Hodges, "The GSM Radio Interface", *British Telecom Technical Journal*, 8(1), January 1990; 31–43.

[HOL-00] H. Holma, A Toskala, *WCDMA for UMTS: Radio Access for 3rd Generation Mobile Communications*, Wiley, Chichester, 2000.

[HUB-00] J.F. Huber, D. Weiler, H. Brand, "UMTS, The Mobile Multimedia Vision for IMT-2000: A Focus on Standardization", *IEEE Communications Magazine*, 38(9), September 2000; 129–136.

[ITU-97] *ITU-R Rec. M.1033-1*, "Technical and Operational Characteristics of Cordless Telephones and Cordless Telecommunication Systems", 1997.

[KLE-98] G. Kleindl, *DECT...An Overview*, October 1998.

[LEE-89] W.C.Y. Lee, *Mobile Cellular Telecommunications Systems*, McGraw-Hill International Editions, Singapore, 1989.

[MAC-93] R.C.V. Macario, *Cellular Radio. Principles and Design*, MacMillan Press Ltd., New York, 1993.

[MCI-99] *Mobile Communications International*, Issue 63, July/August 1999.

[MCI-99b] *The 1999 DECT Supplement*, published by MCI in co-operation with the DECT Forum, January 1999.

[PEE-00] G. Peersman, S. Cvetkovic, P. Griffiths, H. Spear, "The Global System for Mobile Communications Short Message Service", *IEEE Personal Communications*, 7(3), June 2000; 15–23.

[PER-98] C.E. Perkins, *Mobile IP: Design Principles and Practice*, Addison-Wesley, New York, 1998.

[SAL-99] A.K. Salkintzis, "A Survey of Mobile Data Networks", *IEEE Communications Surveys*, 2(3), Third Quarter, 1999.

[SAR-00] B. Sarikaya, "Packet Mode in Wireless Networks: Overview of Transition to Third Generation", *IEEE Communications Magazine*, 38(9), September 2000; 164–172.

[SOL-98] J.D. Solomon, *Mobile IP: The Internet Unplugged*, Prentice-Hall, Englewood Cliffs, NJ, 1998.

[TUT-90] W.H.W. Tuttlebee (Ed.), *Cordless Telecommunications in Europe*, Springer-Verlag, Berlin, 1990.

[UMT-97] *A Regulatory Framework for UMTS*, No. 1 Report from the UMTS Forum, June 1997.

[UMT-98a] "Mobilennium", *UMTS Forum Newsletter*, No. 3, May 1998; 6.

[UMT-98b] *The Path Towards UMTS – Technologies of the Information Society*, No. 2 Report from the UMTS Forum, 1998.

[UMT-98c] *Consideration of Licensing Conditions for UMTS Network Operations*, No. 4 Report from the UMTS Forum, October 1998.

[UMT-98d] *Minimum Spectrum Demand per Public Terrestrial UMTS Operator in the Initial Phase*, No. 5 Report from the UMTS Forum, October 1998.

[UMT-98e] *UMTS/IMT-2000 Spectrum*, No. 6 Report from the UMTS Forum, December 1998.

[UMT-99a] *The Impact of Licence Cost Levels on the UMTS Business Case*, No. 3 Report from the UMTS Forum, February 1999.

[UMT-99b] *Report on Candidate Extension Bands for UMTS/IMT-200 Terrestrial Component*, No. 7 Report from the UMTS Forum, 2nd Edition March 1999.

[UMT-99c] *The Future Mobile Market Global Trends and Developments with a Focus on Western Europe*, No. 8 Report from the UMTS Forum, March 1999.

[UMT-00a] *The UMTS Third Generation Market – Structuring the Service Revenues Opportunities*, No. 9 Report from the UMTS Forum, September 2000.

[UMT-00b] *Shaping the Mobile Multimedia Future – An Extended Vision from the UMTS Forum*, No. 10 Report from the UMTS Forum, September 2000.

[UMT-00c] *Enabling UMTS/Third Generation Services and Applications*, No. 11 Report from the UMTS Forum, October 2000.

[UMT-01] *Naming, Addressing & Identification Issues for UMTS*, No. 12 Report from the UMTS Forum, February 2001.

[YOU-79] W.R. Young, "Advanced Mobile Phone Service: Introduction, Background, and Objectives", *Bell System Technical Journal*, 58(1); 1–14.

[WAL-99] B.H. Walke, *Mobile Radio Networks Networking and Protocols*, Wiley, Chichester, 1999.

2

Mobile Satellite Systems

2.1 Introduction

2.1.1 Current Status

Satellites have been used to provide telecommunication services since the mid-1960s. Since then, key developments in satellite payload technology, transmission techniques, antennas and launch capabilities have enabled a new generation of services to be made available to the public and private sectors. For example, satellite television is currently available in both digital and analogue formats, while global positioning system (GPS) navigation reception is now being incorporated into new car systems [DIA-99].

In a similar time frame to that of terrestrial cellular development, mobile-satellite services have been around since the start of the 1980s, when they were first used to provide communications to the maritime sector. Since then, aeronautical, land-mobile and personal communication services have been introduced.

Satellites are categorised by their orbital type. Specifically, there are four types of orbits that need to be considered: geostationary orbit, highly elliptical orbit, low Earth orbit (LEO) and medium Earth orbit (MEO) (sometimes referred to as intermediate circular orbit). Up until very recently, geostationary satellites had been used as the sole basis for the provision of such services. Over the years, as a geostationary satellite's power and antenna gain characteristics have increased, combined with improvements in receiver technology, it has been possible to decrease the size of the user's terminal to something approaching the dimensions of a briefcase, a small portable computer or a hand-held device.

Significantly, it is now possible to receive via satellite a telephone call virtually anywhere in the world using a hand-held mobile receiver, of roughly a similar dimension to existing cellular mobile phones. In addition to stand-alone satellite receivers, it is also possible to buy dual-mode phones that also operate with a cellular network, such as GSM; simple, alphanumeric pagers are also on the market. These latest developments were initially made possible through the launch of satellite personal communication services (S-PCS), which make use of non-geostationary satellites. This class of satellite can be placed in LEO, at between 750 and 2000 km above the Earth; or MEO at between 10 000 and 20 000 km above the Earth. GLOBALSTAR is a system that exploits the low Earth orbit, while NEW ICO is a MEO system. Recent advances in geostationary satellite payload technology, in particular the use of multi-spot-beam coverage, has enabled this category of orbit to provide hand-held communication facilities.

2.1.2 *Network Architecture*

2.1.2.1 Overview

The basic network architecture of a mobile-satellite access network is shown Figure 2.1.

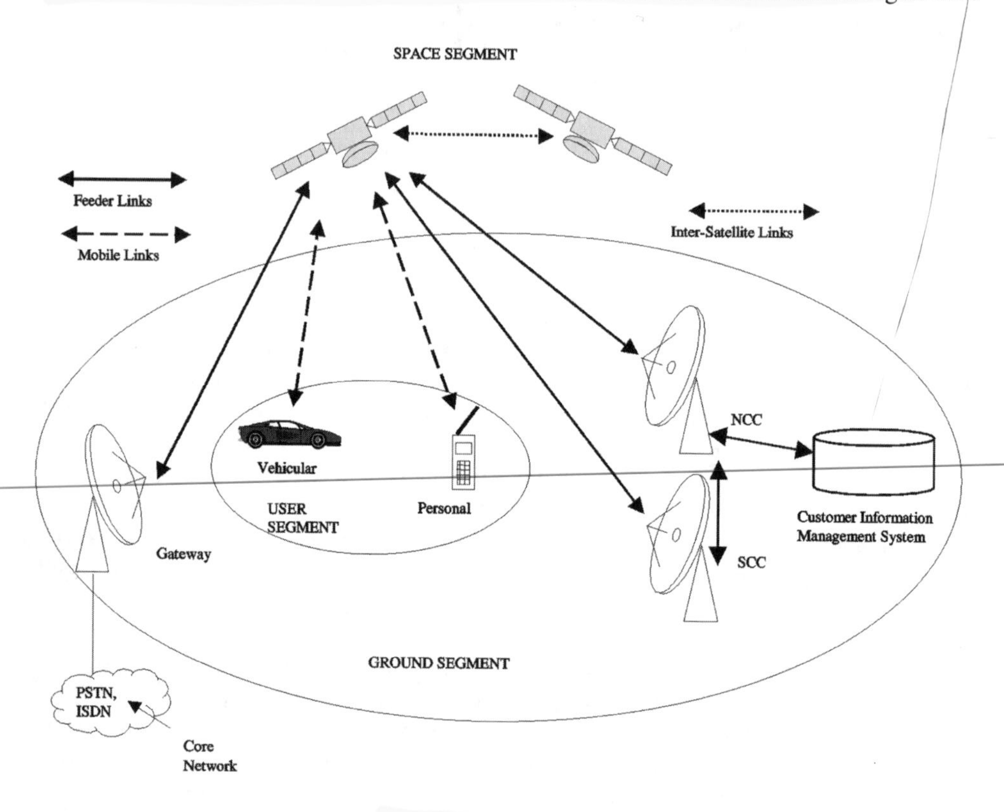

Figure 2.1 Mobile-satellite network architecture.

In its simplest form, the network architecture consists of the three entities: user segment, ground segment and space segment. The roles of each segment are discussed in the following.

2.1.2.2 The User Segment

The user segment comprises of user terminal units. A terminal's characteristics are highly related to its application and operational environment. Terminals can be categorised into two main classes.

- Mobile terminals – Mobile terminals are those that support full mobility during operation. They can be further divided into two categories: mobile personal terminals and mobile group terminals.
  – *Mobile personal terminals* often refer to hand-held and palm-top devices. Other mobile

personal terminal categories include those situated on board a mobile platform, such as a car.

– *Mobile group terminals* are designed for group usage and for installation on board a collective transport system such as a ship, cruise liner, train, bus or aircraft.

- Portable terminals – Portable terminals are typically of a dimension similar to that of a briefcase or lap-top computer. As the name implies, these terminals can be transported from one site to another, however, operation while mobile will not normally be supported.

2.1.2.3 The Ground Segment

The ground segment consists of three main network elements: gateways, sometimes called fixed Earth stations (FES), the network control centre (NCC) and the satellite control centre (SCC).

Gateways provide fixed entry points to the satellite access network by furnishing a connection to existing core networks (CN), such as the public switched telephone network (PSTN) and public land mobile network (PLMN), through local exchanges. A single gateway can be associated with a particular spot-beam or alternatively, a number of gateways can be located within a spot-beam in the case where the satellite coverage transcends national borders, for example. Similarly, a gateway could provide access to more than one spot-beam in cases where the coverage of beams overlap. Hence, gateways allow user terminals to gain access to the fixed network within their own particular coverage region.

Integrating with a mobile network, such as GSM, introduces a number of additional considerations that need to be implemented in the gateway. From a functional point of view, gateways provide the radio modem functions of a terrestrial base transceiver system (BTS), the radio resource management functions of a base station controller (BSC) and the switching functions of a mobile switching centre (MSC) [ETS-99], the latter being connected to the local mobility registers (visitor location registration (VLR)/home location registration (HLR)). Figure 2.2 shows a gateway's internal structure as defined in Ref. [ETS-99]. The RF/IF components and the traffic channel equipment (TCE) together form the gateway transceiver subsystem (GTS). The gateway subsystem (GWS) consists of both the GTS and the gateway station control (GSC).

The NCC, also known as the network management station (NMS) is connected to the

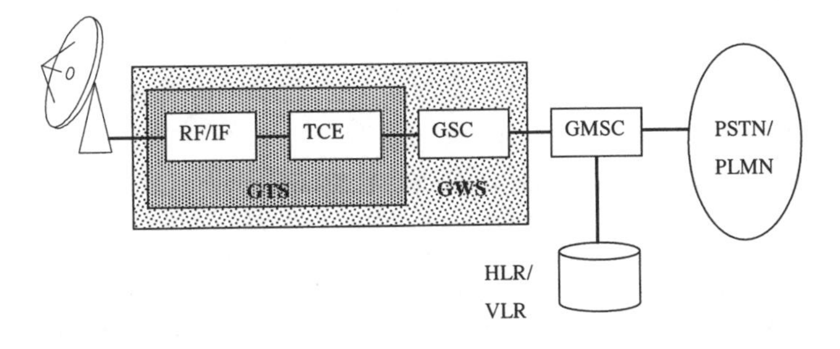

Figure 2.2 Gateway internal structure.

Customer Information Management System (CIMS) to co-ordinate access to the satellite resource and performs the logical functions associated with network management and control. The role of the these two logical functions can be summarised as follows.

- Network management functions: The network management functions include [ETS-99a]:
  - Development of call traffic profiles
  - System resource management and network synchronisation
  - Operation and maintenance (OAM) functions
  - Management of inter-station signalling links
  - Congestion control
  - Provision of support in user terminal commissioning
- Call control functions include:
  - Common channel signalling functions
  - Gateway selection for mobile origination
  - Definition of gateway configurations

The SCC monitors the performance of the satellite constellation and controls a satellite's position in the sky. Call control functions specifically associated with the satellite payload may also be performed by the SCC. The following summarises the functions associated with the SCC.

- Satellite control functions, including:
  - Generation and distribution of satellite ephemera
  - Generation and transmission of commands for satellite payload and bus
  - Reception and processing of telemetry
  - Transmission of beam pointing commands
  - Generation and transmission of commands for inclined orbit operations
  - Performance of range calibration
- Call control functions including provision of real-time switching for mobile-to-mobile calls.

The CIMS is responsible for maintaining gateway configuration data; performing system billing and accounting and processing call detail records.

The NCC, SCC and CIMS can be collectively grouped together into what is known as the *control segment*.

2.1.2.4 The Space Segment

The space segment provides the connection between the users of the network and gateways. Direct connections between users via the space segment is also achievable using the latest generation of satellites. The space segment consists of one or more constellations of satellites, each with an associated set of orbital and individual satellite parameters. Satellite constellations are usually formed by a particular orbital type; hybrid satellite constellations may also be deployed in the space segment. One such example is the planned ELLIPSO network (see later in this chapter), which will use a circular orbit to provide a band of coverage over the Equatorial region and elliptical orbits to cover Northern temperate latitudes. The choice of a space segment's orbital parameters is determined at an early stage in the design by the need to provide a specified guaranteed quality of service (QoS) for a desired region of coverage. In order to provide continuous global coverage, the satellite constellation has to be designed very carefully, taking into account technical and commercial requirements of the network.

In simple, functional terms, a communication satellite can be regarded as a distant repeater, the main function of which is to receive uplink carriers and to transmit them back to the downlink receivers. As a result of advances in technology, communication satellites nowadays contain multi-channel repeaters made up of different components, resembling that of a terrestrial microwave radio relay link repeater. The path of each channel in a multi-channel repeater is called a *transponder*, which is responsible for signal amplification, interference suppression and frequency translation. There are mainly three options for the satellite architecture (see Chapter 5):

- Transparent payload
- On-board processing (OBP) capability
- Inter-satellite links (ISL) within the constellation, or inter-constellation links with other data relay satellites to carry traffic and signalling

The space segment can be shared among different networks. For non-geostationary satellite systems, the space segment can be shared in both time and space [ETSI-93]. Time sharing refers to the sharing of satellite resources among different networks located within a common region at different times. This type of sharing is also applicable to a geostationary satellite system. Space sharing, in contrast, is the sharing of satellite resources among different networks located in different regions. Time and space sharing do not guarantee continuous coverage over a particular area. A non-continuous non-geostationary satellite system coverage provides space sharing among different networks in different areas and time sharing for networks within the same area. Time sharing requires a more efficient co-ordination procedure than that for space sharing. In addition to performing the communication tasks, the space segment can also perform resource management and routing functions and network connectivity using ISL, this being dependent upon the degree of intelligence on board the satellite (see Chapter 6).

The space segment can be designed in a number of ways, depending on the orbital type of the satellites and the payload technology available on board. The use of different satellite orbits to provide complementary services, each optimised for the particular orbital type, is certainly feasible (see Chapter 9 for possible service scenarios). Satellites can be used to connect with each other, through the use of ISL or inter-orbit links (IOL), which when combined with on-board routing facilities, can be used to form a network in the sky. The more sophisticated the space segment, the less reliant it is on the ground network, thus reducing the need for gateways.

Figure 2.3 shows a set of four possible satellite-personal communication network (S-PCN) architectures as identified by European Telecommunications Standards Institute (ETSI) [ETS-96], concentrating on the use of non-geostationary orbit (NGEO) satellites, which in some cases interwork with geostationary satellites (GEO). Here, a global coverage scenario is assumed, whereby a particular gateway is only able to communicate with a satellite providing coverage to one of the parties involved in establishing the mobile call. In this case, mobile-to-

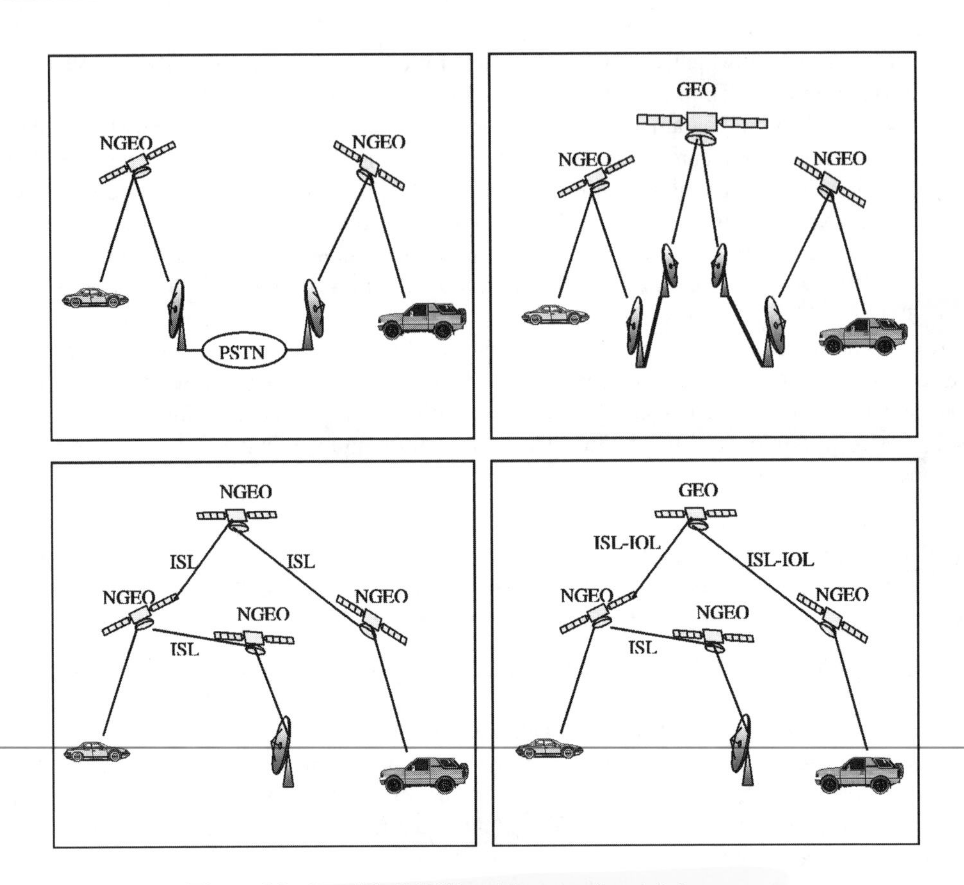

Figure 2.3 Possible S-PCN architectures for global coverage.

mobile calls are considered. Establishing a call between a fixed user and a mobile would require the mobile to form a connection with an appropriately located gateway, as discussed previously.

In option (a), transparent transponders are used in the space segment and the network relies on the ground segment to connect gateways. Satellites do not have the capability to perform ISLs and the delay in a mobile-to-mobile call is equal to at least two NGEO hop-delays plus the fixed network delay between gateways. Option (b) uses a GEO satellite to provide connectivity among Earth stations. As with option (a), no ISL technology is employed. The geostationary satellite is used to reduce the dependency on the terrestrial network, which may otherwise be needed to transport data over long distances. In this option, a mobile-to-mobile call is delayed by at least two NGEO hops plus a GEO hop. Option (c) uses ISLs to establish links with other satellites within the same orbital configuration. The ground segment may still perform some network functions but the need for gateways is reduced. A mobile-to-mobile call may have delays of varying duration depending on the route chosen through the ISL backbone. In the final option (d), a two-tier satellite network is formed through the use of a hybrid constellation. Interconnection between NGEO satellites is established through ISL, as in option (c), and inter-satellite inter-orbit links (IOL) (ISL-IOL) via a data relay geostationary satellite is employed. The mobile-to-mobile call is delayed by two half-NGEO hops plus one NGEO to GEO hop (NGEO-GEO-NGEO). In this configuration, the GEO satellite is directly accessed by an NGEO. To ensure complete global interconnection, three GEO data relay satellites would be required.

While option (a) is applicable to areas of the world where the ground network is fully developed and is able to support S-PCN operation, the other options can be adopted independently of the development of the ground network and its capability of supporting S-PCN operation. In principle, a global network can employ any one or combinations of the four options. A trade-off analysis between the complexity of the network management process, the propagation delay and the cost would have to be carried out before implementation.

2.1.3 Operational Frequency

Mobile-satellite systems now operate in a variety of frequency bands, depending on the type of services offered. Originally, the International Telecommunication Union (ITU) allocated spectrum to mobile-satellite services in the L-/S-bands. As the range of systems and services on offer have increased, the demand for bandwidth has resulted in a greater range of operating frequencies, from VHF up to Ka-band, and eventually even into the V-band. The potential for broadband multimedia communications in the Ka-band has received much attention of late. Experimental trials in the US, Japan and Europe have demonstrated the potential for operating in these bands and, no doubt, this will take on greater significance once the demand for broader bandwidth services begins to materialise. Communications between gateways and satellites, known as feeder links, are usually in the C-band or Ku-band, although recently the broader bandwidth offered by the Ka-band has been put into operation by satellite-PCN operators. Table 2.1 summarises the nomenclature used to categorise each particular frequency band.

Table 2.1 Frequency band terminology

Band	Frequency Range (MHz)
P	225–390
L	390–1550
S	1550–3900
C	3900–8500
X	8500–10900
Ku	10900–17250
Ka	17250–36000
Q	36000–46000
V	46000–56000
W	56000–100000

2.1.4 Logical Channels

2.1.4.1 Traffic Channels

Mobile-satellite networks adopt a similar channel structure to that of their terrestrial counterparts. This is particularly important when considering integration between the respective networks. As an example, the following considers the channels recommended by ETSI under its geo mobile radio (GMR) specifications.

Satellite-traffic channels (S-TCH) are used to carry either encoded speech or user data. The traffic channels in ETSI's GMR-2 specifications are organised to be as close as possible to those of GSM. They are divided into traffic channels and control channels.

Four forms of traffic channels are defined in Ref. [ETS-99b]:

- Satellite full-rate traffic channel (S-TCH/F): Gross data rate of 24 kbps
- Satellite half-rate traffic channel (S-TCH/H): Gross data rate of 12 kbps
- Satellite quarter-rate traffic channel (S-TCH/Q): Gross data rate of 6 kbps
- Satellite eighth-rate traffic channel (S-TCH/E): Gross data rate of 3 kbps

These traffic channels are further categorised into speech traffic channels and data traffic channels. Table 2.2 summarises each category.

Table 2.2 S-TCCH categories

Traffic channel type	Traffic channel listing	Abbreviations
Speech traffic channels[a]	Satellite half-rate traffic channel for enhanced speech	S-TCH/HES
	Satellite half-rate traffic channel for robust speech	S-TCH/HRS
	Satellite quarter-rate traffic channel for basic speech	S-TCH/QBS
	Satellite eighth-rate traffic channel for low-rate speech	S-TCH/ELS
Data traffic channels	Satellite full-rate traffic channel for 9.6 kbps user data	S-TCH/F9.6
	Satellite half-rate traffic channel for 4.8 kbps user data	S-TCH/H4.8
	Satellite half-rate robust traffic channel for 2.4 kbps user data	S-TCH/HR2.4
	Satellite quarter-rate traffic channel for 2.4 kbps user data	S-TCH/Q2.4

[a] The full rate traffic channel defined in GSM is not used for speech over satellite in the GMR specifications.

2.1.4.2 Control Channels

Control channels are used for carrying signalling and synchronisation data. As in GSM, the GMR specifications categorise control channels into broadcast, common and dedicated [ETS-99b]. Table 2.3 summarises the different categories as defined in the GMR specifications.

2.1.5 Orbital Types

One of the most important criteria in assessing the capability of a mobile network is its degree of geographical coverage. Terrestrial cellular coverage is unlikely to ever achieve 100% geographic coverage (as opposed to demographic coverage) and certainly not within the first few years of operation. A satellite provides uniform coverage to all areas within its antenna footprint. This does not necessarily mean that a mobile terminal will be in line-of-sight of the satellite since blockage from buildings, trees, etc. particularly in urban and built-up areas, will curtail signal strength, making communication impossible in certain instances. This is considered further in Chapter 4.

Table 2.3 Satellite control channel categories

Group	Channel name abbreviations	Descriptions
Broadcast (S-BCCCH)	Satellite synchronisation channel (S-SCH)	Carries information on the frame count and the spot-beam ID to the mobile Earth station
	Satellite broadcast control channel (S-BCCH)	Broadcasts general information on a spot-beam by spot-beam basis
	Satellite high margin synchronisation channel (S-HMSCH)	Provides frequency correction and synchronisation reference to the user channel. The channel provides high link margin necessary for user terminal reception (in particular, handset) to allow the user terminal to sufficiently correct for frequency/time alignment
	Satellite high margin broadcast control channel (S-HBCCH)	Provides the same information as the S-BCCH. It provides the high link margin necessary for user terminal reception when it is in a disadvantaged scenario
	Satellite beam broadcast channel (S-BBCH)	Broadcast a slowly changing system-wide information message. This channel is optional
Control (S-CCCH)	Satellite high penetration alerting channel (S-HPACH)	Uses additional link margin to alert users of incoming calls – forward link only
	Satellite paging channels (S-PCH and S-PCH/R[a])	Used to page user terminal – forward link only
	Satellite random access channel (S-RACH)	Used to request access to the system – return link only
	Satellite access grant channels (S-AGCH and S-AGCH/R)	Used for channel allocation – forward link only
Dedicated (S-DCCH)	Satellite standalone dedicated control channel (S-SDCCH)	As in GSM SDCCH
	Satellite slow associated control channel (S-SACCH)	As in GSM SACCH
	Satellite fast associated control channel (S-FACCH)	As in GSM FACCH

[a] R stands for robust.

Geostationary satellites provide coverage over a fixed area and can be effectively used to provide regional coverage, concentrating on a particular service area, or global coverage, by using three or more satellites distributed around the Equatorial plane. The characteristics of a highly elliptical orbit lend itself to coverage of the temperate latitudes of the Northern and Southern Hemispheres. A non-geostationary satellite, on the other hand, provides time-dependent coverage over a particular area, the duration of time being dependent on the altitude of the satellite above the Earth. Multi-satellite constellations are required for continuous global coverage. In the early years of mobile-satellite deployment, geostationary satellites were solely employed for such services. However, by the end of the 1990s, commercial non-geostationary satellite systems were in operation, providing services ranging from store-and-forward messaging to voice and facsimile.

Table 2.4a,b summarises the advantages and drawbacks of each satellite orbit from operational and implementation perspectives, respectively.

The aim of the remainder of this chapter is to present the developments in mobile-satellite technology over the last 20 years, from the initial maritime services to the planned satellite-universal mobile telecommunications system (S-UMTS) systems.

2.2 Geostationary Satellite Systems

2.2.1 General Characteristics

Geostationary satellites have been used to provide mobile communication services, in one form or another, for over 20 years. The geostationary orbit is a special case of the geosynchronous orbit, which has an orbital period of 23 h 56 min 4.1 s. This time period is termed the sidereal day and is equal to the actual time that the Earth takes to fully rotate on its axis. The geosynchronous orbit may have any particular value of inclination angle and eccentricity. These terms are used with others to define the spatial characteristics of the orbit and are discussed further in the following chapter, where orbital design considerations are described. The geostationary orbit has values of 0° for inclination and 0 for eccentricity. This defines the orbit as circular and places it on the Equatorial plane.

With the exception of the polar regions, global coverage can be achieved with a theoretical minimum of three satellites, equally distributed around the Equatorial plane, as shown in Figure 2.4. Satellites orbit the Earth at about 35 786 km above the Equator in circular orbits. Their orbital period ensures that they appear to be stationary in the sky with respect to an observer on the ground. This is particularly advantageous in fixed and broadcast communications, where line-of-sight to the satellite can be guaranteed. The satellite single-hop transmission delay is in the region of 250–280 ms and with the addition of processing and buffering, the resultant delay can exceed 300 ms. This necessitates the use of some form of echo-cancellation when used for voice communications. The ITU specifies a maximum delay of 400 ms for telephony, which can only be achieved using a single-hop via a geostationary satellite. In order to perform direct mobile-to-mobile communications, without the need to perform a double-hop (Figure 2.5), some form of OBP is required on board the satellite in order to perform the call monitoring functions that would otherwise be performed via the ground segment.

Continuous regional or continental coverage can be achieved with a single satellite, although a second satellite is usually deployed to ensure service availability in the case of a satellite failure.

Presently, geostationary satellites are used to provide regional mobile communications in

Table 2.4a Comparison of satellite orbits: operational considerations

	Geostationary orbit	Low Earth orbit	Medium Earth orbit	Highly elliptical orbit
Altitude	35 786 km	750–2000 km	10 000–20 000 km	Apogee: 40 000–50 000 km Perigee: 1000–20 000 km
Coverage	Ideally suited for continuous, regional coverage using a single satellite. Can also be used equally effectively for global coverage using a minimum of three satellites	Multi-satellite constellations of upwards of 30 satellites are required for global, continuous coverage. Single satellites can be used in store-and-forward mode for localised coverage but only appear for short periods of time	Multi-satellite constellations of between 10 and 20 satellites are required for global coverage	Three or four satellites are needed to provide continuous coverage to a region
Visibility	Mobile to satellite visibility decreases with increased latitude of the user. Poor visibility in built-up, urban regions	The use of satellite diversity, by which more than one satellite is visible at any given time, can be used to optimise the link. This can be achieved by either selecting the optimum link or combining the reception of two or more links. The higher the guaranteed minimum elevation angle to the user, the more satellites are needed in the constellation	Good to excellent global visibility, augmented by the use of satellite diversity techniques	Particularly designed to provide high guaranteed elevation angle to satellite for Northern and Southern temperate latitudes

Table 2.4b Comparison of satellite orbits: implementation considerations

	Geostationary orbit	Low Earth orbit	Medium Earth orbit	Highly elliptical orbit
Network complexity	A relatively straightforward network architecture. Satellites appear stationary in the sky, no handover between satellites during a call	The dynamic nature of the satellite orbits introduces a level of complexity into the network. Handover between satellites during a call is required. A large number of gateways may be required to support the global network if inter-satellite link technology is not employed	The motion of the satellites around the Earth will necessitate the need to perform handover between satellites, although not as frequently as that of a LEO. Moreover, the larger coverage offered by a MEO in comparison to a LEO reduces the requirements on the supporting terrestrial network infrastructure	Handover between satellites will need to occur three or four times per day. Otherwise the network complexity is similar to that of a geostationary network
Technology	Has been used to provide mobile services for over two decades. The recent introduction of multi-spot-beam payloads, of the order of 200+ beams per satellite, with the associated on-board processing and routing capabilities represents the next significant technological advancement for this type of orbit	Introduced into service at the end of the last decade with a mixed response with respect to quality of service	Yet to be used for commercial mobile-satellite services. Some significant technological advances are required but not as significant as the LEO solution. The MEO orbit is used for the GPS and GLONASS navigation systems	Has been used to provide TV services to Russia for a number of years

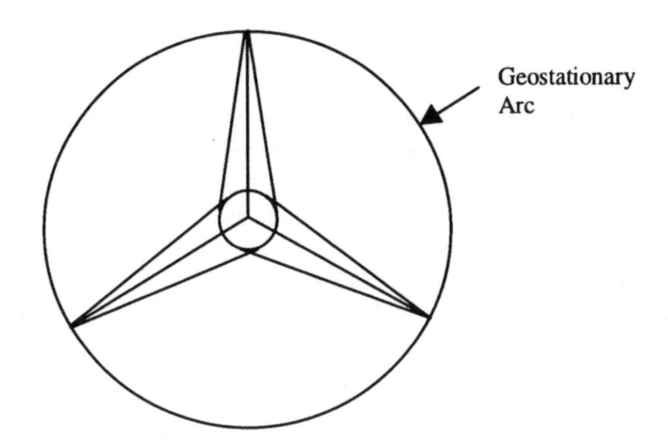

Figure 2.4 Minimum three geostationary satellite configuration.

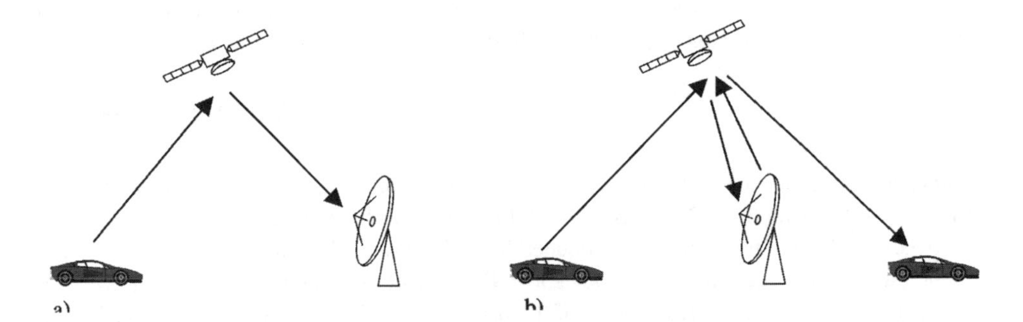

Figure 2.5 (a) Single-hop, (b) double-hop transmissions.

Europe, North America, Australia, the Middle East and South-East Asia. Inmarsat has operated a global mobile system for over 20 years using a configuration of geostationary satellites.

When geostationary satellites were first employed for mobile-satellite services, limitations in satellite effective isotropic radiated power (EIRP) and terminal characteristics placed restrictions on the type of mobile terminals that would be able to operate and the services that could be offered. Consequently, specialised, niche markets were addressed, particularly in the maritime sector. Advances in satellite payload and antenna technologies have resulted in multi-spot-beam coverage patterns being deployed in recent years. This has resulted in an increased satellite EIRP and subsequent reduction in a mobile terminal's dimensions and increased data rate, such that integrated services digital network (ISDN) compatibility can now be offered. Hand-held terminals, similar to digital cellular, are now appearing on the market. Moreover, the cost of producing mobile-satellite receivers and the air-time charges associated with their use has fallen significantly in recent years, thus increasing the marketability of these products. Geostationary satellites now have the capabilities to address a number of market sectors, including aeronautical, land-mobile and the professional business traveller, requiring office-type service availability in remote or developing areas of the world.

2.2.2 *Inmarsat*

2.2.2.1 System

Inmarsat was founded in 1979 to serve the maritime community, with the aim of providing ship management and distress and safety applications via satellite. Commercial services began in 1982, and since then, Inmarsat's range of delivered services has broadened to include land and aeronautical market sectors. Inmarsat was formed on the basis of a joint co-operative venture between governments. Each government was represented by a signatory, usually the national telecommunications provider. By the start of the 1990s, Inmarsat had 64 member countries. In April 1999, Inmarsat became a limited company with its headquarters based in London.

The Inmarsat system consists of three basic elements.

- The Inmarsat space segment, which consists of geostationary satellites deployed over the Atlantic (East (AOR-E) and West (AOR-W)), Pacific (POR) and Indian Ocean regions (IOR).
- Land Earth stations (LES), that are owned by telecommunication operators and provide the connection to the terrestrial network infrastructure. Presently there are about 40 LESs deployed throughout the world, with at least one in each of the satellite coverage areas.
- Mobile Earth stations, which provide the user with the ability to communicate via satellite.

Inmarsat started service by leasing satellite capacity from Comsat General of three MARISAT spacecraft, located at 72.5° East, 176.5° East and 106.5° West, respectively.

Between 1990 and 1992, Inmarsat launched four of its own INMARSAT-2 satellites. They provided a capacity equivalent to about 250 INMARSAT-A circuits (See section 2.2.2.2), which is roughly 3–4 times the capacity of the original leased satellites. The satellites have a payload comprising two transponders supporting space to mobile links in the L-/S-bands (1.6 GHz for the uplink, 1.5 GHz for the downlink) and Space to Earth links in the C-/S-bands (6.4 GHz for the uplink, 3.6 GHz for the downlink). The satellites had a launch mass of 1300 kg, which reduced to 800 kg in orbit. The satellites transmit global beams with an EIRP of 39 dBW at the L-band.

The next phase in the development of the space segment was with the launch of the INMARSAT-3 satellites. Significantly, these satellites employ spot-beam technology to increase EIRP and frequency re-use capabilities. Each INMARSAT-3 satellite has a global beam plus five spot-beams. The satellites offer a spot-beam EIRP of up to 48 dBW, eight times the power of the INMARSAT-2 global beams. Bandwidth and power can be dynamically allocated between beams in order to optimise coverage according to demand. This has a significant bearing on the type of services that Inmarsat can now offer and also on the equipment that can be used to access the network. In addition to the communication payload, INMARSAT-3 satellites also carry a navigation payload to enhance the GPS and GLONASS satellite navigation systems (see Chapter 9).

Presently Inmarsat employs four operational INMARSAT-3 satellites and six spares, comprising three INMARSAT-3 and three INMARSAT-2 satellites. Three other Inmarsat satellites are being offered for lease capacity. The satellite configuration is listed in Table 2.5.

Table 2.5 Inmarsat satellite configuration

Region	Operational	Spare
AOR-W	INMARSAT-3 F4 (54° W)	INMARSAT-2 F2 (98° W)
		INMARSAT-3 F2 (15.5° W)
AOR-E	INMARSAT-3 F2 (15.5° W)	INMARSAT-3 F5 (25° E)
		INMARSAT-3 F4 (54° W)
IOR	INMARSAT-3 F1 (64° E)	INMARSAT-2 F3 (65° E)
POR	INMARSAT-3 F3 (178° E)	INMARSAT-2 F1 (179° E)

The world-wide coverage provided by the Inmarsat system is illustrated in Figure 2.6.

2.2.2.2 Inmarsat Services

Maritime and Land-Mobile Inmarsat offer a wide range of services through a family of Inmarsat systems.

In 1982, the INMARSAT-A system was the first to be introduced into service under the commercial name STANDARD-A. Transportable, portable and maritime modes of operation are available. Transportable terminals are about the size of one or two suitcases, depending on the manufacturer, and weigh between 20 and 50 kg. The terminal operates with a parabolic antenna of about 1 m in diameter, with a 36-dBW EIRP and a G/T of -4 dBK^{-1}. Usually, users can select the route to make the call, in terms of the available satellite and LES. INMARSAT-A voice services occupy the band 300–3000 Hz using single channel per carrier frequency modulation (SCPC/FM). Voice activation and demand assignment techniques are used to increase the efficiency of the satellite resource. BPSK modulation is used for data at rates of up to 19.2 kbit/s and facsimile services at a rate of up to 14.4 bit/s. There is also the possibility to increase data transmission to 64 kbit/s, in which case quadrature phase shift keying (QPSK) modulation is employed. A terminal requests a channel to establish a call by transmitting a 4.8-kbit/s BPSK modulated signal using ALOHA (see Chapter 5). INMARSAT-A operates in the 1636.5–1645 MHz transmit band and 1535–1543.5 MHz receive band. Voice channels operate with a 50-kHz spacing, while data channels are separated by 25 kHz. INMARSAT-A terminals are no longer produced.

INMARSAT-B was introduced into service in 1993, essentially to provide a digital version of the INMARSAT-A voice service. INMARSAT-B terminals are available in transportable, portable and maritime versions, just like INMARSAT-A. The system incorporates voice activation and active power control to minimise satellite EIRP requirements. Terminals operate at 33, 29 or 25 dBW, with a G/T of -4 dBK^{-1}. Voice is generated at 16 kbit/s using adaptive predictive coding (APC), which is then 3/4-rate convolutional coded, increasing the channel rate to 24 kbit/s. The signal is modulated using offset-QPSK. Data are transmitted at rates of between 2.4 and 9.6 kbit/s, while facsimile is transmitted at up to 9.6 kbit/s, using offset-QPSK modulation. INMARSAT-B high-speed data (HSD) services offer 64 kbit/s digital communications to maritime and land users, and provide the capability to connect to the ISDN via an appropriately connected LES. A terminal requests a channel to establish a call by transmitting a 24-kbit/s offset-QPSK modulated signal using ALOHA.

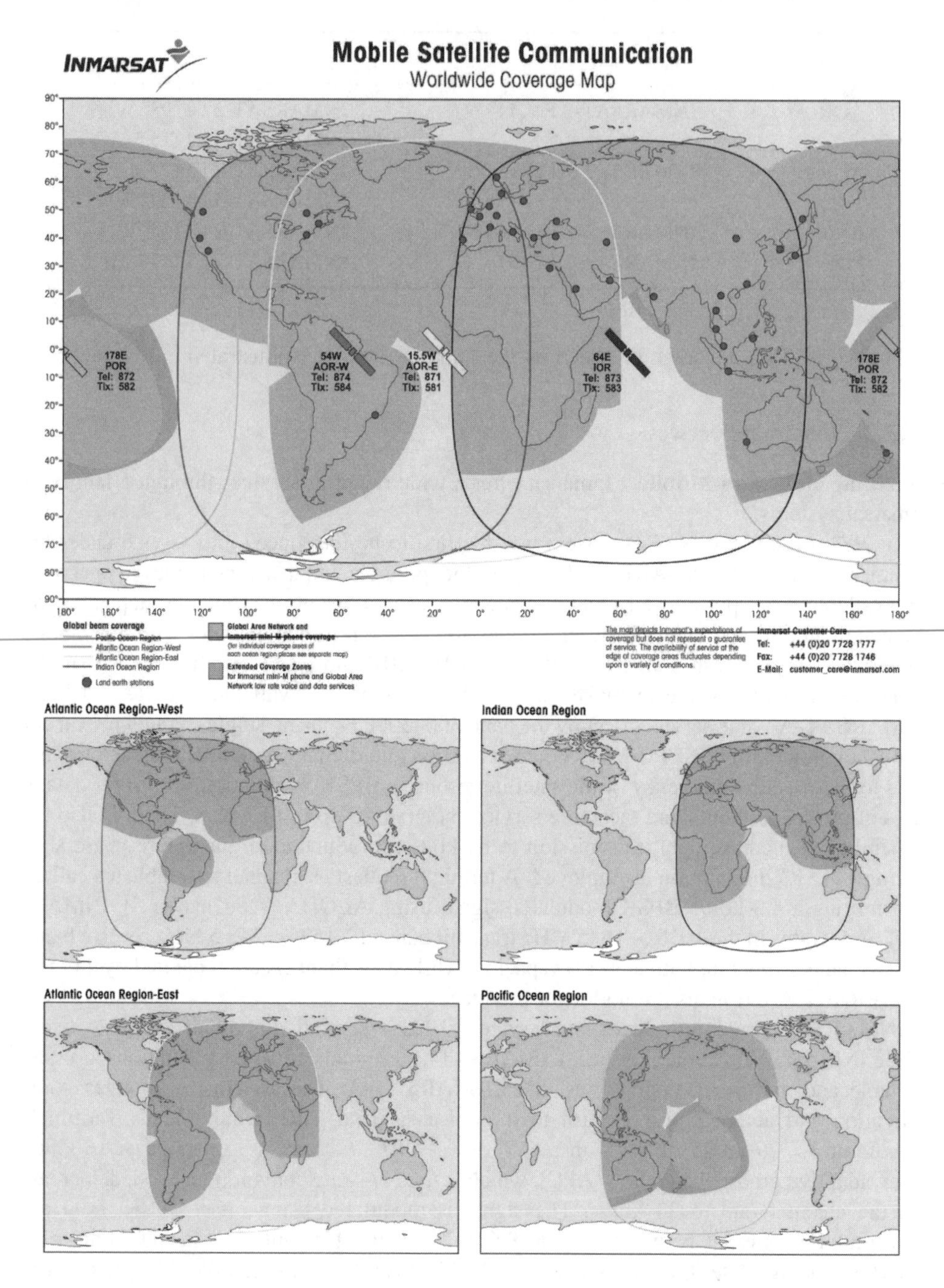

Figure 2.6 Inmarsat service coverage (courtesy of Inmarsat).

Channels are assigned using a BPSK TDM channel. INMARSAT-B operates in the 1626.5–1646.5 MHz transmit band and the 1525–1545 MHz receive band.

INMARSAT-C terminals provide low data rate services at an information rate of 600 bit/s. Half-rate convolutional coding, of constraint length 7, results in a transmission rate of 1200 bit/s. Signals are transmitted using BPSK modulation within a 2.5-kHz bandwidth. Terminals are small, lightweight devices that typically operate with an omni-directional antenna. Terminals operate with a G/T of -23 dBK^{-1} and an EIRP in the range 11–16 dBW. The return request channel employs ALOHA BPSK modulated signals at 600 bit/s. Channels are assigned using a TDM BPSK modulated signal. The system provides two-way store and forward messaging and data services, data reporting, position reporting and enhanced group call (EGC) broadcast services. The EGC allows two types of broadcast to be transmitted: SafetyNET, which provides the transmission of maritime safety information; and FleetNET, which allows commercial information to be sent to a specified group of users. Terminals can be attached to vehicular or maritime vessels and briefcase type terminals are also available. INMARSAT-C operates in the 1626.5–1645.5 MHz (transmit) and 1530.0–1545.0 MHz (receive) bands, using increments of 5 kHz.

INMARSAT-M was introduced into commercial service in December 1992 with the claim of being the first personal, portable mobile-satellite phone [INM-93]. The system provides 4.8 kbit/s telephony using improved multi-band excitation coding (IMBE), which after 3/4-rate convolutional coding increases to a transmission rate of 8 kbit/s. Additionally, 2.4 kbit/s facsimile and data services (1.2–2.4 kbit/s) are also provided. INMARSAT-M operates in maritime and land mobile modes. Maritime terminals operate with an EIRP of either 27 or 21 dBW and a G/T of -10 dBK^{-1}. Land mobile terminals operate with an EIRP of either 25 or 19 dBW and a G/T of -12 dBK^{-1}. The return request channel employs slotted-ALOHA BPSK modulated signals at 3 kbit/s. Channels are assigned using a TDM BPSK modulated signal. INMARSAT-M maritime operates in the 1626.5–1646.5 MHz (transmit) and 1525.0–1545.0 MHz (receive) bands, with a channel spacing of 10 kHz. The land mobile version operates in the bands 1626.5–1660.5 MHz (transmit) and 1525.0–1559.0 MHz (receive) bands, again with a channel spacing of 10 kHz.

The INMARSAT MINI-M terminal exploits the spot-beam power of the INMARSAT-3 satellites to provide M-type services but using smaller terminals than those of the INMARSAT-M. Terminals are small, compact devices, about the size of a lap-top computer, weighing less than 5 kg. Vehicular and maritime versions are also available, as are rural-phone versions, which require an 80-cm dish.

Other systems offered by Inmarsat include the INMARSAT-D+, which is used to store and display messages of up to 128 alphanumeric characters. Typical applications include personal messaging, supervisory control and data acquisition (SCADA) and point-to-multipoint broadcasting. The INMARSAT-E system is used to provide global maritime distress alerting services via Inmarsat satellites.

Aeronautical Inmarsat provides a range of aeronautical services with approximately 2,000 aircraft now fitted with aero terminals. As with the maritime and mobile sectors, aero terminals come in a range of terminal types, catered for particular market needs. The MINI-M AERO, based on the land mobile equivalent, is aimed at small aircraft users and offers a single channel for telephone calls, fax and data transmissions.

The AERO-C is the aeronautical equivalent of the INMARSAT-C terminal, and enables low data rate store-and-forward text or data messages to be sent or received by an aircraft. The AERO-H offers multi-channel voice, fax and data communications at up to 10.5 kbit/s, anywhere within the global beam. AERO-H operates in the bands 1530–1559 MHz (transmit) and 1626.5–1660.5 MHz (receive). The AERO-H+ is an evolution of the AERO-H, and operates primarily in the spot-beam coverage areas provided by the INMARSAT-3 satellites and can switch to the global beam when outside of spot-beam coverage.

The AERO-I system also exploits the spot-beam, capabilities of the INMARSAT-3 satellites, and is aimed at the short and medium haul aircraft markets. AERO-I provides up to seven channels per aircraft Earth station. Packet-data services are also available via the global beam. The AERO-L provides low speed data communications at 600 kbit/s, and is mainly used for air traffic control, operational and administration procedures.

Global Access Network (GAN) Inmarsat launched the GAN at the end of 1999. The aim of GAN is to provide mobile-ISDN and mobile-Internet protocol (IP) services. The services supported by GAN are 64 kbit/s HSD services, 4.8 kbit/s voice using the advanced multi-band excitation coding algorithm and analogue voice-band modem services. Terminals nominally operate at 25 dBW, with a G/T of 7 dBK^{-1}. The channel rates are at 5.6 and 65.2 kbit/s with channel spacing of 5 and 40 kHz, respectively. Terminals operate in the 1626.5–1660.5 MHz (transmit) and 1525–1559 MHz (receive) bands.

Terminals are lap-top like, weighing about 4 kg, and connection to the satellite is via two- or three-panel antennas. Manufacturers tend to provide the option of adding a DECT base station (BS) to the modem unit, operating in the 1880–1900 MHz band. This allows the terminals to operate with a DECT telephone, providing the benefits of cordless operation, as shown in Figure 2.7.

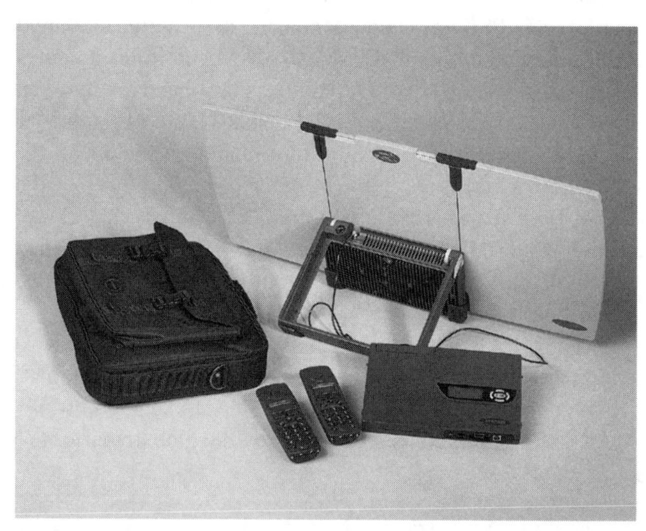

Figure 2.7 An example GAN terminal (courtesy of Nera Telecommunications).

Project Horizons In December 1999, Inmarsat's Board of Directors approved the next phase of development of the space segment with the decision to proceed with a request for tender for the $1.4 billion INMARSAT-4 satellites. The fourth-generation of satellites will comprise two in-orbit satellites plus one ground spare. The satellites will be located at 54° West and 64° East and each satellite will weigh 3 tonnes, three times the weight of INMARSAT-3 satellites. The satellites will be designed to support services of data rates in the range 144–432 kbit/s and will provide complementary services to those of the terrestrial UMTS/IMT-2000 network. This will be known as the broadband GAN (BGAN) [FRA-00]. Both circuit and packet-switched services will be supported on the network. The user-terminals are likely to be similar to the lap-top terminals that are used for the GAN services. Aeronautical, maritime and remote FES will also be supported. The satellite payload will comprise 200 narrow spot-beams with an EIRP of 67 dBW, covering land and the main aeronautical and maritime routes; 19 overlay wide spot-beams, providing 56 dBW, and a global beam of 39 dBW. The satellites will operate in the 1.5/1.6 GHz bands and are expected to be in service by the end of 2004, two years after the introduction of terrestrial-UMTS services.

2.2.3 EUTELSAT

2.2.3.1 EUTELTRACS

EUTELTRACS is a fleet-management system that is used to send/receive text messages to/from vehicles via a geostationary satellite. Introduced as Europe's first commercial land mobile-satellite service, the system also provides a position reporting service, allowing the tracking of vehicles to be performed. EUTELTRACS operates in the Ku-band and is based on a centralised network architecture, organised around a single hub station operated by the European Telecommunication Satellite Organisation (EUTELSAT) [VAN-97]. The network consists of five elements: the hub Earth station, space segment, the service provider network management centre (SNMC), the dispatch terminal and the mobile communications terminal (MCT), which is mounted on the vehicle. The network architecture is shown in Figure 2.8.

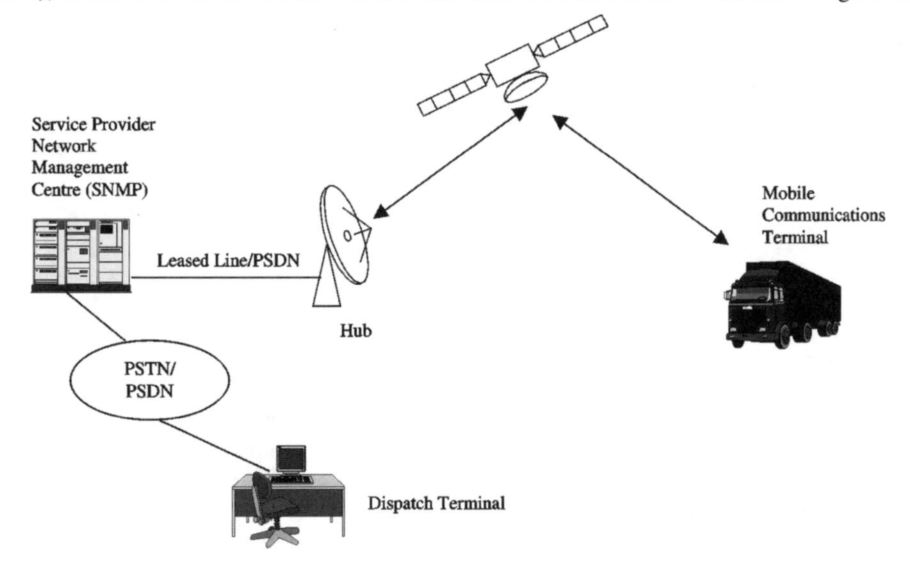

Figure 2.8 EUTELTRACS network architecture.

The hub station controls satellite access and provides network management and the service billing capabilities. Customers send and receive messages from a dispatch terminal, which is connected to the hub station via an SNMC. The dispatch terminal is a PC, which runs proprietary software associated with the operation of the system. The SNMC is connected directly to the hub via a leased line or public switched digital network (PSDN) and is used to maintain a record of transactions with the customer. Connection between the SNMC and the dispatch terminal is either via a PSTN, PSDN or leased-line connection. Vehicles communicate with the hub station using an MCT. EUTELTRACS is a closed group service, that is to say, messages are only provided between the end-user and the associated fleet of vehicles.

EUTELTRACS employs a time division multiplex (TDM) scheme in the forward link, that is from the hub station to the MCT, the signal being spread over a 2-MHz bandwidth in order to avoid causing interference to other satellites within proximity and also to counteract multipath fading. The system employs two data rates known as 1X and 3X, the choice of which is dependent upon the transmission environment. The 1X data rate is at 4.96 kbit/s, which is subject to half-rate Golay encoding and modulated using BPSK. The 3X data rate, at a basic rate of 14.88 kbit/s is 3/4-rate encoded, prior to QPSK modulation. Hence, the basic symbol rate in both cases is 9920 symbols/s.

On the return link, a 1/3-rate convolutional encoder of constraint length nine, in conjunction with Viterbi decoding, is employed. After encoding, data are block interleaved to protect the information from the burst-error channel introduced by the mobile environment. The output of the interleaver is then fed into a 32-ary frequency shift keying (FSK) scheme, which is used to map five symbols onto a FSK signal. This FSK signal is then combined with direct spreading sequence (DSS) minimum shift keying (MSK) modulation at a rate of 1 MHz. The signal is then spread over the 36 MHz bandwidth offered by the satellite transponder by applying a frequency hopping sequence using a pseudo-random sequence known to both the hub station and the mobile. Further information on this technique can be found in Chapter 5. As with the forward link, two data rates are available, either 1X at 55 bit/s, corresponding to a single 32-ary FSK symbol, or 3X at 165 bit/s corresponding to three 32-ary FSK symbols.

EUTELTRACS employs a store-and-forward payload to ensure that data is received correctly. On the forward link, the mobile transmits an acknowledgement (ACK) message upon correct reception of a packet. If no ACK is received, the hub station continues to transmit the same packet intermittently in order to take into account the varying conditions of the propagation environment. This takes place for up to 12 times in 1 h, before terminating the message. Similarly, if an ACK is not received on the return link, the mobile retransmits the same packet up to 50 times before terminating the message [COL-95].

The system is essentially based on Qualcomm's OMNITRACS, which has been operating in the US since 1989 [JAL-89].

2.2.3.2 EMSAT

EUTELSAT also offers a mobile voice and data service under the commercial name EMSAT. The services are specifically: 4.8 kbit/s voice, Group 3 Fax at 2.4 kbit/s, data transfer at 2.4 kbit/s, messaging in 44 bits per packet and positioning using an integrated GPS card. These services are available over Europe and the Mediterranean basin and are provided in the L-band using the European mobile-satellite (EMS) payload on-board ITALSAT F-2.

2.2.4 *Asia Cellular Satellite, THURAYA and Other Systems*

There are now a number of systems deployed that exploit geostationary satellites to provide mobile services at the L-band. These regionally deployed systems focus on providing mobile services to regions of the world that are either sparsely populated or under-served by terrestrial mobile communications. Such systems are presently deployed over Australia (OPTUS) [NEW-90], Japan (N-Star) [FUR-96], North America (MSAT) [JOH-93], South-East Asia (ACeS), India, North Africa and the Middle East (THURAYA).

The Asia cellular satellite (ACeS) system provides services to a region bounded by Japan in the East, Pakistan in the West, North China in the North and Indonesia in the South [NGU-97]. The area is covered by a total of 140 spot-beams in the L-band plus a single coverage beam in the C-band. The region encompasses approximately 3 billion people and is an area in which terrestrial communication facilities are not so prevalent. Indeed, Asia represents one of the major market opportunities for satellite communications [HU-96] and the regional concentration in demand makes it perfectly suited for geostationary satellites.

The first ACeS satellite, GARUDA-1, was launched into orbit on February 12, 2000 using a Proton launch vehicle. The satellite has a designed operational life of 12 years and can support at least 11 000 simultaneous voice channels with a 10-dB link margin. To provide coverage at the L-band, the satellite employs separate transmit and receive antennas, both of 12 m diameter. The satellite payload provides on-board switching and routing of calls, allowing single-hop mobile-to-mobile calls to be performed.

The network comprises an NCC, satellite control facility (SCF), user terminals and regional gateways. The NCC and SCF are co-located at Batam Island, Indonesia, sharing a 15.5-m tracking antenna. The NCC performs the control and management functions of the network, such as call set-up and clear down and resource management. The NCC also incorporates the ACeS CIMS, the primary role of which is to handle the billing system. Initially, regional gateways will be located in Indonesia, Philippines, Taiwan and Thailand. The gateways manage a subset of the system resources provided by the NCC. As with the GSM network, ACeS subscribers are registered with a home gateway and can roam to other gateways by registering as a visitor. Gateways perform the local billing and access to the core terrestrial network. The NCC and the gateways operate in the C-/S-bands, specifically 6425–6725 MHz (Earth to Space) and 3400–3700 MHz (Space to Earth).

User terminals, which provide fax, voice and data services, operate in the L-/S-bands, specifically the bands 1626.5–1660.5 MHz (Earth to Space) and 1525.0–1559.0 MHz (Space to Earth). Terminals can be broadly categorised as mobile, hand-held or fixed. Mobile and hand-held terminals allow dual-mode operation with the GSM network. Since the ACeS radio interface is based on that of GSM, it should allow the future upgrading of the network, along similar lines to that of the terrestrial mobile network, to provide packet-oriented services based on general packet radio service (GPRS) or enhanced data rates for GSM evolution (EDGE) technologies.

Future launches of the GARUDA satellite are planned to enable coverage to extend towards mid-Asia and Europe.

The other major geostationary satellite system to address the mobile needs of the Middle East and Asian markets is the THURAYA system, which commenced service in 2001 with a scheduled life-span of 12 years. The THURAYA-1 satellite, which was launched on October 20, 2000 by Sea Launch, is located at 44° East. The satellite provides coverage over a region bounded by −20° West to 100° East and 60° North to −2° South. A satellite can support up to 13 750 simultaneous calls with a call blocking probability of 2%. The THURAYA-2 satellite acts as an on-ground spare.

Like ACeS, the THURAYA system has been designed to be compatible with the GSM network. The dual-mode hand-held terminals are of a similar dimension to current GSM phones, as shown in Figure 2.9. Vehicular dual-mode phones are also supported, as are satellite-mode only fixed terminals and payphones. The mobile link operates in the L-/S-bands, specifically the 1626.5–1660.5 MHz (Earth to Space) and 1525.0–1559.0 MHz (Space to Earth) bands. The feeder links operate in the C-/S-bands, 6425.0–6725.0 MHz (Earth to Space) and 3400–3625.0 MHz (Space to Earth) bands. Coverage at the L-band is provided by a 12.25 × 16 m mesh transmit-receive reflector and a 128-element dipole L-band feed array. The digital beam forming capabilities on-board the satellite provide between 250 and 300 spot-beams over the coverage area. Coverage at the C-band is achieved using a 1.27-m round dual-polarised shaped reflector. The multiple access scheme employed is FDMA/TDMA and QPSK is used to modulate the signals. The network supports voice, and fax and data at the rates 2.4, 4.8 and 9.6 kbit/s.

The THURAYA system aims to attract somewhere in the region of 1 750 000 subscribers to the network.

Figure 2.9 THURAYA mobile terminal (courtesy of Boeing Satellite Systems Inc.).

2.3 Little LEO Satellites

2.3.1 Regulatory Background

So called "little LEO" systems aim to provide non-voice, low bit rate mobile data and messaging services, including e-mail, remote monitoring and utility meter reading, on a global basis using a constellation of non-geostationary satellites. As their name suggests, these satellites operate in the low Earth orbit, in the region of 700–2000 km above the Earth's surface. Services operate in near real-time or in store-and-forward mode, depending on the degree of network coverage available. The degree of network coverage is determined both by the satellite constellation and the availability of the terrestrial network infrastructure that supports the satellite constellation. For example, a satellite may only be able to download its data when passing over a certain coverage area, corresponding to the location of a satellite gateway connected to the terrestrial network infrastructure.

The term "little LEO" arises in comparison with the non-geostationary satellites that are used to provide satellite-PCN services, which tend to be larger and more sophisticated in design. It should be noted, however, that unlike little LEOs, satellite-PCN services are not particularly constrained to the LEO orbit.

In order to be able to provide a service within a particular frequency band in the US, an operating license has to be awarded by the US Federal Communications Commission (FCC) office of Engineering and Technology. Without this license, services could not be provided over the US, seriously limiting the potential market opportunities of any prospective operator. In 1990, the FCC received filings for a little LEO operating license from three potential operators, namely Orbital Services Corporation (ORBCOMM), STARSYS global positioning system (STARSYS) and Volunteers in Technical Assistance (VITA).

Operating frequencies for this form of communication were first made available at the World Administrative Radio Conference in 1992, [RUS-92] as shown in Table 2.6. (Note: The ITU allocates frequencies to services in shared frequency bands on either a primary or secondary status. A primary status service is guaranteed interference protection from secondary status services. When more than one service is allocated primary status within a common

Table 2.6 Little LEO frequency allocations below 500 MHz

Frequency (MHz)	Status	Direction
137–137.025	Primary	Space to Earth
137.025–137.175	Secondary to meteorological-satellite	Space to Earth
137.175–137.825	Primary	Space to Earth
137.825–138	Secondary to meteorological-satellite	Space to Earth
148–148.9	Primary	Earth to space
149.9–150.05	Primary land mobile-satellite only	Uplink
312–315	Secondary	Uplink
387–390	Secondary	Space to Earth
399–400.05	Primary	Earth to space
400.15–401	Primary	Space to Earth
406–406.1	Primary	Earth to space

band, the service operators must co-ordinate their transmissions so as not to cause mutual interference. Secondary status services are not guaranteed immunity from interference from primary status services sharing the same band. In allocating frequencies, the ITU divides the world into three regions, where: Region 1 corresponds to the Americas; Region 2 corresponds to Europe, Africa, and former the Soviet Union; Region 3 corresponds to Australasia.)

In October 1994, the FCC awarded a license to ORBCOMM to operate 36 satellites. This was followed in 1995 by the decision to allow VITA to operate a single satellite on a non-commercial basis and STARSYS to operate a 24-satellite constellation, which was subsequently returned in 1997.

Following a second round of applications, additional licences were awarded in 1998 to: LEO ONE™ for 48 satellites; E-SAT for six satellites; Final Analysis for a 26-satellite system; ORBCOMM to increase the number of its satellites from 36 to 48, and VITA to increase its number of satellites from one to two.

ORBCOMM™ became the first little LEO system to enter commercial service on November 30, 1998.

2.3.2 *ORBCOMM™*

ORBCOMM is a system developed by Orbital Sciences Corporation in collaboration with Teleglobe Industries and Technology Resources Industries.

ORBCOMM presently operates with a constellation of 36 satellites, with FCC approval to operate 48 satellites in the future. The constellation is arranged as follows:

- Three satellite planes, inclined at 45°, with eight satellites per plane, at an altitude of 825 km.
- Two planes at inclination angles of 70° and 108°, comprising of two satellites per plane, placed 180° apart, at an altitude of 780 km.
- And eight satellites in the Equatorial plane.

Each ORBCOMM MICROSTAR™ satellite weighs in the region of 43 kg, costing approximately $1.2 million to manufacture.

The ORBCOMM system provides user data rates of up to 2.4 kbit/s on the uplink and 4.8 kbit/s on the downlink, with a provision to increase this to 9.6 kbit/s. Symmetric differential phase shift keying (SDPSK) with raised cosine filtering is employed on both up and down links. The subscriber/satellite link operates in the bands 148–149.9 MHz on the uplink and 137–138 MHz on the downlink. The satellite also transmits a beacon signal at 400.1 MHz.

Apart from satellites, the ORBCOMM network comprises the following: subscriber communicators (SC); a NCC; and gateways. The NCC, located in the US, provides global management of the network, including monitoring the performance of the ORBCOMM satellites. The network architecture is shown in Figure 2.10.

Subscriber communicators are small, compact devices, weighing about 1 kg with a transmit power of 5 W and a nominal EIRP of 7 dBW. Manufacturers are free to develop their own designs but they must be type approved by ORBCOMM.

Each gateway consists of two distinct elements: a gateway control centre (GCC); and one or more gateway Earth stations (GES). GESs are connected to a GCC via an ISDN line. A GES provides the connection to a satellite. Operating with a gain in the region of 15 dB and a transmission power of 200 W, each GES provides 57.6 kbit/s TDMA links to the satellite at

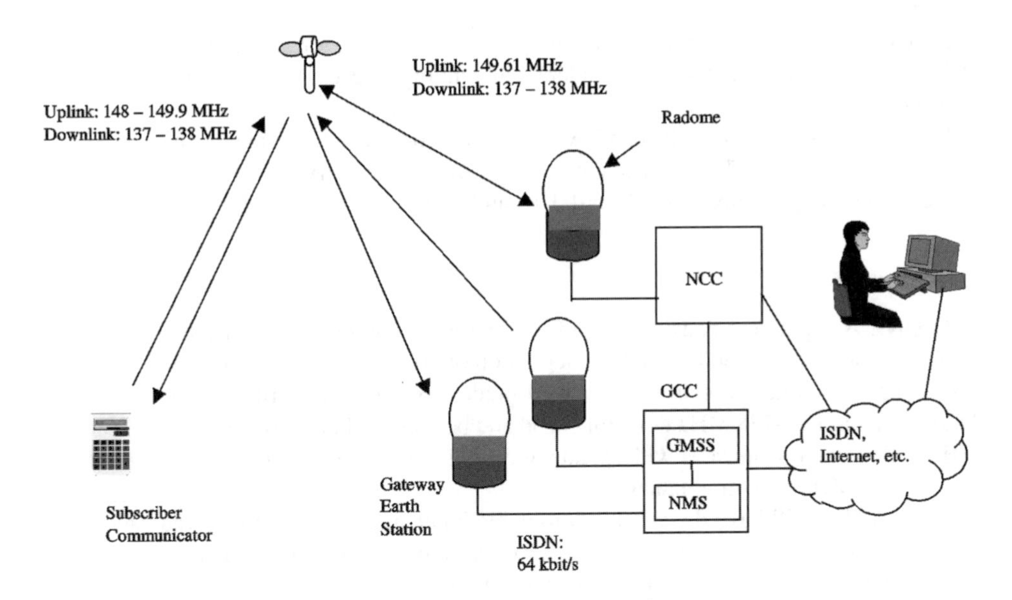

Figure 2.10 ORBCOMM network architecture.

a transmission frequency of 149.61 MHz. The satellite transmits a 3-W signal at the same data rate as the uplink in the 137–138 MHz band. Offset-QPSK modulation is applied in both directions. Each satellite can communicate with up to 16 GESs located within its coverage area.

The GCC provides the connection to the fixed network (ISDN, Internet, etc.). A GCC is located in each country in which a service agreement is in place. It comprises the gateway message switching system (GMSS) and the network management system. A critical component of the GMSS is the ORBCOMM message switch, which manages all SCs and GESs within its service area. This includes managing the handover between GESs when a satellite moves from one GES coverage area to another [MAZ-99].

2.3.3 E-SAT™

E-SAT is a six-satellite constellation that will orbit the Earth using polar orbits at an altitude of 800 km. The satellite constellation will be deployed in two orbital planes. Each satellite weighs approximately 130 kg. The E-SAT system is targeted at servicing on a store-and-forward basis, industrial data acquisition equipment, such as electric, gas and water meters and road traffic monitoring devices. The system aims to provide a global service.

E-SAT terminals will operate in the 148–148.55 MHz band for uplink communications in a direct sequence – spread spectrum multiple access (DS-SSMA) mode, providing 800 bit/s of useful information. BPSK will be employed to modulate the spreading code with the symbols and the spreading chips will use MSK to modulate the carrier. Terminals will transmit using 4.9 W of power, resulting in a peak EIRP of 5.4 dBW. Each E-SAT satellite can support up to 15 simultaneous terminal transmissions. Each terminal transmits in 450-ms bursts, containing 36 bytes of user data plus synchronisation and error detection and correcting codes.

The 137.0725–137.9275 MHz band will be used for inbound feeder downlinks to the gateway, service downlinks for terminal command and polling, and telemetry links for the satellite orbits. Service transmissions will operate continuously in TDM mode, at a data rate of 4.8 kbit/s. The same method of modulation and multiple access technique will be used as in the uplink.

The 148.855–148.905 MHz band will be used for the telecommand channel and the outbound feeder link. The network comprises of a Telemetry, Tracking and Command (TT& C) and SCC, located in Guildford, UK, and a gateway located in Spitzbergen, Norway.

2.3.4 LEO ONE™

LEO ONE is a planned constellation of 48 satellites, comprising six satellites per orbital plane, at an altitude of 930 km, and an inclination of 50° to the Equator [GOL-99]. LEO ONE is designed to operate in store-and-forward mode, providing subscribers with data rates of 24 kbit/s in the 137–138 MHz downlink band and between 2.4 and 9.6 kbit/s in the 148–149.9 MHz uplink band. Gateways will operate at 50 kbit/s in the 149.5–150.05 MHz uplink and 400.15–401 MHz downlink bands.

Each satellite will have the capability to demodulate and decode all received packets, and then store on-board for later transmission or to directly transmit to a gateway ground station in view. Each satellite payload has four transmitters and 15 receivers.

Initially, LEO ONE plans to have two gateways in the Continental US (CONUS) plus one in Alaska. Further gateways will be deployed outside of the US once service agreements are in place. Gateways provide connection to terrestrial CNs.

In addition to gateways, the LEO ONE network also comprises a constellation management centre, which provides TT & C and the network management centre, which provides the control of the communication network, including performance monitoring and subscriber validation.

The satellite constellation is expected to be fully deployed by 2002.

2.3.5 Other Systems

FAISAT™ is a planned 30-satellite network that will be developed by Final Analysis. The constellation will consist of 26 operational satellites plus four in-orbit spares. At the time of writing, there is a proposal under consideration by the FCC to increase the constellation to 36 satellites. At present, 24 of the satellites will be equally divided into six orbital planes at 66° inclination, including one spare satellite per plane. The remaining two satellites will be divided into polar orbits, inclined at 83°. Satellites will be placed in orbit 1000 km above the Earth. Should the constellation be increased to 36 satellites, the inclination angle will decrease to 51°, with six satellites per inclined orbital plane.

The non-profit organisation Volunteers in Technical Assistance has been promoting the use of non-geostationary satellite technology for the provision of communication services to the developing world, since the start of the 1980s. VITA utilises two satellites, which are deployed in polar orbits.

2.4 Satellite-Personal Communication Networks (S-PCN)

2.4.1 General Characteristics

As illustrated in the previous chapter, the start of the 1990s marked a key point in the development of mobile communications. Analogue cellular technology was making significant market inroads and new digital cellular services, such as GSM, were starting to appear on the high street. Although at the time, it would have been virtually impossible to predict the phenomenal market take-up that occurred over the next 10 years, it was clear that there was a substantial potential market for mobile communication services. Furthermore, the new cellular technologies that were being introduced, although continental in nature, could still be considered to be regional on a global basis. The enthusiastic global take-up of systems like GSM and cmdaOne was far from a reality.

The start of the 1990s also denoted the next significant phase in the evolution of mobile-satellite communications, with several proposals for non-geostationary satellite systems. The previous section has already discussed the developments in "little LEO" satellite technologies. At about the same time, a number of sophisticated non-geostationary systems, aimed primarily at the provision of voice services, were also proposed. The services that these satellites offered came under the general description of S-PCNs. Unlike terrestrial mobile network developments, these new satellite systems were virtually all initiated by American organisations, with limited European industrial involvement.

The primary aim of S-PCN is to provide voice and low data rate services, similar to those available via terrestrial cellular networks, using hand-held phones via satellites in either the LEO or MEO orbits. Satellites in the MEO are located in the region of 10 000–20 000 km

Table 2.7 Allocation of mobile-satellite service frequencies in the L-/S-bands

Frequency (MHz)	Status	Direction	Region
1492–1525	Primary	Space to Earth	Region 2
1525–1530	Primary	Space to Earth	Region 2/Region 3
1610–1626.5	Primary	Earth to space	World-wide
1613.8–1626.5	Secondary	Space to Earth	World-wide
1626.5–1631.5	Primary	Earth to space	Region 2/Region 3
1675–1710	Primary	Earth to space	Region 2
1930–1970	Secondary	Earth to space	Region 2
1970–1980	Primary	Earth to space	Region 2
1980–2010	Primary: intended for IMT-2000 satellite component	Earth to space	World-wide
2120–2160	Secondary	Space to Earth	Region 2
2160–2170	Primary	Space to Earth	Region 2
2170–2200	Primary: intended for IMT-2000 satellite component	Space to Earth	World-wide
2483.5–2500	Primary	Space to Earth	World-wide
2500–2520	Primary	Space to Earth	World-wide
2670–2690	Primary	Earth to space	World-wide

above the Earth with an orbital period of about 6 h. Satellites in the low Earth orbit are termed "big LEO" and orbit the Earth roughly every 90 min. These satellites are deployed in the region of 750–2000 km above the Earth. Due to the effects of the Van Allen radiation belt, satellites in the LEO orbit usually have an expected life-time of about 5–7 years. Satellites in the MEO orbit are not so affected and have an operational life-time of about 10–12 years.

S-PCN networks operate in the L- and S-bands. Table 2.7 shows the allocation of frequencies to the mobile-satellite service between 1 and 3 GHz. As was noted in Chapter 1, WRC 2000 also made these frequencies available to satellite-UMTS/IMT-2000 services.

S-PCNs provide global service coverage by making use of multi-satellite constellations. The number of satellites in a constellation is a function of the altitude of the orbit and the service availability characteristics, usually defined in terms of minimum user-to-satellite elevation angle. The design of non-geostationary satellite constellations will be addressed in the following chapter. Although these types of networks can provide communications to satellite-only hand-held terminals, systems also incorporate dual-mode phones, which operate with terrestrial as well as satellite mobile networks. Dual-mode phones operate primarily in the terrestrial mode, for example GSM, and only use the satellite connection when there is no terrestrial channel available. In this respect, it can be seen that the satellite service is complementary to the terrestrial networks.

The establishment of an S-PCN requires a significant up-front investment with an associated high degree of financial risk. For example, the IRIDIUM system cost somewhere in the region of $5.8 billion to get off the ground. Needless to say, a significant market take-up of the services on offer is required in order to sustain commercial viability. One of the original markets identified for S-PCN was widely known as the international business traveller. The idea here was that business users in Europe using GSM, for example, when travelling abroad would rely on a S-PCN to provide their communication capabilities. The possibility of a European businessperson being able to use their GSM phone in China, for example, by the end of the 1990s was not considered possible. At the start of the 1990s this may have proved a significant S-PCN market, however, the roll-out and world-wide take-up of terrestrial networks, such as GSM, which is now virtually a world standard, has severely diminished the size of this market. This scenario will be further discussed in Chapter 8, where the satellite market is investigated in detail. Other markets that were considered included the provision of telephony services to developing regions of the world, where fixed telephone lines were unavailable.

Table 2.7 summarises the key characteristics of the operational and planned S-PCN.

2.4.2 *IRIDIUM™*

Motorola announced its intention to develop the IRIDIUM system in 1990, with the initial aim of launching a commercial service in 1996. In January 1995, IRIDIUM gained an operating license from the US FCC. Similar regional license agreements with telecom operators around the world then followed over the next few years. On November 1, 1998, IRIDIUM became the first S-PCN to enter into commercial service. This followed a concerted 2-year launch campaign, between May 5, 1997 and June 12, 1999, during which time 88 satellites were successfully launched into low Earth orbit, including the launch of 72 satellites in the first year of the campaign. Three different types of launcher were used to deploy the satellites into orbit: Boeing Space Systems Delta II, which was used in 11 launches to deploy 55 satellites; Russia's Khrunichev Proton, which was used in three launches to deploy 21 satellites; and the Chinese Great Wall Long March 2C/SD, which deployed 14 satellites in seven launches. Further information on these satellite launchers can be found in Ref. [MAR-98].

From a satellite payload perspective, IRIDIUM is probably the most sophisticated of all the first generation S-PCNs. Originally intended as a 77-satellite system, hence its name (the atomic number of the element IRIDIUM is 77), early in its design it was decided to reduce the constellation to 66 satellites. These satellites are equally divided into six polar orbital planes, inclined at 86.4°. The satellites orbit the Earth at an altitude of approximately 780 km. Each satellite weighs 689 kg and has an anticipated life-span of between 5 and 8 years.

Connection to the terrestrial network is via a world-wide network of gateways, comprising 12 in total, placed strategically around the globe in 11 countries. Gateway and TT & C links operate in the Ka-band, specifically 19.4–19.6 GHz on the downlink and 29.1–29.3 GHz on the uplink. Unlike other LEO S-PCN systems, IRIDIUM satellites have a degree of OBP, which enables calls to be routed via ISL. This reduces the dependency on the terrestrial network to perform the routing of calls, hence the relatively few number of gateways needed to provide a global service. ISL operates in the Ka-band at frequencies in the range 23.18–23.38 GHz. At the time of writing, no other commercial mobile-satellite system operates using inter-satellite link technology. Each satellite can transmit up to 48 spot-beams, a number of which are systematically deactivated as a satellite approaches the North and South poles in order to limit inter-satellite interference.

The IRIDIUM system provides full-duplex voice, data and facsimile services at 2.4 kbit/s. Data rates of 10 kbit/s are anticipated. QPSK is employed to modulate signals, with FDMA/TDMA being used as the multiple access technique. Mobile transmissions operate in the 1616–1626.5 MHz band.

As was noted earlier, IRIDIUM began service in November 1998, more than a year ahead of its nearest market rival, GLOBALSTAR. Despite this early market lead, the take-up of IRIDIUM services by subscribers was significantly lower than predicted. The lack of available hand-sets was widely cited as one of the reasons for the poor level of interest in the network from potential customers. On August 13, 1999, IRIDIUM filed for Chapter 11 bankruptcy protection in the US. As it turned out, IRIDIUM was not the only S-PCN provider to take such action.

IRIDIUM, unable to find a buyer for its network, ceased service provision on March 17, 2000. However, this was not the end of the IRIDIUM constellation. On December 12, 2000, IRIDIUM Satellite LLC announced the acquisition of the IRIDIUM network, with a re-alignment of the target markets towards US government and industrial clients, including the maritime, aviation, oil and gas and forestry sectors, with services starting early in 2001.

2.4.3 GLOBALSTAR™

Loral Space and Communications and Qualcomm launched the concept of GLOBALSTAR at a similar time to that of IRIDIUM. GLOBALSTAR gained an operating license from the US FCC in November 1996. GLOBALSTAR's first launch of four satellites occurred in May 1998, eventually completing its deployment of 48 satellites plus four spares early in the year 2000. Two types of launcher were used to deploy the satellites: Boeing Space Systems Delta

II, which was used in seven launches to deploy 28 satellites; and the Startsem Rocket Soyuz-Ikar, which was used in six launches to deploy 24 satellites.

GLOBALSTAR is a 48-satellite system, deployed equally into eight orbital planes, inclined at 52° to the Equator, with a 45° ascending node separation between planes. The satellites orbit the Earth at a low Earth orbit altitude of 1414 km. In this respect, GLOBALSTAR does not provide total global coverage but is restricted to the areas between 70° latitudes North and South of the Equator.

Unlike IRIDIUM, GLOBALSTAR satellites are transparent, implying that no OBP and routing of signals are performed. Each satellite utilises a 16-spot-beam coverage pattern on the forward and reverse links. Since satellites are transparent, a communication link between a mobile user and the fixed network can only be established when both the mobile terminal and a gateway are simultaneously in view of a satellite. Due to the relatively small coverage area offered by a LEO satellite, a large number of terrestrial gateways are necessary in order to ensure global service availability if ISL technology is not employed. In GLOBALSTAR's case, connection to the terrestrial network is achieved through the use of more than 100 gateways, distributed throughout the world. Each gateway serves an operator area of approximately 500 km radius and can serve more than one country, provided appropriate service agreements are in place. Local GLOBALSTAR service providers operate gateways. A service provider buys from GLOBALSTAR an exclusive right to provide GLOBALSTAR services. The service provider is responsible for obtaining regulatory approval to operate its network and to set the retail structure, in terms of service costs, handset price, etc. for the service. A gateway comprises three to four antennas and a switching station and is used to provide connection to the local fixed and mobile networks.

Gateway links operate in the C-band, specifically 6.875–7.055 GHz on the downlink and 5.091–5.250 GHz on the uplink. In addition to the gateways, the GLOBALSTAR ground segment comprises ground operation control centres (GOCCs), which are responsible for the allocation of satellite resources to the gateways; and the satellite operation control centre (SOCC), which manages the space segment. The gateways, GOCCs and the SOCC are connected together via the GLOBALSTAR Data Network.

Mobile to satellite links are in the S-band, specifically 1610–1626.5 MHz on the uplink and 2483.5–2500 MHz on the downlink. The bandwidth is divided into 13 FDM channels, each 1.23 MHz wide. The GLOBALSTAR air interface is based on the cmdaOne (IS-95) solution discussed in the previous chapter [SCH-00]. The GLOBALSTAR voice service is provided through an adaptive 0.6–9.6 kbit/s (2.2 kbit/s on average) voice codec. Data transmissions are supported over a voice channel at a basic rate of 2.4 kbit/s. QPSK is employed to modulate signals and CDMA is used as the multiple access technique.

In the forward direction, each gateway has access to 128 CDMA channels, each derived from one row of a 128 × 128 Walsh Hadamard matrix, with the row numbers going from 0 to 127. One of these channels, Walsh function 0, is used to transmit a pilot signal, comprising all zeros. A synchronisation channel is used to provide mobile terminals with important control information such as the gateway identifier, the system time and the assigned paging channel. Like IS-95, the SYNC channel always transmits at a data rate of 1.2 kbit/s.

The remaining 126 channels are available for traffic, with up to the first seven of these being available for paging. The paging channel operates at 4.8 kbit/s.

As with IS-95, GLOBALSTAR supports two different rate sets, although the data rates are different from that of the cellular network. Rate Set 1 supports 2.4 and 4.8 kbit/s, while Rate

Set 2 operates at either 2.4, 4.8 or 9.6 kbit/s. Data is grouped into 20-ms frames, which are then encoded using a half-rate convolutional encoder with a constraint length of 9. In order to ensure a constant rate of 9.6 kbit/s for Rate Set 1 or 19.2 kbit/s for Rate Set 2, repetition of lower rate code bits is performed prior to interleaving and spreading by a pseudo-random sequence derived from the long code and a long code mask, which is unique to each terminal. This 9.6 kbit/s or 19.2 kbit/s signal is then multiplexed with power control bits, which are applied at 50 bit/s. The Walsh code assigned to the user's traffic channel is then used to spread the signal before being spread by an outer PN sequence generator of length 288 chips, at a rate of 1200 outer PN chip/s. The outer PN sequence, which identifies the satellite, modulates the I and Q pseudo-noise inner sequences, which have a length of 2^{10} chips, and a chip rate of 1.2288 Mchip/s. Baseband filtering is then applied prior to modulating the I and Q components onto a CDMA channel. The forward link transmission scheme for Rate Set 1 is illustrated in Figure 2.11.

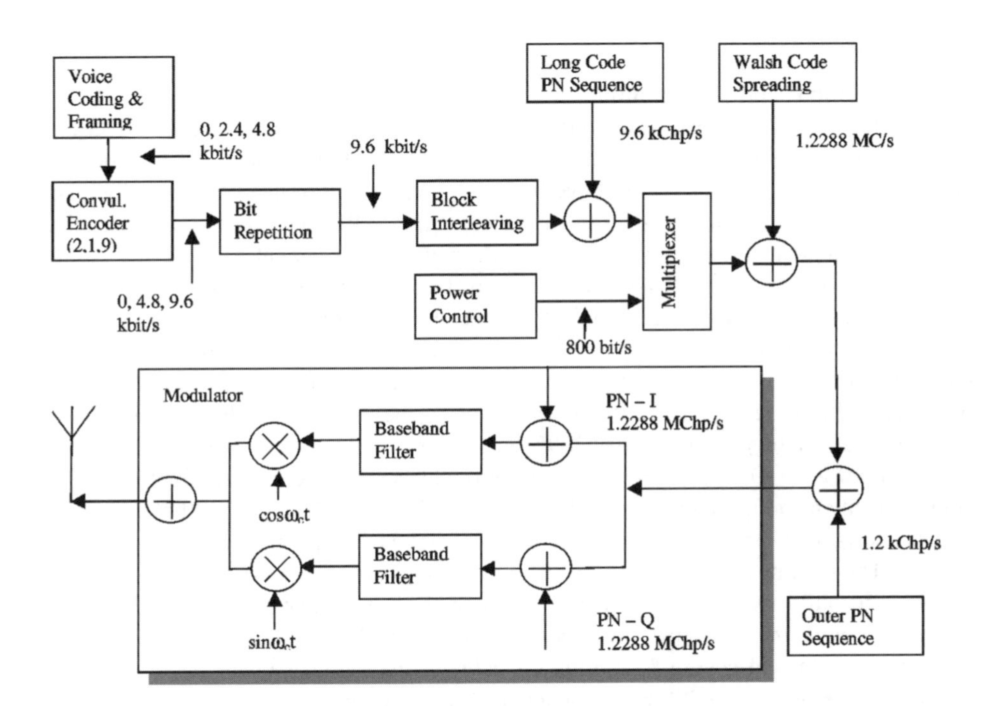

Figure 2.11 GLOBALSTAR forward traffic modulation for Rate Set 1.

In the reverse direction, the traffic channel supports 2.4, 4.8 and 9.6 kbit/s. Data are organised into 20-ms frames. For Rate Set 1, this is then encoded using a half-rate convolutional encoder with a constraint length of 9. Orthogonal Walsh modulation is then performed, by which a block of six code symbols is used to generate a sequence of 64 bits, corresponding to a row of a 64 × 64 Walsh Hadamard matrix. Each mobile transmits a different Walsh code, enabling the base station to identify the transmitter. The long code, which repeats itself after $2^{42}-1$ chips, and long code mask, which is unique to each terminal and contains the mobile's electronic serial number, is then used to spread the signal at 1.2288 Mchip/s. Offset-QPSK is used to modulate the carrier. This involves delaying the quadrature channel by half a chip. A quadrature pair of pseudo-noise (PN) sequences is then used to spread the signal at 1.2288 Mchip/s with a periodicity of $2^{15}-1$ chips. Baseband filtering is then applied to ensure that the components of the modulation remain within the channel before modulating the in-phase and quadrature signals onto the CDMA channel.

A catastrophic launch failure in September 1998, in which 12 GLOBALSTAR satellites were lost, contributed to the delay of the launch of GLOBALSTAR, which was initially aimed to start service in November 1998. Eventually GLOBALSTAR began commercial trials in November 1999, with initial availability over North and South America, China, Korea and parts of Europe. Commercial services began in spring 2000, however, as with IRIDIUM, commercial deployment of GLOBALSTAR has been anything but smooth, with only 55,000 subscribers as of July 2001.

2.4.4 NEW ICO™

2.4.4.1 Commercial Background

In September 1991, Inmarsat announced its strategy for its future evolutionary development of mobile-satellite communications, under the heading Project-21 [LUN-91]. The culmination of this strategy was to have been the introduction of hand-held satellite phones in the 1998–2000 time frame under the service name INMARSAT-P. It was recognised that in order to implement this new type of service, a new space segment architecture would be required. During the early 1990s, Inmarsat evaluated a number of possibilities for the INMARSAT-P space segment, including enhanced geostationary satellites, LEO solutions and a MEO solution. These investigations subsequently led to the identification of a MEO constellation as the optimum solution and the eventual establishment of ICO Global Communications Ltd in order to finance the development of the new system. ICO Global Communications was established in January 1995, as a commercial spin-off of Inmarsat.

In addition to ICO, the American organisation TRW also proposed to exploit the MEO solution using a configuration of satellites, named ODYSSEY. The ODYSSEY constellation was to consist of 12 satellites, equally divided into three orbital planes, inclined at 55° to the equator. The satellites were to be placed 10 600 km above the Earth. The FCC awarded TRW a license to build its MEO system in 1995, with the caveat that building of the first two satellites should commence by November 1997. ODYSSEY was predicted to start service in 1999, at an estimated cost of $3.2 billion. Unable to find another major investor willing to support the project, ODYSSEY was abandoned in December 1997. Also during this time, ICO and TRW had been involved in a patent dispute over the rights to exploit the MEO architecture. TRW had been awarded a patent by the US Patent Office in July 1995 following the filing of a patent in May 1992 [PAT-95]. With the abandonment of the ODYSSEY project, all legal actions were dropped and TRW took a 7% stake in ICO.

In August 1999, 2 weeks after IRIDIUM filed for Chapter 11 bankruptcy protection in the US, ICO Global Communications followed suite. This was after failing to secure sufficient funding to implement the next stage of development, the upgrade of its terrestrial network to allow the provision of high-speed Internet services via its satellites. The investment of $1.2 billion in ICO Global Communications by TELEDESIC was announced in November 1999.

TELEDESIC, although primarily aiming to provide Internet-in-the-Sky™ services to fixed users via its constellation of non-geostationary satellites, also foresees the possibility of offering services to the maritime and aeronautical sectors.

TELEDESIC's system began life as an 840 + 84 spare satellite constellation placed in the LEO orbit at 700 km above the Earth [STU-95]. The large number of satellites were necessary to guarantee a minimum 40° elevation angle to the satellite. Following a design review, the planned constellation was revised to a multi-satellite LEO configuration comprising 288 satellites, equally divided into 12 orbital planes, at an altitude of 1400 km. Satellites will be connected via ISL. TELEDESIC, which will be a packet-switched network, plans to offer user-data rates of 2 Mbps on the uplink and 64 Mbps on the downlink. In this respect, the TELEDESIC network is claimed to be the space equivalent of the terrestrial fibre network. The system will operate in the Ka-band frequencies allocated for non-geostationary fixed satellite services, specifically 18.8–19.3 GHz on the downlink and 20.6–29.1 GHz on the uplink. TELEDESIC anticipates starting operation in 2004, although the links with NEW ICO could further delay its launch.

On the May 17, 2000, NEW ICO, formerly ICO Global Communications, successfully emerged from bankruptcy protection.

2.4.4.2 System Aspects

NEW ICO's constellation comprises ten satellites, equally divided into two orbital planes, inclined at 45° to the Equator, with a 180° separation between ascending nodes. Each orbital plane also contains one spare satellite. The satellites will orbit the Earth at 10 390 km altitude, each satellite providing a capacity of 4500 voice circuits. Each satellite has an anticipated life span of approximately 12 years, which is about 5 years longer than its LEO counterparts, and roughly equal to that of the latest geostationary satellites. Figure 2.12 illustrates an artist's impression of the NEW ICO constellation and a constituent satellite.

Since NEW ICO operates using a MEO constellation, the degree of coverage offered by each satellite is significantly greater than its LEO counterparts. Indeed, each ICO satellite covers approximately 30% of the Earth's surface. A user on the ground would normally see two satellites and sometimes up to as many as four. This enables what is known as path diversity to be implemented, whereby the link to the satellite can be selected to optimise reception. Furthermore, as a consequence of the large coverage area, the number of gateways required to support access to the ground segment is not as great as that of GLOBALSTAR, for example.

NEW ICO's ground network is referred to as ICONET, access to which is gained through one of 12 inter-connected satellite access nodes (SANs) distributed throughout the world. Each SAN comprises five tracking antennas for communicating with the space segment, and associated switching equipment and databases. The NEW ICO ground segment is based on the GSM network architecture, incorporating home location and visitor location registers, and authentication centres. Connection to the fixed network is via a GSM based MSC. An inter-working function is used to automatically convert IS-41 mobility management functions to those of GSM. SANs perform the selection of optimum satellite for a link, and the setting up and clearing down of connections. Six of the 12 SANs also have TT & C facilities. SANs are controlled by the network management centre, located in London, which is responsible for the overall management of the resource. SANs communicate with satellites in the C-band,

Figure 2.12 NEW ICO satellite and constellation (courtesy of NEW ICO).

specifically in the 5150–5250 MHz band on the uplink and the 6975–7075 MHz band on the downlink.

NEW ICO has chosen to operate its mobile links in the frequency bands allocated to the IMT-2000 satellite component. Specifically, in the 1985–2015 MHz band on the uplink and the 2170–2200 MHz band on the downlink. QPSK with 40% raised cosine filtering is used to modulate voice services. For data services, GMSK modulation is applied on the mobile uplink, with either BPSK or QPSK used in the downlink, depending on the operating environment. TDMA is employed as the multiple access scheme. Each TDMA frame, which is of 40 ms duration and comprises six time-slots, is transmitted at 36 kbit/s. TDMA frames are divided into 25-kHz channels. Each slot is made up of 4.8 kbit/s of user information plus 1.2 kbit/s of in-band signalling and framing information [GHE-99].

NEW ICO will initially provide telephony, bi-directional SMS, data and Group 3 fax services. The output of the NEW ICO voice coder provides a basic rate of 3.6 kbit/s, which then has 3/4-rate convolution coding applied, resulting in a net rate of 4.8 kbit/s. Data services, at a basic rate of 2.4 kbit/s are coded using a half-rate convolution encoder, resulting in a net rate of 4.8 kbit/s. Higher data rates can be achieved by using multi-slots within a time frame. High-speed circuit-switched data services at 38.4 kbit/s are planned for vehicular and semi-fixed terminals. NEW ICO plans to commence service in 2004.

2.4.5 CONSTELLATION COMMUNICATIONS™

CONSTELLATION COMMUNICATIONS is a low Earth orbit satellite system that is targeting developing regions of the world [JAN-99]. CONSTELLATION obtained an operating license from the US FCC in July 1997, for its first phase introduction of services.

The satellite configuration comprises 11 satellites plus one orbital spare, deployed in the Equatorial plane. The satellites are designed to orbit the Earth at an altitude of 2000 km, providing a belt of coverage between the 23° N and S latitudes. The satellites each transmit 24 spot-beams and have a simple transparent payload. Each satellite weighs 500 kg with a 2-kW power rating.

CONSTELLATION's ground segment will initially comprise 12 gateways, an SCC and an NCC. Gateways provide connection to the fixed and mobile networks. Like GLOBALSTAR, CONSTELLATION sells satellite capacity to gateway operators, who in turn establish agreements with local service providers. Each gateway operates with three antennas in the bands 5091–5250 MHz on the uplink and 6924–7075 MHz on the downlink. Selected gateways on the network also have TT & C equipment installed.

The SCC is connected directly to the selected gateways and provides the necessary TT & C commands. The NCC is connected to all the gateways and allocates space segment resources and collects billing information.

CONSTELLATION aims to provide voice, data and facsimile services to hand-held, vehicular and fixed terminals and public payphone units. Voice is coded at a basic rate of 4 kbit/s. Following the addition of signalling information and half-rate convolution coding, voice is transmitted in the gateway to subscriber (outbound) direction at 9.0 kbit/s and 9.33 kbit/s in the subscriber to gateway (inbound) direction. Outbound signals are modulated using QPSK, while offset-QPSK is applied on the inbound channel. Data services are provided at basic rates of 2.4, 4.8 and 9.6 kbit/s. CONSTELLATION has selected CDMA as its multiple access technique, at a rate of 432 kchipps/s (outbound) and 298.69 kchipps/s (inbound) per I and Q channel, respectively.

Once the initial deployment of satellites has been established, the next phase in CONSTELLATION's implementation schedule is planned to begin with the expansion from regional to global services. This will be achieved through a constellation of 35 satellites and seven spares deployed into seven circular orbits, inclined at 62° to the Equator. The satellites will orbit the Earth at an altitude of just under 2000 km. This new proposal is under consideration by the FCC at the time of writing and is planned for operation in 2003.

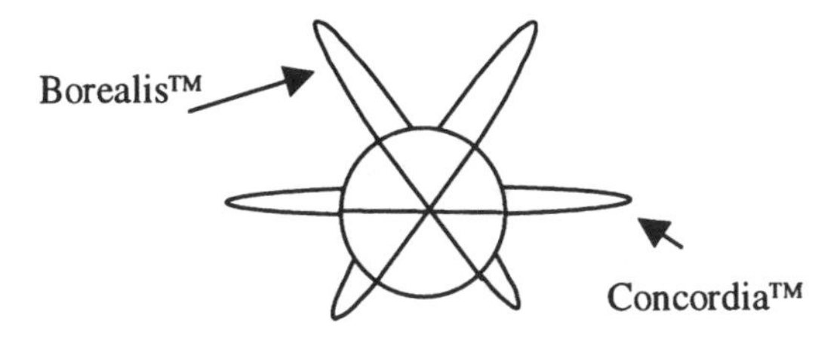

Figure 2.13 ELLIPSO orbital characteristics.

Table 2.8 S-PCN characteristics: satellite and orbit

	IRIDIUM	GLOBALSTAR	NEW ICO	CONSTELLATION	ELLIPSO
Orbit altitude (km)	780	1414	10 390	2000	Bore: 7605, Conc.: 8050
Type	LEO	LEO	MEO	LEO	Hybrid LEO/HEO
Launch mass (kg)	689	450	2750	500	650
Number of satellites	66	48	10	11	Bore:10, Conc.: 7
Satellites/plane	11	6	5	11	Bore: 5, Conc.: 7
ISL	Yes, 23.18–23.38 GHz	No	No	No	No
OBP	Yes	No	No	No	No
No. of spot-beams	48	16		24	61

Table 2.9 S-PCN characteristics: services

	IRIDIUM	GLOBALSTAR	NEW ICO	CONSTELLATION	ELLIPSO
Launch date	1 Nov. 1998, terminated 17 Mar. 2000 Re-launched 2001	Spring 2000	2004	2001	Not available
Trans. Rate	2.4 kbit/s	0.6–9.6 bit/s (voice) 2.4 kbit/s (data)	4.8 kbit/s (voice) 2.4–9.6 kbit/s (hand-held data) 8.0–38.4 kbit/s vehicular/ semi-fixed data)	4 kbit/s (voice) 2.4, 4.8 and 9.6 kbit/s (data)	2.4 kbit/s (voice), up to 28.8 kbit/s (data)
Services	Voice, fax, data	Voice, fax, data, SMS	Voice, bi-directional SMS, fax, Internet access, high speed circuit switched data	Voice, data, fax	Voice, e-mail, Internet access, fax, data, push-to-talk, global positioning
Coverage	Global	Global within bounds ±70° latitude	Global	Initially, regional within bounds ±23° latitude. Global coverage aimed for 2003 by incorporating new satellites into constellation	Global above 50° South

Table 2.10 S-PCN characteristics: radio interface

		IRIDIUM	GLOBALSTAR	NEW ICO	CONSTELLATION	ELLIPSO
Mobile (MHz)	↑	1616–1626.5	1610–1626.5	1985–2015	2483.5–2500	1610–1626.5
Mobile (MHz)	↓	1616–1626.5	2483.5–2500	2170–2200	1610–1626.5	2483.5–2500
Feeder (GHz)	↑	29.1–29.3	5.091–5.250	5.150–5.250	5.091–5.250	15.45–15.65
Feeder (GHz)	↓	19.4–19.6	6.875–7.055	6.975–7.075	6.924–7.075	6.875–7.075
Multiple access		FDMA/TDMA	CDMA	FDMA/TDMA	CDMA	W-CDMA
Modulation		QPSK	QPSK	GMSK uplink, BPSK/QPSK downlink	QPSK outbound, O-QPSK inbound	Not available

2.4.6 ELLIPSO™

ELLIPSO is the only S-PCN to employ satellites placed in elliptical orbit as part of its constellation [DRA-97]. Like CONSTELLATION, ELLIPSO is targeting developing regions of the world and has designed its constellation accordingly.

ELLLPSO employs a unique approach to its satellite deployment, which was recognised with the award a US Patent in December 1996 [PAT-96] to Mobile Communications Holding Industry. The space segment comprises two distinct orbital configurations, termed BOREALIS™ and CONCORDIA™. Once operational, services can be provided independently by each orbit. The BOREALIS configuration comprises ten satellites, deployed in two elliptical orbits, inclined at 116.6° to the Equator. The apogee and perigee of the elliptical orbit are at 7605 and 633 km above the Earth, respectively. This orbital configuration is used to provide coverage to northern temperate latitudes. The CONCORDIA orbital configuration comprises seven satellites deployed in the Equatorial plane, at an altitude of 8050 km. Figure 2.13 illustrates the ELLIPSO constellation. The satellites provide coverage between 50° North and 50° South, with each satellite providing 61 spot-beams using receive and transmit planar array antennas. The ELLIPSO satellites are based on Boeing's GPS satellites. Ariane Space will launch the satellites. Satellites, which employ transparent payloads, have an anticipated lifetime of between 5 and 7 years.

The terrestrial segment initially comprises 12 gateways, which provide the connection with local terrestrial networks. Significantly, ELLIPSO aims to deploy an IP CN to connect their gateways, which is in agreement with the development of 3G CNs described in the previous chapter.

ELLIPSO plans to deliver voice, Internet access and e-mail facilities to a variety of terminal types, including hand-held, vehicular, public payphone and residential. Voice services will be delivered at 2.4 kbit/s and data at up to 28.8 kbit/s. The user terminal will transmit in the 1610–1621.5 MHz band and receive in the 2483.5–2500 MHz band. Terminals will be based on 3G technology and will use a derivative of W-CDMA for the multiple access scheme.

In March 2001, NEW ICO and ELLIPSO announced their intention to collaborate on technical, regulatory, business and financial issues.

The characteristics of the S-PCNs discussed in this section are summarised in Tables 2.8–2.10.

References

[COL-95] J.-N. Colcy, G. Hall, R. Steinhäuser, "Euteltracs: The European Mobile Satellite Service", *Electronics & Communications Engineering Journal*, 7(2), April 1995; 81–88.

[DIA-99] M. Diaz, "Integrating GPS Receivers into Consumer Mobile Electronics", *IEEE Multimedia Magazine*, 6(4), October–December; 88–90.

[DRA-97] J.E. Draim, D. Castiel, W.J. Brosius, A.G. Helman, "ELLIPSO™ – An Affordable Global, Mobile Personal Communications System", *Proceedings of 5th International Mobile Satellite Conference*, Pasadena, 16–18 June 1997; 153–158.

[ETS-93] ETSI Technical Report, "Satellite Earth Stations & Systems (SES), Possible European Standardisation of Certain Aspects of S-PCN, Phase 2: Objectives and Options for Standardisation", *ETSI TR DTR/SES-05007*, September 1993.

[ETS-96] ETSI Technical Report "Satellite Earth Stations & Systems (SES), Phase 2: Objectives and Options for Standardisation", *ETSI TR DTR/SES-00002*, June 1996.

[ETS-99] ETSI Technical Specification "GEO-Mobile Radio Interface Specifications: GMR-2 General System Description", *ETSI TS101377-01-03*, GMR-2 01.002, 1999.

[ETS-99a] ETSI Technical Specification "GEO-Mobile Radio Interface Specifications: Network Architecture", *ETSI TS101377-03-2*, GMR-2 03.002, 1999.

[ETS-99b] ETSI Technical Specification "GEO-Mobile Radio Interface Specifications: Network Architecture", *ETSI TS101377-05-02*, GMR-2 05.002, 1999.

[FRA-00] A. Franchi, A. Howell, J. Sengupta, "Broadband Mobile via Satellite: Inmarsat BGAN", *Proceedings of IEE Colloquium on Broadband Satellite Systems*, London, 16–17 October 2000; 23/1–23/7.

[FUR-96] K. Furukawa, Y. Nishi, M. Kondo, T. Veda, Y. Yasuda, "N-STAR Mobile Communication Satellite System", *Proceedings of IEEE Global Telecommunications Conference*, London, November 1996; 390–395.

[GHE-99] L. Ghedia, K. Smith, G. Titzer, "Satellite PCN – the ICO System", *International Journal of Satellite Communications*, 17(4), July–August 1999; 273–289.

[GOL-99] E. Goldman, "Little LEOs Serve an Unmet Demand: LEO One System Architecture Optimized to meet Market Requirements", *International Journal of Satellite Communications*, 17(4), July–August 1999; 225–242.

[HU-96] Y.F. Hu, R.E. Sheriff, "Asia, the Dominant Future Market for Mobile-Satellite Communications", *Proceedings of International Conference on Communications Technology*, Beijing, 5–7 May 1996; 301–304.

[INM-93] *Project 21: The Development of Personal Mobile Satellite Communications*, Inmarsat, March 1993.

[JAN-99] J.D. Jancso, B. Kraeselsky, "The Constellation LEO Satellite System: A Wide-Area Solution to Telecom Needs in Underserved Areas Worldwide", *International Journal of Satellite Communications*, 17(4), July – August 1999; 257–271.

[JAL-89] I.M. Jacobs, "An Overview of OmniTRACS: the First Operational Mobile Ku-Band Satellite Communications System", *Space Communications*, 7(1), December 1989; 25–35.

[JOH-93] G.A. Johanson, N.G. Davies, W.R.H. Tisdale, "The American Mobile Satellite System: Implementation of a System to Provide Mobile Satellite Services in North America", *Space Communications*, 11(2), 1993; 121–128.

[LUN-91] *Project 21: A Vision for the 21st Century*, Statement by O. Lundberg, Director General, Inmarsat, September 12 1991.

[MAR-98] G. Maral, M. Bousquet, *Satellite Communication Systems*, 3rd edition, Wiley, Chichester, 1998.

[MAZ-99] S. Mazur, "A Description of Current and Planned Location Strategies within the ORBCOMM Network", *International Journal of Satellite Communications*, 17(4), July–August 1999; 209–223.

[NEW-90] W. Newland, "AUSSAT Mobilesat System Description", *Space Communications*, 8(1), December 1990; 37–52.

[NGU-97] N.P. Nguyen, P.A. Buhion, A.R. Adiwoso, "The Asia Cellular Satellite System", *Proceedings of 5th International Mobile Satellite Conference*, Pasadena, 16–18 June 1997; 145–152.

[PAT-95] *Medium-Earth-Altitude Satellite-Based Cellular Telecommunications System*, United States Patent No. 5,433,726, Filed May 28, 1992, Awarded 18 July 1995.

[RUS-92] C.M. Rush, "How WARC'92 Will Affect Mobile Services", *IEEE Communications Magazine*, 30(10), October 1992; 90–96.

[SCH-00] L. Schiff, A. Chockalingam, "Signal Design and System Operation of GLOBALSTAR™ versus IS-95 CDMA – Similarities and Differences", *Wireless Networks*, 6(1), 2000; 47–57.

[STU-95] M.A. Sturza, "Architecture of the TELEDESIC Satellite System", *Proceedings of 4th International Mobile Satellite Conference*, Ottawa, 6–8 June 1995; 212–218.

[VAN-97] L. Vandebrouck, "EUTELSAT Development Plans in Mobile Satellite Communications", *Proceedings of 5th International Mobile Satellite Conference*, Pasadena, 16–18 June 1997; 499–502.

3

Constellation Characteristics and Orbital Parameters

3.1 Satellite Motion

3.1.1 Historical Context

In 1543, the Polish Canon Nicolas Copernicus wrote a book called *On the Revolutions of the Heavenly Spheres*, which for the first time placed the Sun, rather than the Earth, as the centre of the Universe. According to Copernicus, the Earth and other planets rotated around the Sun in circular orbits. This was the first significant advancement in astronomy since the Alexandrian astronomer Ptolemy in his publication *Almagest* put forward the geocentric universe sometime during the period 100–170 AD. Ptolemy theorised that the five known planets at the time, together with the Sun and Moon, orbited the Earth.

From more than 20 years of observational data obtained by the astronomer Tycho Brahe, Johannes Kepler discovered a minor discrepancy between the observed position of the planet Mars and that predicted using Copernicus' model. Kepler went on to prove that planets orbit the Sun in elliptical rather than circular orbits. This was summarised in Kepler's three planetary laws of motion. The first two of these laws were published in his book *New Astronomy* in 1609 and the third law in the book *Harmony of the World* a decade later in 1619.

Kepler's three laws are as follows, with their applicability to describe a satellite orbiting around the Earth highlighted in brackets.

- First law: the orbit of a planet (satellite) follows an elliptical trajectory, with the Sun (gravitational centre of the Earth) at one of its foci.
- Second law: the radius vector joining the planet (satellite) and the Sun (centre of the Earth) sweeps out equal areas in equal periods of time.
- Third law: the square of the orbital period of a planet (satellite) is proportional to the cube of the semi-major axis of the ellipse.

While Kepler's laws were based on observational records, it was sometime before these laws would be derived mathematically. In 1687, Sir Isaac Newton published his breakthrough work *Principia Mathematica* in which he formulated the Three Laws of Motion:

Law I: every body continues in its state of rest or uniform motion in a straight line, unless impressed forces act upon it.

Law II: the change of momentum per unit time is proportional to the impressed force and takes place in the direction of the straight line along which the force acts.
Law III: to every action, there is always an equal and opposite reaction.

Newton's first law expresses the idea of inertia.

The mathematical description of the second law is as follows:

$$\mathbf{F} = m\frac{\mathrm{d}^2\mathbf{r}}{\mathrm{d}t^2} = m\ddot{\mathbf{r}} \tag{3.1}$$

where $\mathbf{F}$ is the vector sum of all forces acting on the mass m; $\ddot{\mathbf{r}}$ is the vector acceleration of the mass.

In addition to the Three Laws of Motion, Newton stated the "two-body problem" and formulated the Law of Universal Gravitation:

$$\mathbf{F} = Gm_1m_2\frac{1}{r^2}\frac{\mathbf{r}}{r} \tag{3.2}$$

where $\mathbf{F}$ is the vector force on mass m_1 due to m_2 in the direction from m_1 to m_2; $G = 6.672 \times 10^{-11}$ Nm/kg^2 is the Universal Gravitational Constant; r is the distance between the two bodies; $\mathbf{r}/r$ is the unit vector from m_1 to m_2.

The Law of Universal Gravitation states that the force of attraction of any two bodies is proportional to the product of their masses and inversely proportional to the square of the distance between them. The solution to the two-body problem together with Newton's Three Laws of Motion are used to provide a first approximation of the satellite orbital motion around the Earth and to prove the validity of Kepler's three laws.

3.1.2 *Equation of Satellite Orbit – Proof of Kepler's First Law*

The solution to the two-body problem is obtained by combining equations (3.1) and (3.2). In the formulation, the centre of the Earth is the origin in the co-ordinate system and the radius vector $\mathbf{r}$ is defined as positive in the direction away from the origin. Re-expressing equations (3.1) and (3.2) to describe the force acting on the satellite of mass m due to the mass of the Earth, M:

$$\mathbf{F}_m = -GmM\frac{\mathbf{r}}{r^3} = -m\mu\frac{\mathbf{r}}{r^3} \tag{3.3}$$

where $\mu = GM = 3.9861352 \times 10^5$ km^3/s^2 is Kepler's constant.

The negative sign in equation (3.3) indicates that the force is acting towards the origin. Equation (3.1) and (3.3) gives rise to:

$$\frac{\mathrm{d}^2\mathbf{r}}{\mathrm{d}t^2} = -\mu\frac{\mathbf{r}}{r^3} \tag{3.4}$$

The above equation represents the *Law of Conservation of Energy* [BAT-71].

Cross multiplying equation (3.4) with $\mathbf{r}$:

$$\mathbf{r} \times \frac{\mathrm{d}^2\mathbf{r}}{\mathrm{d}t^2} = -\mu\mathbf{r} \times \frac{\mathbf{r}}{r^3} \tag{3.5}$$

Since the cross product of any vector with itself is zero, i.e. $\mathbf{r} \times \mathbf{r} = 0$, hence:

$$\mathbf{r} \times \frac{\mathrm{d}^2\mathbf{r}}{\mathrm{d}t^2} = 0 \tag{3.6}$$

Consider the following equation:

$$\frac{\mathrm{d}}{\mathrm{d}t}\left[\mathbf{r} \times \frac{\mathrm{d}\mathbf{r}}{\mathrm{d}t}\right] = \frac{\mathrm{d}\mathbf{r}}{\mathrm{d}t} \times \frac{\mathrm{d}\mathbf{r}}{\mathrm{d}t} + \mathbf{r} \times \frac{\mathrm{d}^2\mathbf{r}}{\mathrm{d}t^2} \tag{3.7}$$

From equation (3.6) and from the definition of vector cross product, the two terms on the right hand side of equation (3.7) are both equal to zero. It follows that:

$$\frac{\mathrm{d}}{\mathrm{d}t}\left[\mathbf{r} \times \frac{\mathrm{d}\mathbf{r}}{\mathrm{d}t}\right] = 0 \tag{3.8}$$

Hence,

$$\mathbf{r} \times \frac{\mathrm{d}\mathbf{r}}{\mathrm{d}t} = \mathbf{h} \tag{3.9}$$

where $\mathbf{h}$ is a constant vector and is referred to as the orbital areal velocity of the satellite.

Cross multiplying equation (3.4) by $\mathbf{h}$ and making use of equation (3.9):

$$\frac{\mathrm{d}^2\mathbf{r}}{\mathrm{d}t^2} \times \mathbf{h} = -\frac{\mu}{r^3}\mathbf{r} \times \mathbf{h} = \frac{\mu}{r^3}\mathbf{r} \times \left[\mathbf{r} \times \frac{\mathrm{d}\mathbf{r}}{\mathrm{d}t}\right] \tag{3.10}$$

By making use of the rule for vector triple product: $a \times (b \times c) = (a{\cdot}c)b - (a{\cdot}b)c$, the rightmost term of equation (3.10) can be expressed as:

$$\frac{\mu}{r^3}\mathbf{r} \times \left[\mathbf{r} \times \frac{\mathrm{d}\mathbf{r}}{\mathrm{d}t}\right] = -\frac{\mu}{r^3}\left[\left(\mathbf{r}\cdot\frac{\mathrm{d}\mathbf{r}}{\mathrm{d}t}\right)\mathbf{r} - (\mathbf{r}\cdot\mathbf{r})\frac{\mathrm{d}\mathbf{r}}{\mathrm{d}t}\right] \tag{3.11}$$

Since

$$\mathbf{r}\cdot\frac{\mathrm{d}\mathbf{r}}{\mathrm{d}t} = 0$$

this implies

$$\frac{\mu}{r^3}\mathbf{r} \times \left[\mathbf{r} \times \frac{\mathrm{d}\mathbf{r}}{\mathrm{d}t}\right] = \mu\frac{\mathrm{d}}{\mathrm{d}t}\left[\frac{\mathbf{r}}{r}\right] \tag{3.12}$$

Comparing (3.10) with (3.12) gives:

$$\frac{\mathrm{d}^2\mathbf{r}}{\mathrm{d}t^2} \times \mathbf{h} = \mu\frac{\mathrm{d}}{\mathrm{d}t}\left[\frac{\mathbf{r}}{r}\right] \tag{3.13}$$

Integrating (3.13) with respect to t:

$$\frac{\mathrm{d}\mathbf{r}}{\mathrm{d}t} \times \mathbf{h} = \mu\frac{\mathbf{r}}{r} + \mathbf{c} \tag{3.14}$$

Taking the dot product of (3.14) with $\mathbf{r}$ gives:

$$\mathbf{r}\cdot\left[\frac{\mathrm{d}\mathbf{r}}{\mathrm{d}t}\times\mathbf{h}\right]=\mathbf{r}\cdot\left[\mu\frac{\mathbf{r}}{r}+\mathbf{c}\right] \tag{3.15}$$

By making use of the rule for scalar triple product, equation (3.15) becomes:

$$\mathbf{h}\cdot\left[\mathbf{r}\times\frac{\mathrm{d}\mathbf{r}}{\mathrm{d}t}\right]=\frac{\mu}{r}\mathbf{r}\cdot\mathbf{r}+\mathbf{c}\cdot\mathbf{r} \tag{3.16}$$

Substituting (3.9) into (3.16) gives

$$\mathrm{h}^2=\mu\mathrm{r}+\mathrm{rc}\cos\vartheta \tag{3.17}$$

where ϑ is the angle between vectors **c** and **r** and is referred to as the true anomaly in the satellite orbital plane.

By expressing:

$$c=\mu e \tag{3.18}$$

Hence:

$$r=\frac{h^2/\mu}{1+e\cos\vartheta} \tag{3.19}$$

Equation (3.19) is the general polar equation for a conic section with focus at the origin. For $0\leq e<1$, the equation describes an ellipse and the semi-latus rectum, p, is given by:

$$p=\frac{h^2}{\mu}=a(1-e^2) \tag{3.20}$$

where a and e are the semi-major axis and the eccentricity of the ellipse, respectively.

This proves Kepler's first law. Figure 3.1 shows the satellite orbital plane.

3.1.3. Satellite Swept Area per Unit Time – Proof of Kepler's Second Law

Referring to Figure 3.2, a satellite moves from M to N in time Δt, the area swept by the position vector **r** is approximately equal to half of the parallelogram with sides **r** and $\Delta\mathbf{r}$, i.e.

$$\Delta\mathbf{A}=\frac{1}{2}\mathbf{r}\times\Delta\mathbf{r} \tag{3.21}$$

Then the approximate area swept out by the radius vector per unit time is given by:

$$\frac{\Delta\mathbf{A}}{\Delta t}=\frac{1}{2}\mathbf{r}\times\frac{\Delta\mathbf{r}}{\Delta t} \tag{3.22}$$

Hence, the instantaneous time rate of change in area is:

$$\frac{\mathrm{d}\mathbf{A}}{\mathrm{d}t}=\lim_{\Delta t\to 0}\frac{1}{2}\mathbf{r}\times\frac{\Delta\mathbf{r}}{\Delta \mathrm{t}}=\frac{1}{2}\mathbf{r}\times\frac{\mathrm{d}\mathbf{r}}{\mathrm{d}t} \tag{3.23}$$

Substituting equation (3.9) into (3.23) gives:

$$\frac{\mathrm{d}\mathbf{A}}{\mathrm{d}t}=\frac{\mathbf{h}}{2} \tag{3.24}$$

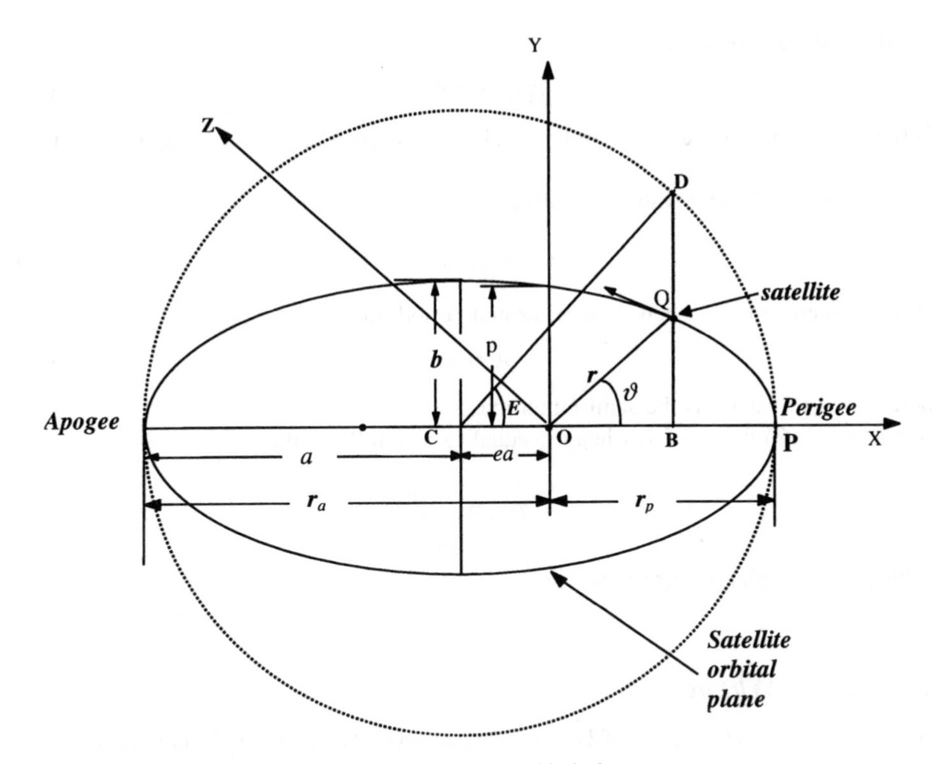

Figure 3.1 Satellite orbital plane.

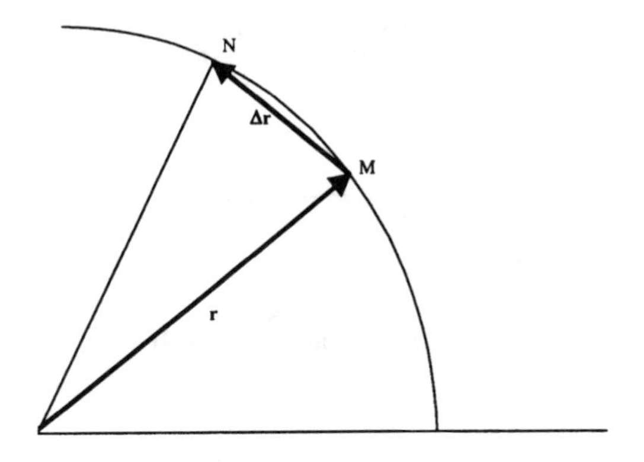

Figure 3.2 Area swept by the radius vector per unit time.

Since **h** is a constant vector, it follows that the satellite sweeps out equal areas in equal periods of time. This proves Kepler's second law.

3.1.4 The Orbital Period – Proof of Kepler's Third Law

From equation (3.20),

$$h = \sqrt{\mu a(1 - e^2)} \tag{3.25}$$

At the perigee and apogee,

$$r_p v_p = r_a v_a = h \tag{3.26}$$

where v_p and v_a are the velocities of the satellite at the perigee and the apogee, respectively.

Integrating (3.25) with respect to t from $t = 0$ to $t = t_1$ gives:

$$A = t_1 \sqrt{\mu a(1 - e^2)} \tag{3.27}$$

When t is equal to T, where T is the orbital period, then:

$$A = \pi ab \tag{3.28}$$

where $b = a(1 - e^2)^{1/2}$ is the semi-minor axis.

Equating (3.27) with (3.28) when t is equal to T, it follows that:

$$T = 2\pi \sqrt{\frac{a^3}{\mu}} \tag{3.29}$$

This proves Kepler's Third Law.

3.1.5 Satellite Velocity

Using the Law of Conservation of Energy in equation (3.4) and taking its dot product with $\boldsymbol{v}$, where $\boldsymbol{v}$ is the satellite velocity, gives:

$$\frac{d^2\mathbf{r}}{dt^2} \cdot \boldsymbol{v} = -\mu \frac{\mathbf{r}}{r^3} \cdot \boldsymbol{v} \tag{3.30}$$

From equation (3.30):

$$\frac{d\boldsymbol{v}}{dt} \cdot \boldsymbol{v} = -\frac{\mu}{r^3} \mathbf{r} \cdot \frac{d\mathbf{r}}{dt} \tag{3.31}$$

For any two vectors **a** and **b**

$$\frac{d}{dt}(\mathbf{a} \cdot \mathbf{b}) = \mathbf{a} \cdot \frac{d\mathbf{b}}{dt} + \frac{d\mathbf{a}}{dt} \cdot \mathbf{b}$$

It follows that

$$\frac{d(\boldsymbol{v} \cdot \boldsymbol{v})}{dt} = \frac{dv^2}{dt} = 2 \frac{d\boldsymbol{v}}{dt} \cdot \boldsymbol{v} \tag{3.32}$$

and

$$\frac{dr^2}{dt} = 2\mathbf{r} \cdot \frac{d\mathbf{r}}{dt} \tag{3.33}$$

Substituting (3.32) and (3.33) into (3.31) gives:

$$\frac{1}{2} \frac{dv^2}{dt} = -\frac{\mu}{2r^3} \frac{dr^2}{dt} \tag{3.34}$$

Integrating (3.34) with respect to t:

$$\frac{1}{2}v^2 = \frac{\mu}{r} + k \tag{3.35}$$

where k is a constant.

From equation (3.26) and evaluating k at the perigee gives

$$\begin{aligned} k &= \frac{1}{2}v_{\mathrm{p}}^2 - \frac{\mu}{r_{\mathrm{p}}} \\ &= \frac{1}{2}\left(\frac{h}{r_p}\right)^2 - \frac{\mu}{r_p} \\ &= \frac{1}{2}\frac{\mu a(1-e^2)}{a^2(1-e^2)} - \frac{\mu}{a(1-e)} \\ &= -\frac{\mu}{2a} \end{aligned} \tag{3.36}$$

Hence

$$\frac{1}{2}v^2 - \frac{\mu}{r} = -\frac{\mu}{2a} \tag{3.37}$$

It follows that

$$v^2 = \mu\left(\frac{2}{r} - \frac{1}{a}\right) \tag{3.38}$$

It follows from Figure 3.1 that the respective velocities, v_{p} and v_{a} at the perigee and apogee are:

$$v_{\mathrm{p}} = \sqrt{\frac{\mu}{a}\left[\frac{a(1+e)}{a(1-e)}\right]} = \sqrt{\frac{\mu r_{\mathrm{a}}}{a r_{\mathrm{p}}}} \tag{3.39}$$

$$v_{\mathrm{a}} = \sqrt{\frac{\mu}{a}\left[\frac{a(1-e)}{a(1+e)}\right]} = \sqrt{\frac{\mu r_{\mathrm{p}}}{a r_{\mathrm{a}}}} \tag{3.40}$$

where $r_{\mathrm{a}} = a(1 + e)$ is the apogee radius and $r_{\mathrm{p}} = a(1 - \mathrm{e})$ is the perigee radius.

3.2 Satellite Location

3.2.1 Overview

In order to design a satellite constellation for world-wide or partial coverage, a satellite's location in the sky has to be determined. A satellite's position can be identified with different co-ordinate systems, the choice being dependent upon the type of application. For example, radio communication engineers prefer to use look angles, specified in terms of azimuth and elevation, for antenna pointing exercises. The most commonly used co-ordinate systems are described in the following sections.

3.2.2 Satellite Parameters

A set of six orbital parameters is used to fully describe the position of a satellite in a point in space at any given time:

- Ω: the right ascension of ascending node, the angle in the equatorial plane measured counter-clockwise from the direction of the vernal equinox direction to that of the ascending node;
- i: inclination angle of the orbital plane measured between the equatorial plane and the plane of the orbit;
- ω: argument of the perigee, the angle between the direction of ascending node and direction of the perigee;
- e: eccentricity ($0 \leq e < 1$);
- a: semi-major axis of the elliptical orbit;
- ϑ: true anomaly.

The first three parameters, Ω, i and ω define the orientation of the orbital plane. They are used to locate the satellite with respect to the rotating Earth. The latter three parameters e, a and ν define the orbital geometrical shape and satellite motion; they are used to locate the satellite in the orbital plane. Figure 3.3 shows the orbital parameters with respect to the Earth's equatorial plane. The co-ordinate system is called the geocentric-equatorial co-ordinate system, which is used to locate the satellite with respect to the Earth. In this co-ordinate system, the centre of the Earth is the origin, O, and the xy-plane coincides with the equatorial plane. The z-axis coincides with the Earth's axis of rotation and points in the direction of the North Pole, while the x-axis points to the direction of the vernal equinox. The points at which

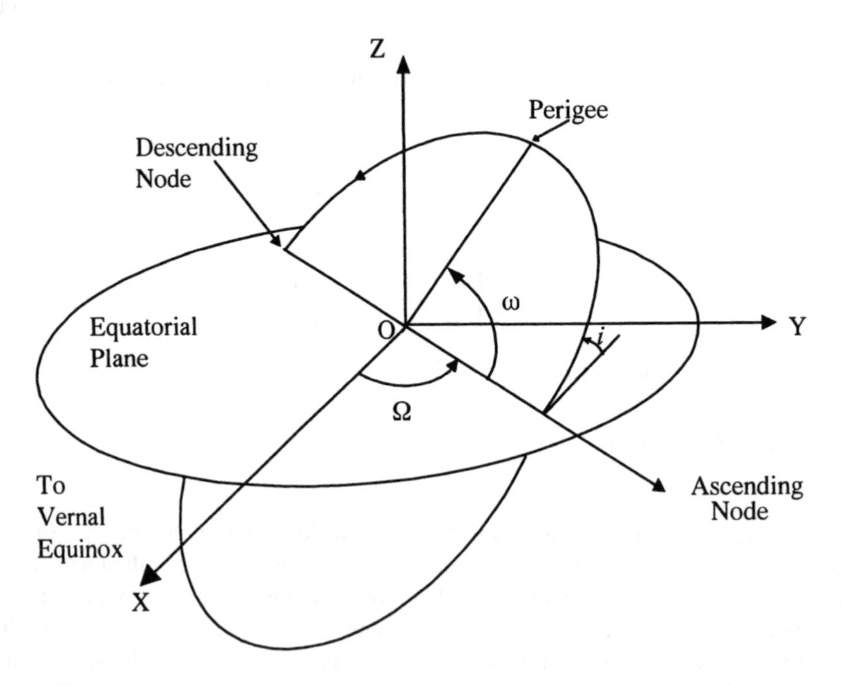

Figure 3.3 Satellite parameters in the geocentric-equatorial co-ordinate system.

the satellite moves upward and downward through the equatorial plane are called *ascending node* and *descending node*, respectively.

In addition to defining the location of a satellite in space, it is important to determine the direction at which an Earth station's antenna should point to the satellite in order to communicate with it. This direction is defined by the look angles – the elevation and azimuth angles – in relation to the latitude and the longitude of the Earth station. The following sections discuss the location of the satellite with respect to the different co-ordinate systems. Note: the formulation of the satellite location outlined in the following sections assumes that the Earth is a perfect sphere.

3.2.3 Satellite Location in the Orbital Plane

The location of a satellite in its orbit at any time t is determined by its true anomaly, ϑ, as shown in Figure 3.4. In the figure, the orbit is circumscribed by a circle of radius equal to the semi-major axis, a, of the orbit. O is the centre of the Earth and is the origin of the co-ordinate system. C is the centre of the elliptical orbit and the centre of the circumscribed circle. E is the eccentric anomaly.

Refer back to Figure 3.1 and consider the quadrant containing the points P, B, O, C and D as shown in Figure 3.4. In order to locate a satellite's position at any time t, the angular velocity, ϖ, and the mean anomaly, M, have to be found. By using the perigee as the reference point, the mean anomaly is defined as the arc length (in radians) that a satellite would have traversed at time t after passing through the perigee at time t_0 had it proceeded on the circumscribed circle with the same angular velocity. The angular velocity is obtained from (3.29) and is given by:

$$\varpi = \frac{2\pi}{T} = \sqrt{\frac{\mu}{a^3}} \tag{3.41}$$

Referring to Kepler's Third Law, the area swept out by the radius vector at time t after the

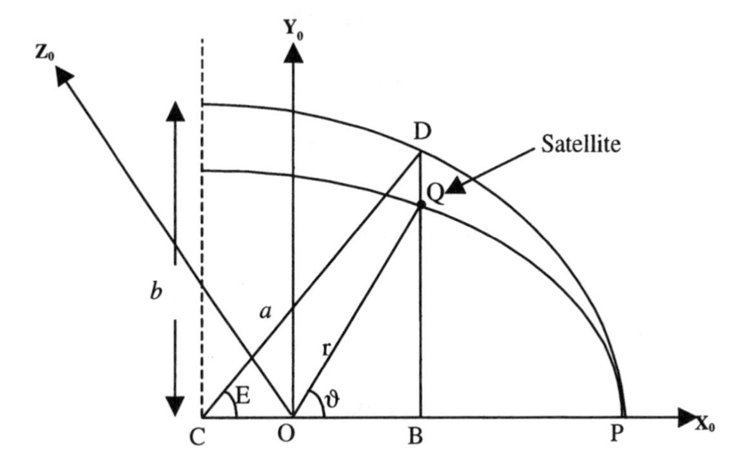

Figure 3.4 Satellite location with respect to an orbital plane co-ordinate system.

satellite moves through the perigee at time t_0 is given by:

$$\delta A = A\frac{t - t_0}{T} \tag{3.42}$$

Substituting (3.41) into (3.42):

$$\delta A = A\frac{\varpi(t - t_0)}{2\pi} = A\frac{M}{2\pi} \tag{3.43}$$

From equation (3.43)

$$M = \varpi(t - t_0) \tag{3.44}$$

From Figure 3.4, it can be seen that the area (CPD) is equal to $(a^2E/2)$ and area (CDB) is equal to $(a^2\cos E\sin E/2)$, this implies:

$$\text{Area (DBP)} = a^2[E - \cos E\sin E]/2 \tag{3.45}$$

Since DB/QB $= b/a$
then Area (QBP) $= (b/a)$Area (DBP) $= ab[E - \cos E\sin E]/2$
and Area (OQB) $=$ (OB)(QB)/2 $= (a\cos E - ae)(b\sin E)/2 = ab\sin E(\cos E - e)/2$
Therefore

$$\text{Area (OQP)} = \text{Area (OQB)} + \text{Area(QBP)} = \frac{ab}{2}[E - e\sin E] = \delta A \tag{3.46}$$

Equating (3.43) and (3.46) gives:

$$M = E - e\sin E \tag{3.47}$$

Equation (3.47) is known as Kepler's equation. In order to find E from (3.47), a numerical approximation method has to be used. Referring to Figure 3.4,

$$\begin{aligned} \text{QB} &= r\sin\vartheta \\ &= (b/a)(a\sin E) \\ &= b\sin E \\ &= a(1 - e^2)^{1/2} sinE \end{aligned} \tag{3.48}$$

and

$$\begin{aligned} \text{OB} &= r\cos\vartheta \\ &= a\cos E - ae \end{aligned} \tag{3.49}$$

Adding QB^2 and OB^2 from equations (3.48) and (3.49) gives:

$$r = \sqrt{a^2(1 - e\cos E)^2} \tag{3.50}$$

and

$$\vartheta = 2\tan^{-1}\left[\left(\frac{1 + e}{1 - e}\right)^{1/2}\tan\frac{E}{2}\right] \tag{3.51}$$

The location of the satellite in the orbital plane is then given as:

$$\begin{bmatrix} x_0 \\ y_0 \\ z_0 \end{bmatrix} = \begin{bmatrix} r\cos\vartheta \\ r\sin\vartheta \\ 0 \end{bmatrix} \tag{3.52}$$

3.2.4 Satellite Location with Respect to the Rotating Earth

Given the three orbital parameters, Ω, i, and ω, as shown in Figure 3.3, the location of the satellite in the geocentric-equatorial system in terms of the orbital plane –ordinates system is given by [PRA-86]:

$$\begin{bmatrix} x_f \\ y_f \\ z_f \end{bmatrix} = \begin{bmatrix} \cos\Omega\cos\omega - \sin\Omega\cos i\sin\omega & -\cos\Omega\sin\omega - \sin\Omega\cos i\cos\omega & \sin\Omega\sin i \\ \sin\Omega\cos\omega + \cos\Omega\cos i\sin\omega & -\sin\Omega\sin\omega + \sin\Omega\cos i\cos\omega & -\cos\Omega\sin i \\ \sin i\sin\omega & \sin i\cos\omega & \cos i \end{bmatrix} \begin{bmatrix} x_0 \\ y_0 \\ z_0 \end{bmatrix} \tag{3.53}$$

where the first matrix on the right hand side of equation (3.53) is called the rotation matrix. The geocentric-equatorial co-ordinate system (x_f, y_f, z_f) represents a fixed Earth system. In order to account for the rotation of the Earth, a transformation of co-ordinates from the fixed Earth system to a rotating Earth system is required. The relationship between the rotating Earth system and the geocentric-equatorial system is shown in Figure 3.5.

In this figure, the Earth is rotating at an angular velocity ϖ_e. If T_e is the elapsed time since the x_f-axis and the x_r-axis last coincided, then the location of the satellite in the rotating Earth system is related to the geocentric-equatorial system as follows [NAS-63]:

$$\begin{bmatrix} x_r \\ y_r \\ z_r \end{bmatrix} = \begin{bmatrix} \cos(\varpi_e T_e) & \sin(\varpi_e T_e) & 0 \\ -\sin(\varpi_e T_e) & \cos(\varpi_e T_e) & 0 \\ 0 & 0 & 1 \end{bmatrix} \begin{bmatrix} x_f \\ y_f \\ z_f \end{bmatrix} \tag{3.54}$$

The value of $\varpi_e T_e$ at any time t in minutes is given by:

$$\varpi_e T_e = \alpha_{g,o} + 0.25068447t \text{ (degrees)} \tag{3.55}$$

where $\alpha_{g,o} = 99.6909833 + 36000.7689\tau + 0.00038707\tau^2$ (degrees); $\tau = (\mathrm{JD} - 2415020)/36525$ (Julian centuries); $\mathrm{JD} = 2415020 + (\mathrm{Y} - 1899) \times 365 + \mathrm{Int}[(\mathrm{Y} - 1899)/4] + M_m + (D_m - D) + [(h - 12)]/24$.

JD is the Julian date in year Y, on day D of month m and at h hours (using pm/am convention). JD is calculated using the reference Julian Day 2415020, which is at noon on December 31, 1899. M_m is the number of days between month m and December, the value of which for different m is tabulated in Table 3.1. D_m is the number of days in month m. In the case of a leap year, $D_m = 29$ in February. $\alpha_{g,o}$ is the right ascension of the Greenwich meridian at 0 h UT at Julian date JD. τ is the elapsed time in Julian centuries between 0 h UT on Julian day JD and noon UT on January 1, 1900.

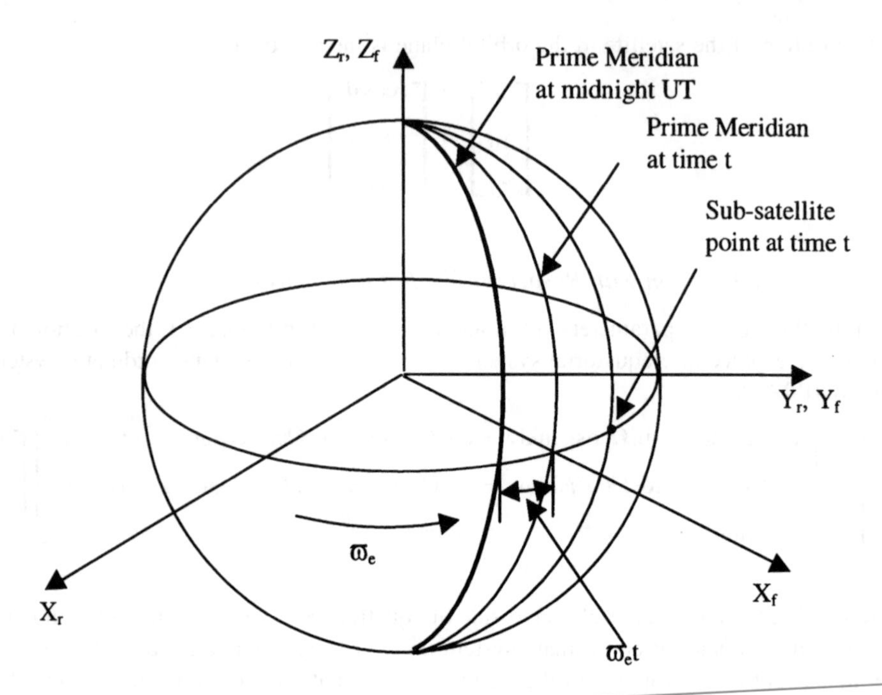

Figure 3.5 The rotating Earth system with respect to the fixed Earth system [PRI-93].

3.2.5 Satellite Location with Respect to the Celestial Sphere

Refer to Figure 3.6 and consider the location of a satellite in its celestial sphere. The position of the satellite can be located by its right ascension angle, α, and its declination angle, δ, on the celestial. From the spherical triangle ASN and using the law of sines,

Table 3.1 Value of M_m for Julian dates calculation

Month	M_m
January	334 (335 if it is a leap year)
February	306
March	275
April	245
May	214
June	184
July	153
August	122
September	92
October	61
November	31
December	0

$$sin(\alpha - \Omega) = \frac{cosi}{cos\delta} sin(\omega + \vartheta) \tag{3.56}$$

From the spherical triangle ABS and using the law of cosines:

$$cos(\omega + \vartheta) = \cos\delta\cos(\alpha - \Omega) \tag{3.57}$$

Eliminating cosδ from equation (3.56) and (3.57) gives:

$$\tan(\alpha - \Omega) = \cos i \tan(\omega + \vartheta) \tag{3.58}$$

Thus,

$$\alpha = \tan^{-1}[\cos i \tan(\omega + \vartheta)] + \Omega \tag{3.59}$$

By the law of sines,

$$\sin\delta = \sin i \sin(\omega + \vartheta) \tag{3.60}$$

From equations (3.57) and (3.60), it follows that

$$\delta = tan^{-1}[\tan i \sin(\alpha - \Omega)] \tag{3.61}$$

3.2.6 Satellite Location with Respect to Satellite-Centred Spherical Co-ordinates

Satellite-centred spherical co-ordinates locate a satellite in relation to an Earth station's latitude, L_g, and relative longitude, l_g, as shown in Figure 3.7.

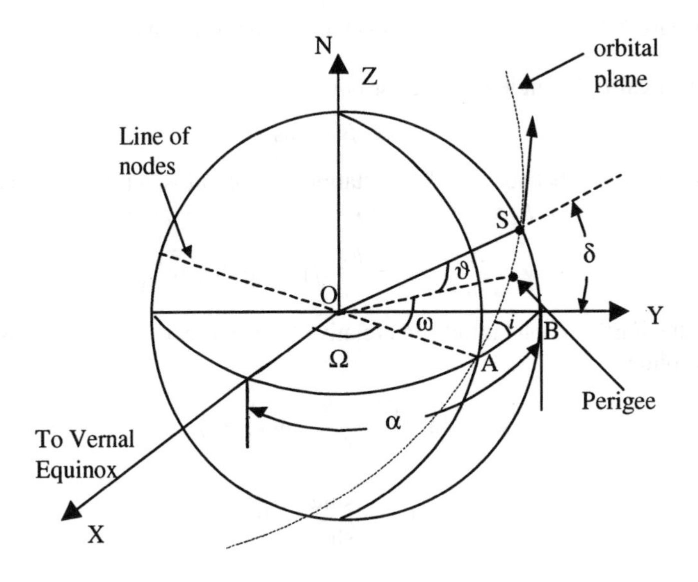

Figure 3.6 Celestial sphere co-ordinate system.

Using the law of sines in triangle GSM′,

$$\sin\beta = \frac{\text{GM}}{R} = \frac{R_\text{E}}{R}\sin L_\text{g} \tag{3.62}$$

Similarly,

$$\sin\alpha = \frac{R_\text{E}}{R}\frac{\cos L_\text{g}\sin l_\text{g}}{\cos\beta} \tag{3.63}$$

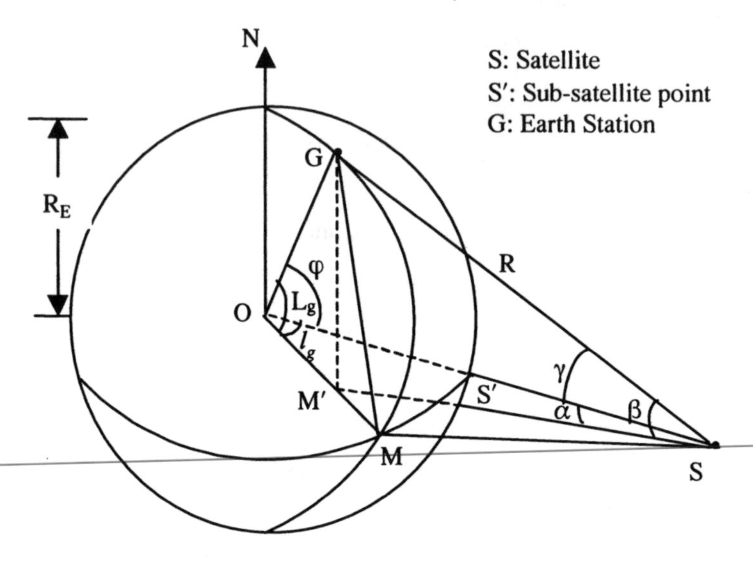

Figure 3.7 Satellite location in satellite-centred spherical co-ordinates.

From the spherical right-angle triangle formula,

$$\cos\alpha\cos\beta = \cos\gamma \tag{3.64}$$

γ is called the *tilt angle* between the Earth station and the sub-satellite point. The tilt angle can be expressed in terms of L_g and l_g as follows:

$$\sin\gamma = \frac{R_\text{E}}{R}\sin\varphi = \frac{R_\text{E}}{R}\sqrt{1 - \cos^2 L_\text{g}\cos^2 l_\text{g}} \tag{3.65}$$

By analogy, the Earth station latitude and relative longitude can also be expressed in terms of γ and β as follows:

$$\varphi = \sin^{-1}\left(\frac{h + R_\text{E}}{R_\text{E}}\sin\gamma\right) - \gamma \tag{3.66}$$

$$\sin L_\text{g} = \frac{\sin\varphi}{\sin\gamma}\sin\beta \tag{3.67}$$

$$\cos l_\text{g} = \frac{\cos\varphi}{\cos L_\text{g}} \tag{3.68}$$

The above equations are normally used to determine the antenna pointing angles and in calculating the antenna gain toward a specified Earth station.

3.2.7 Satellite Location with Respect to the Look Angles

3.2.7.1 Definitions

Radio communications engineers are more familiar with the elevation and azimuth angles as shown in Figure 3.8. The azimuth angle, ξ, is the angle measured North Eastward from the geographic North at the Earth station, G, to the sub-satellite point. The sub-satellite point, S', is defined as the point where the line joining the centre of the Earth, O, and the satellite meets the Earth's surface. The elevation angle, θ, is the angle measured upward from this tangential plane at the Earth station to the direction of the satellite.

In Figure 3.8 (and Figure 3.7), the angle φ is called the *central angle* or the *coverage angle* at the centre of the Earth, O, formed by lines OG and OS, where G denotes the Earth station and S is the satellite. The angle γ is called the *tilt angle* or the *nadir angle* at the satellite, formed by the lines GS and OS. L_g and l_g represent the latitude and the relative longitude (i.e. relative to the longitude of the sub-satellite point) of the Earth station, respectively and L_s is the latitude of the satellite. Note: in this co-ordinate system, Northern latitudes and Eastern longitudes are regarded as positive. Furthermore, the latitude of the sub-satellite point is the same as the declination angle, δ, of the satellite with respect to the geocentric-equatorial co-ordinate system, that is $L_s = \delta$. The slant range R of the satellite is its distance from the Earth station.

3.2.7.2 Elevation Angle

In order to calculate the elevation angle θ, the central angle φ and the slant range R have to be determined. The co-ordinates of G in Figure 3.8 are related to its Northern latitude and Eastern longitude by the following:

$$\begin{bmatrix} x_g \\ y_g \\ z_g \end{bmatrix} = \begin{bmatrix} R_E \cos L_g \sin l_g \\ R_E \cos L_g \cos l_g \\ R_E \sin L_g \end{bmatrix} \tag{3.69}$$

where R_E is the radius of the Earth (6378 km).

By analogy, the co-ordinates of the sub-satellite point, S$'$, are related to its Northern latitude and Eastern longitude by the following:

$$\begin{bmatrix} x_s \\ y_s \\ z_s \end{bmatrix} = \begin{bmatrix} 0 \\ R_E \cos L_s \cos l_s \\ R_E \sin L_s \end{bmatrix} \tag{3.70}$$

Consider triangle OS$'$G, the distance, d, between G and S$'$ can be found using the law of cosines:

$$d^2 = 2R_E^2(1 - \cos\varphi) \tag{3.71}$$

d can also be obtained from the following equation:

$$d^2 = (x_g - x_s)^2 + (y_g - y_s)^2 + (z_g - z_s)^2 \tag{3.72}$$

Substituting (3.69) and (3.70) into (3.72) and then equating (3.71) and (3.72) gives:

$$\cos\varphi = \cos L_g \cos L_s \cos l_g + \sin L_g \sin L_s \tag{3.73}$$

Now consider triangle GOS and using the law of cosines again, the slant range, R, is given by

$$R = \sqrt{R_E^2 + (R_E + h)^2 - 2R_E(R_E + h)\cos\varphi} \tag{3.74}$$

where h is the altitude of the satellite above the Earth's surface.

By applying the law of sines:

$$\frac{R_e + h}{\sin(90° + \theta)} = \frac{R}{\sin\varphi} \tag{3.75}$$

Substituting (3.67) into (3.68) and expressing θ in terms of φ gives:

$$\cos\theta = \frac{\sin\varphi}{\sqrt{1 + \left(\frac{R_E}{R_E + h}\right)^2 - 2\left(\frac{R_E}{R_E + h}\right)\cos\varphi}} \tag{3.76}$$

3.2.7.3 Azimuth Angle

The calculation of the azimuth angle, ξ, is slightly more complicated than that for the elevation angle due to the relative position of the sub-satellite point with respect to the Earth station and the hemisphere on which the sub-satellite point and the Earth station are located. For the time being, consider Figure 3.8 in which the sub-satellite point is in the direction East of the Earth station.

Referring to the spherical triangle GNS′, $\angle GNS' = l_g$ and $\text{arc}(GS') = \varphi$. Using the law of sines:

$$\frac{\sin\xi}{\sin(90° - L_s)} = \frac{\sin l_g}{\sin\varphi} \tag{3.77}$$

Since $L_s = \delta$, hence

$$\xi = \sin^{-1}\left[\frac{\cos\delta \sin l_g}{\sin\varphi}\right] \tag{3.78}$$

However, the above equation for ξ has not taken into account the relative position of the sub-satellite point with respect to the Earth station. Table 3.2 summarises the actual azimuth angle when this is taken into account.

3.2.7.4 Minimum Elevation Angle – Visibility

The condition for a satellite to be visible from an Earth station is $\theta \geq 0°$. From Figure 3.8, for the condition to be met

Table 3.2 Value of the azimuth angle ξ with respect to the relative position of the sub-satellite point

	Sub-satellite point (S′) position w.r.t. to the Earth station (G)	Azimuth angle (ξ)
Northern Hemisphere	S′ Northwest of G	$360° - \xi$
	S′ Southwest of G	$360° - \xi$
	S′ Northeast of G	ξ
	S′ Southeast of G	ξ
Southern Hemisphere	S′ Northwest of G	$180° + \xi$
	S′ Southwest of G	$180° + \xi$
	S′ Northeast of G	$180° - \xi$
	S′ Southeast of G	$180° - \xi$

$$R_{\mathrm{E}} + h \geq \frac{R_{\mathrm{E}}}{\cos\varphi} \tag{3.79}$$

This implies that for the condition of satellite visibility to be met,

$$\cos\varphi \leq \frac{R_{\mathrm{E}}}{R_{\mathrm{E}} + h} \tag{3.80}$$

From a geometrical point of view, the minimum condition that the satellite is visible from an earth station is $\theta = 0°$. In practise, however, the minimum value of θ, termed the minimum elevation angle, $\theta_{\min}$, should be high enough to avoid any propagation factors such as

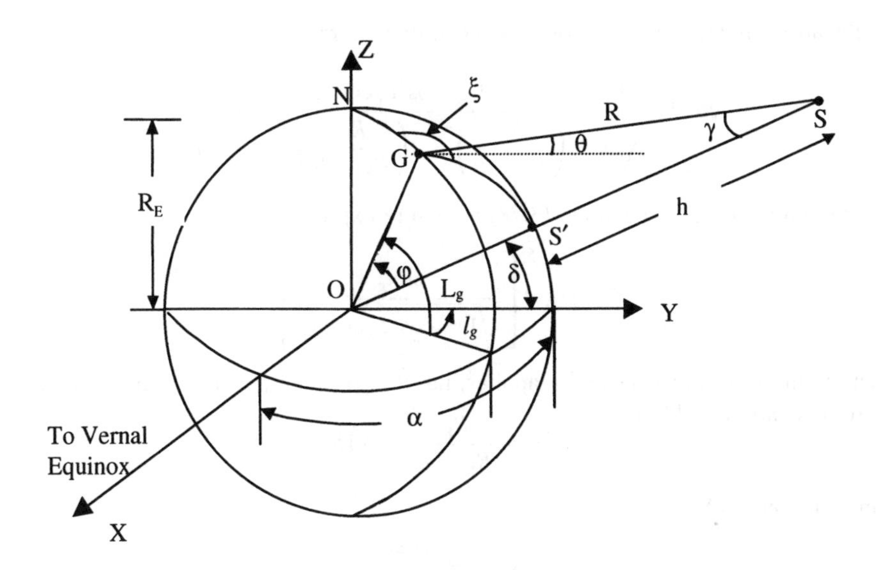

Figure 3.8 Satellite location with respect to the look angle.

shadowing. It is then convenient to express φ in terms of $\theta_{\min}$. Normally, a minimum elevation angle in the range 5–7° is used.

From Figure 3.8, it can be shown from triangle SGO that:

$$\varphi = \cos^{-1}\left[\frac{R_E \cos\theta_{\min}}{R_E + h}\right] - \theta_{\min} \tag{3.81}$$

Equation (3.81) can be used to calculate φ when $\theta_{\min}$ instead of γ is known as in equation (3.66).

3.2.7.5 Tilt Angle

Referring to triangle OSG in Figure 3.8, the tilt angle, γ, measured at the satellite from the sub-satellite point to the Earth station is given by:

$$\sin\gamma = \frac{R_E}{R}\cos\theta \tag{3.82}$$

Equation (3.81) can be used to calculate γ when $\theta_{\min}$ instead of φ is known as in equation (3.65).

3.2.8 Geostationary Satellite Location

For geostationary orbits, the inclination angle $i = 0°$, eccentricity $e = 0$ and, since the satellite is placed in the equatorial plane, the satellite's latitude, $L_s = 0°$. Furthermore, for geostationary satellites, $R_E + h = 42164$ km. Bearing this in mind, the central angle, φ, in equation (3.81) can be rewritten as:

$$\cos\varphi = \cos L_g \cos l_g \tag{3.83}$$

The elevation angle, θ, in equation (3.76) can then be re-expressed as:

$$\cos\theta = \sqrt{\frac{1 - \cos^2 L_g \cos^2 l_g}{1 + \left(\frac{R_E}{R_E + h}\right)^2 - 2\left(\frac{R_E}{R_E + h}\right)\cos L_g \cos l_g}} \tag{3.84}$$

The azimuth angle, ξ, in equation (3.78) is then re-expressed as:

$$\xi = \sin^{-1}\left[\frac{\sin l_g}{\sqrt{1 - \cos^2 L_g \cos^2 l_g}}\right] \tag{3.85}$$

When the limit of optical visibility applies, i.e. $\theta = 0°$, then the central angle, φ, can be found from equation (3.73) as:

$$\cos\varphi = \frac{R_E}{R_E + h} = 0.151 \tag{3.86}$$

From equation (3.68),

$$\cos L_g = \frac{\cos\varphi}{\cos l_g} \tag{3.87}$$

If the satellite and the Earth station are on the same meridian, it follows that $l_g = 0$. Thus, the maximum latitude, $L_{g,max}$ for which the satellite is visible can be obtained from the following equation:

$$\cos L_{g,max} = \cos\varphi = 0.151 \tag{3.88}$$

This implies that $L_{g,max} = 81.3°$.

3.3 Orbital Perturbation

3.3.1 General Discussion

The orbital equations derived in the previous section are based on two basic assumptions:

- The only force that acts upon the satellite is that due to the Earth's gravitational field;
- The satellite and the Earth are considered as point masses with the shape of the Earth being a perfect sphere.

In practise, the above assumptions do not hold. The shape of the Earth is far from spherical. In addition, apart from the gravitational force due to the Earth, the satellite will also experience gravitational fields due to other planets, and more noticeably, those due to the Sun and the Moon. Other non-gravitational field related factors including the solar radiation pressure and atmospheric drag also contribute to the satellite orbit perturbing around its elliptical path.

In the past, techniques have been derived to include the perturbing forces in the orbital description. By assuming that the perturbing forces cause a constant drift, to the satellite's position from its Keplerian orbit, which varies linearly with time, the satellite's position in terms of the six orbital parameters (see Section 3.2.1) at any instant of time t_1, can be written as:

$$\left[\Omega_0 + \frac{d\Omega}{dt}\delta t,\ i_0 + \frac{di}{dt}\delta t,\ \omega_0 + \frac{d\omega}{dt}\delta t,\ e_0 + \frac{de}{dt}\delta t,\ a_0 + \frac{da}{dt}\delta t,\ \nu_0 + \frac{d\nu}{dt}\delta t\right]$$

where (Ω_0, i_0, ω_0, e_0, a_0, ν_0) are the satellite's orbital parameters at time t_0; d()/dt is the linear drift in the orbital parameter with respect to time; and δt is $(t_1 - t_0)$.

In order to counteract the perturbation effect, station-keeping of the satellite has to be carried out periodically during the satellite's mission lifetime.

3.3.2 Effects of the Moon and the Sun

Gravitational perturbation is inversely proportional to the cube of the distance between two bodies. Hence, the effect of the gravitational pull from planets, other than the Earth, has a more significant effect on geostationary satellites than that on Low Earth Orbit (LEO) satellites. Among these planets, the effect of the Sun and the Moon are more noticeable. Although the mass of the Sun is approximately 30 times that of the Moon, the perturbation effect of the Sun on a geostationary satellite is only half that of the Moon due to the inverse cube law. The lunar–solar perturbation causes a change in the orbital inclination. The rate of change in a geostationary orbital inclination due to the Moon is described by the following formula [AGR-86]:

$$\left(\frac{di}{dt}\right)_{\text{moon}} = \frac{3}{4}\frac{\mu_{\text{m}}}{\text{h}}\frac{r^2}{r_{\text{m}}^3}\sin(\Omega_{\text{sat}} - \Omega_{\text{moon}})\sin i_{\text{m}}\cos i_{\text{m}} \tag{3.89}$$

where $\mu_{\text{m}} = 4902.8$ km^3/s^2 is the gravitational constant of the Moon; $r_{\text{m}} = 3.844 \times 10^5$ km is the Moon's orbital radius; $r = 42164$ km is the radius of a geostationary satellite orbit; $h = r\nu = 129640$ km^2/s is the angular momentum of the orbit.

Similarly, the perturbation effect in the inclination due to the Sun is expressed as:

$$\left(\frac{di}{dt}\right)_{\text{sun}} = \frac{3}{4}\frac{\mu_{\text{s}}}{h}\frac{r^2}{r_{\text{s}}^3}\sin(\Omega_{\text{sat}} - \Omega_{\text{sun}})\sin i_{\text{s}}\cos i_{\text{s}} \tag{3.90}$$

where $\mu_{\text{s}} = 1.32686 \times 10^{11}$ km^3/s^2 is the gravitational constant of the Sun. $r_{\text{s}} = 1.49592 \times 10^8$ km is the distance of the satellite from the Sun.

In both (3.89) and (3.90), the orbital radius of the geostationary satellite is assumed to be 42164 km. Hence, the value of the angular momentum, h, is 129640 km^2/s. The total perturbation effect due to the Sun and Moon together on the orbital inclination is given by [BAL-69]:

$$\left(\frac{di}{dt}\right)_{\text{total}} = \sqrt{(A + B\cos\Omega)^2 + (C\sin\Omega)^2}\ \text{deg/year} \tag{3.91}$$

where $A = 0.8457$, $B = 0.0981$, $C = -0.090$ and Ω is the right ascension of the ascending node of the lunar orbit in the ecliptic plane and can be expressed as:

$$\Omega = -\frac{2\pi}{18.613}(T - 1969.244)\ \text{rad/year} \tag{3.92}$$

where T is the date in years.

From equations (3.89)–(3.92), it can be shown that the lunar perturbation causes a cyclic variation between 0.48° and 0.67° in the orbital plane inclination, whereas the Sun causes a steady change of about 0.27° per year. The cyclic variation from the Moon is due to the fact that the orbit of the Moon is also affected by the Sun's gravitational pull. The resultant perturbation causes a change in the inclination of about 0.75–0.95° per year to a geostationary satellite orbit. This has the effect of causing the satellite to follow an apparent figure of eight ground trace at the rate of one oscillation per sidereal day. In order to correct the change in inclination, an increment in the satellite velocity, $\Delta\nu$, is required. $\Delta\nu$ is related to the change in the orbital inclination, Δi, by the following equation

$$\Delta\nu = 2\nu\sin\frac{\Delta i}{2} \tag{3.93}$$

By expressing (3.93) in terms of the satellite mission lifetime, T, $\Delta\nu$ can *be* re-expressed as:

$$\Delta\nu = \nu\sin\frac{\Delta i}{\Delta t}\frac{\pi}{180°}T \tag{3.94}$$

Assuming an average inclination drift of 0.85° per year for a geostationary satellite orbit due to the lunar-solar perturbation, a velocity increment of about 50 m/s per year is required if T is assumed to take the value of 10 years. If left uncorrected, the change in inclination can be up to 15° in about 27 years after the year of launch. Although this well exceeds the satellite's lifetime, it demonstrates the severity of the problem. The velocity increment is performed by giving the spacecraft a velocity impulse in the North direction to rotate the orbital plane through an angle of $\Delta i = 2\delta i$, where δi is the maximum allowable inclination. It has been shown in [BAL-69] that the optimum time between manoeuvres is given by:

$$T_{\Delta \nu} = \frac{2\delta i}{\mathrm{d}i/\mathrm{d}t} \tag{3.95}$$

If the rate of change in orbital inclination $\mathrm{d}i/\mathrm{d}t$ is 0.85° per year and $\delta i = \pm 0.025°$, $T_{\Delta \nu} \approx$ 21 days.

3.3.3 *Effects of the Oblate Earth*

The Earth is an ellipsoid with the equatorial radius being 21 km more than the polar radius. This non-spherical nature of the Earth increases its gravitational potential. Instead of being represented by a simple formulae $G(r) = -\mu/r$, the gravitation field of the Earth is represented in terms of the Legendre Polynomial of degree n [WOL-61]:

$$G(r, \theta) = -\frac{\mu}{r}\left[1 - \sum_{n=2}^{\infty} J_n \left(\frac{R_E}{r}\right)^n P_n(\cos\theta)\right] \tag{3.96}$$

where r is the distance from the centre of the Earth; $R_E = 6378.137$ km is the equatorial radius of the Earth; θ is the co-latitude; J_n is the harmonic coefficient of the Earth of degree n; and P_n is the Legendre Polynomial of degree n.

The effect of the non-spherical shape of the Earth causes the ascending node Ω, to drift Westward for direct orbits ($i < 90°$).

$$\dot{\Omega} = -\frac{3}{2}\left(\frac{2\pi}{T}\right)\left(\frac{R_E}{a}\right)^2 \frac{J_2}{(1-e^2)^2} \cos i \text{ deg/day} \tag{3.97a}$$

or

$$\dot{\Omega} = -\frac{9.964}{(1-e^2)^2}\left(\frac{R_E}{a}\right)^{3.5} \cos i \text{ deg/day} \tag{3.97b}$$

The negative sign indicates that the node drifts Westward for direct orbits ($i < 90°$) and Eastward for retrograde orbits ($i > 90°$). For polar orbits ($i = 90°$), the ascending node remains unchanged.

Apart from the drift in the ascending node, the argument of the perigee will also rotate either forward or backward. The rate of rotation in the argument of the perigee is given by the following equation:

$$\dot{\omega} = -\frac{3}{4}\left(\frac{2\pi}{T}\right)\left(\frac{R_E}{a}\right)^2 \frac{J_2}{(1-e^2)^2} (5\cos^2 i - 1) \text{ deg/day} \tag{3.98a}$$

or

$$\dot{\omega} = -\frac{4.982}{(1-e^2)^2}\left(\frac{R_E}{a}\right)^{3.5} (5\cos^2 i - 1) \text{ deg/day} \tag{3.98b}$$

At $i = 63.4°$ or 116.6°, ω remains constant.

Another effect due to the oblate Earth is on the period of node-to-node revolution, T_N, which will differ slightly from the ideal Keplerian period for a perfectly spherical Earth. The change in the node-to-node period of revolution is [KAL-63]:

$$\frac{\Delta T_{\mathrm{N}}}{T} = -\frac{3}{8}\left(\frac{R_{\mathrm{E}}}{a}\right)^2 \frac{J_2}{(1-e^2)^2}(7\cos^2 i - 1) \tag{3.99}$$

where a is the semi-major axis and T is the mean Keplerian period.

However, the oblateness of the Earth sometimes can be used for good cause. For Earth resource and surveillance missions where constant illumination conditions are desirable, a Sun-synchronous orbit can be used to make use of the advantage that the ascending node drifts Eastward at 0.9856° per day. This is the rate at which the Earth orbits around the Sun. In this case, the orientation of the orbital plane with respect to the Earth–Sun line remains fixed and a constant illumination condition can be met.

3.3.4 *Atmospheric Drag*

Atmospheric drag affects the rate of the decay of an orbit and the satellite lifetime as a result of the drag force from the atmosphere on the satellite. This is due to the frictional force and heat generated on a satellite caused by collision with the atoms and ions present in the atmosphere. It has a more prominent effect on LEO satellites below 800 km. The drag force on the satellite is expressed as [JEN-62]:

$$\mathbf{D} = -\frac{1}{2} C_{\mathrm{D}} A \rho v \boldsymbol{v} \tag{3.100}$$

where C_{D} is the drag coefficient; A is the cross-sectional area; ρ is the atmospheric density; and v is the satellite velocity.

By rewriting equation (3.4) in terms of the unit vectors $\mathbf{e}_{\mathrm{r}}$ and $\mathbf{e}_{\nu}$, it can be shown that equation (3.4) can be expressed in component form:

$$\left.\begin{aligned} \ddot{r} - r\dot{\nu} &= -\frac{\mu}{r^2} \\ \frac{1}{r}\frac{\mathrm{d}}{\mathrm{d}t}\left(r^2\dot{\nu}\right) &= r\ddot{\nu} + 2\dot{r}\dot{\nu} = 0 \end{aligned}\right\} \tag{3.101}$$

Taking into account the atmospheric drag, equation (3.101) becomes:

$$\left.\begin{aligned} \ddot{r} - r\dot{\nu} &= -\frac{\mu}{r^2} - B\rho v\dot{r} \\ r\ddot{\nu} + 2\dot{r}\dot{\nu} &= -B\rho v r\dot{\nu} \end{aligned}\right\} \tag{3.102}$$

where $B = (C_{\mathrm{D}}A)/2m$ is called the *ballistic coefficient* and m is the mass of the satellite.

For a circular orbit, the orbital decay causes no change on the shape of the orbit, i.e. it will remain circular. However, for an elliptical orbit, the orbital shape will become more circular.

3.4 Satellite Constellation Design

3.4.1 *Design Considerations*

In designing a satellite constellation, a major consideration is to provide the specified coverage area with the fewest number of satellites. When the elevation angle is equal to 0°, the instantaneous coverage area of a satellite is at its maximum. Any point located within this coverage area will be within the geometric visibility to the satellite. However, close to zero elevation angle is not operable due to the high blocking and shadowing effects, as will be discussed in the following chapter. This leads to the concept of minimum elevation angle. The minimum elevation angle is defined as the elevation angle required for the instantaneous coverage area to be within the 'radio-frequency visibility'. For a given minimum elevation angle, the only factor affecting the coverage area is the satellite altitude. Figure 3.9 shows a typical circle of coverage by a satellite at an altitude *h*.

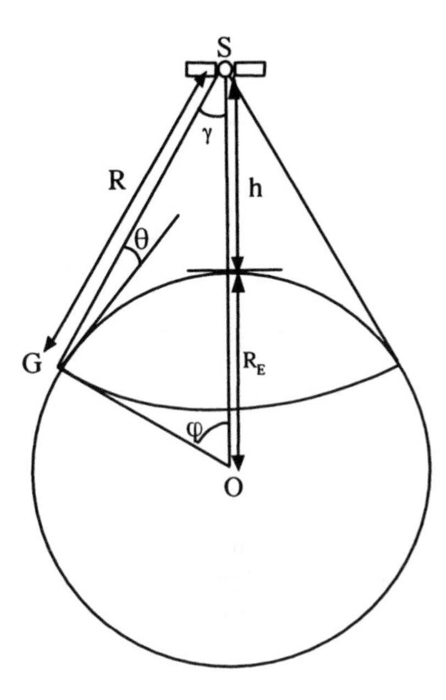

Figure 3.9 Coverage area by a satellite at an altitude *h*.

While a single geostationary satellite can provide continuous coverage, a constellation of satellites is required for non-geostationary orbits. The choice of constellation depends on a number of factors:

1. The elevation angle used should be as high as possible. This is particularly important for mobile-satellite services. With a high elevation angle, the multipath and shadowing problem can be reduced resulting in better link quality. However, there is a trade-off between the elevation angle used and the dimension of the service area.
2. The propagation delay should be as low as possible. This is especially the case for real-time services. This poses a constraint on the satellite altitude.
3. The battery consumption on board the satellite should be as low as possible.
4. Inter- and intra-orbital interference should be kept within an acceptable limit. This poses a requirement on the orbital separation.
5. The regulatory issues governing the allocation of orbital slots for different services and to different countries.

For an optimal constellation of satellites, the most efficient plan is to have the satellites equally spaced within a given orbital plane and the planes equally spaced around the equator. The coverage obtained by successive satellites in a given orbital plane is described by a ground swath or street of coverage as shown in Figure 3.10. Total Earth coverage is achieved by overlapping ground swaths of different orbital planes.

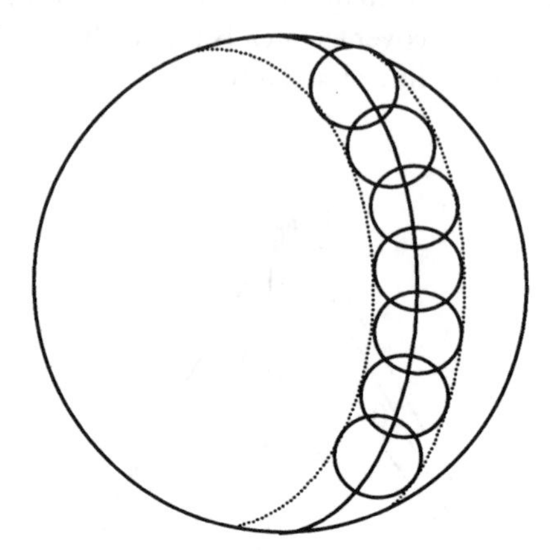

Figure 3.10 Ground swath coverage.

The total number of satellites in a constellation is given by $N = ps$, where p is the number of orbital planes and s is the number of satellites per plane.

Another point of consideration in the design of a satellite constellation is the number of satellites being visible at any one time within a coverage area in order to support certain applications or to provide a guaranteed service.

3.4.2 Polar Orbit Constellation

A polar orbit constellation will usually result in the provision of single satellite coverage near the equatorial region, while the concentration of satellites near the polar caps is significantly higher. The problem of designing orbit constellations to provide continuous single-satellite coverage was first addressed by Lüders [LUD-61]. Lüder's approach was later extended such that satellites were placed in orbital planes which have a common intersection, for example in the polar region. The orbital plane separation and satellite spacing were then adjusted in order to minimise the total number of satellites required. Beste [BES-78] subsequently derived another method for polar constellation design for both single coverage and triple coverage by selecting orbital planes in such a way that a more uniform distribution of satellites over the Earth was obtained. Later on, Adams and Rider [ADA-87] also derived another optimisation technique for designing such a constellation. The geometry used in optimising a polar orbit constellation is shown in Figure 3.11.

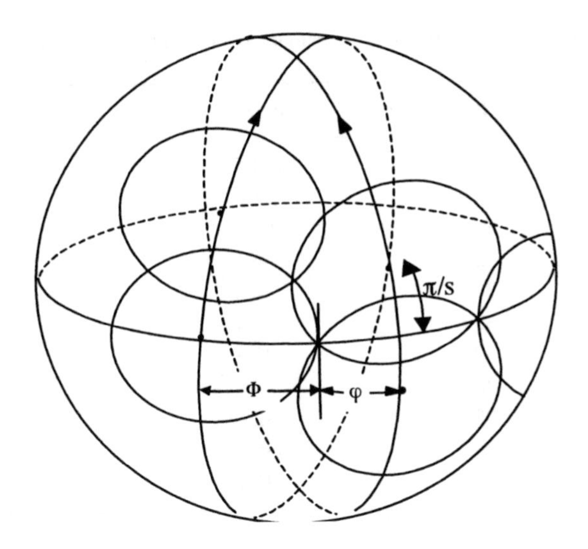

Figure 3.11 Coverage geometry for polar orbit optimisation.

3.4.2.1 Beste Approximations to Polar Orbit Constellation Design

Global Coverage In Figure 3.11, φ is the central angle as defined in Section 3.2. When optimal phasing of orbital planes is considered for global coverage, the point of intersection of overlapping circles of coverage coincides with the boundary of a circle coverage in the adjacent plane, as shown in Figure 3.11. Since the satellites are uniformly distributed in a given orbital plane, the phase separation between two consecutive satellites within an orbital plane is given by $2\pi/s$. By applying spherical trigonometry, the angular half-width, Φ, of the ground swath with single satellite coverage is:

$$\cos\varphi = \frac{\cos\Phi}{\cos\left(\frac{\pi}{s}\right)} \tag{3.103}$$

Satellites in adjacent planes travel in the same direction. The inter-plane angular separation between adjacent planes is equal to $\alpha = (\Phi + \varphi)$. However, satellites in the first and last planes rotate in opposite directions. Because of this counter-rotation effect, the angular separation between the first and last planes is smaller than that between adjacent planes. At the equator, Φ has to satisfy the following condition:

$$(p-1)(\Phi+\varphi)+2\varphi=\pi \tag{3.104}$$

where p is the number of orbital planes.

It follows that

$$\alpha = \Phi + \varphi \geq \pi/p \tag{3.105}$$

and that the angular separation between the first and the last planes is equal to 2φ.

Table 3.3 Requirements of p, s, Φ and φ for single coverage of the entire Earth [BES-78]

p	s	Φ (°)	$\psi = \Phi + \varphi$ (°)	$ps\Omega/4\pi$
2	3	66.7	104.5	1.81
2	4	57.6	98.4	1.86
3	4	48.6	69.3	2.03
3	5	42.3	66.1	1.95
3	6	38.7	64.3	1.97
4	6	33.6	49.4	2.00
4	7	30.8	48.3	1.97
4	8	28.9	47.6	1.99
5	8	25.7	38.6	1.98
5	9	24.2	38.1	1.97
5	10	23.0	37.7	1.99

From equations (3.104) and (3.105), Beste has computed the values of p and s for single coverage of the entire Earth as shown in Table 3.3.

In Table 3.3, Ω is referred to as the solid angle and is equal to $\Omega = 2\pi(1 - \cos\Phi)$. From the results shown in Table 3.3, Beste suggested that the number of satellites required for single-satellite global coverage can be approximated by:

$$N_{B-1} = ps \approx 4/(1 - \cos\Phi), \quad 1.3p < s < 2.2p \tag{3.106}$$

Where N_{B-1} denotes the total number of satellites required to provide global single-satellite coverage as derived by Beste.

Full Coverage Beyond a Specified Latitude Beste has extended the analysis of entire Earth coverage to partial coverage for latitudes beyond a specified value, L, as shown in Figure 3.12.

Beste has shown that in order to provide coverage beyond a specified latitude, L, the constraint specified by equation (3.104) for global coverage becomes:

$$(p - 1)(\Phi + \varphi) + 2\varphi = \pi\cos L \tag{3.107}$$

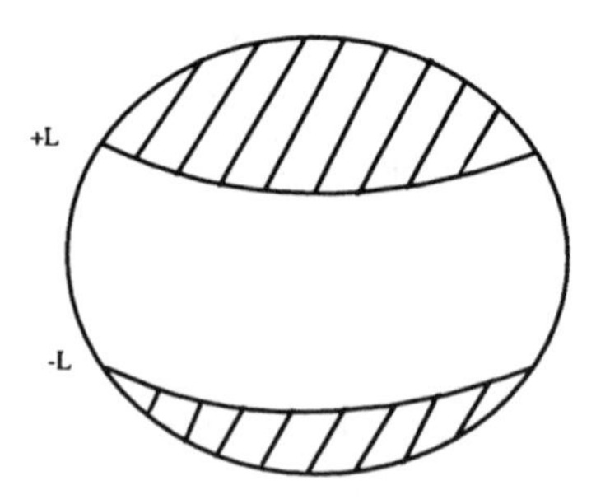

Figure 3.12 Coverage area beyond a specified latitude value.

The total number of satellites as identified in equation (3.106) can be generalised as follows:

$$N_{B-1,L} = ps = 4\cos L/(1 - \cos\Phi), \ 1.3p < s\cos L < 2.2p \tag{3.108}$$

Triple Coverage Beste continued to extend his analysis for triple coverage by using an iterative method. His approximation for triple coverage is as follows:

$$N_{B-3,L} = ps = 11\cos L/(1 - \cos\Phi), \ 1.3p < s\cos L < 2.2p \tag{3.109}$$

Table 3.4 records the results as computed by Beste for triple coverage of the entire Earth.

Table 3.4 Requirements of p, s, Φ and φ for triple coverage of the entire Earth [BES-78]

p	s	Φ (°)	$\alpha = \Phi + \varphi$ (°)	$ps\Omega/4\pi$
3	4	80.7	64.5	5.03
3	5	70.3	62.3	4.98
3	6	63.9	60.3	5.04
3	7	61.1	60.0	5.42
4	6	56.4	46.9	5.36
4	7	52.2	45.9	5.42
4	8	48.3	45.4	5.35
5	8	43.7	36.9	5.54
5	9	41.1	36.6	5.55
5	10	38.8	36.2	5.52
6	9	37.5	30.8	5.59
6	10	35.8	30.5	5.66

3.4.2.2 Adams and Rider Approximation to Polar Orbit Constellation Design

Using similar geometry to that shown in Figure 3.11, Adams and Rider [ADA-87] arrived at a different expression for the total number of satellites, $N_{A-R,}$, required for providing multiple satellite coverage by optimisation techniques using the method of Lagrange multipliers. The exact geometry for the derivation of Adam and Riders approximations is shown in Figure 3.13.

In Adam and Riders approach, polar constellation design for multiple coverage beyond a specified latitude, ϕ, is analysed. From Figure 3.13,

$$\varsigma = \phi + \frac{\pi}{s} \tag{3.110}$$

By applying spherical geometry in spherical triangle NGO:

$$\cos\phi = \cos\left(\frac{\pi}{2} - \varsigma\right)\cos\left(\frac{\pi}{2} - \phi\right) + \sin\left(\frac{\pi}{2} - \varsigma\right)\sin\left(\frac{\pi}{2} - \phi\right)\cos\Phi \tag{3.111}$$

After manipulation, Φ can be obtained as follows:

$$\cos\Phi = \frac{\cos\phi - \sin\varsigma\sin\phi}{\cos\varsigma\cos\phi} \tag{3.112}$$

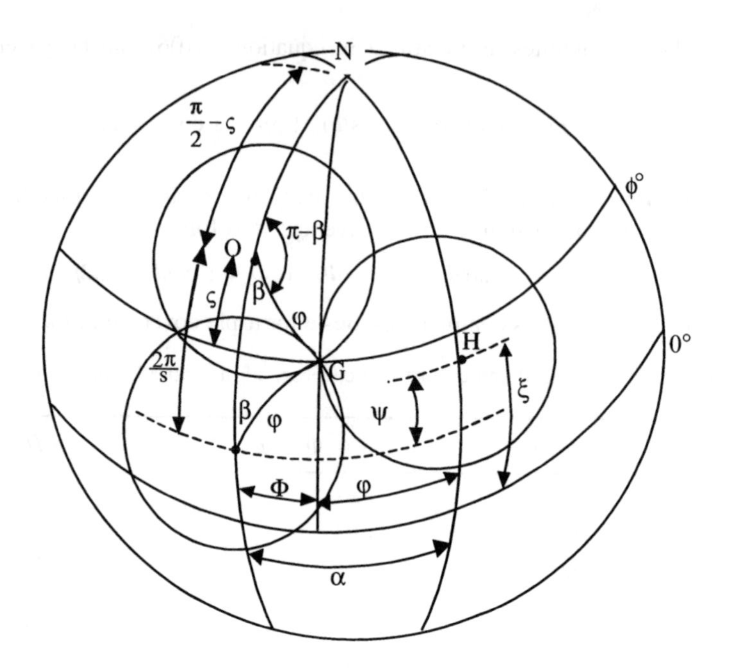

Figure 3.13 Single geometry for single satellite coverage above latitude ϕ.

In the same figure, for a given ψ, $(\pi/2 - \varsigma) + 2\pi/s = (\pi/2 - \xi) + \psi$, this implies:

$$\xi = \varsigma + \psi - \frac{2\pi}{s} \tag{3.113}$$

For optimum phasing, $\psi = \varsigma/2$. By applying spherical trigonometry to the spherical triangle NGH,

$$\cos\phi = \cos\left(\frac{\pi}{2} - \xi\right)\cos\left(\frac{\pi}{2} - \phi\right) + \sin\left(\frac{\pi}{2} - \xi\right)\sin\left(\frac{\pi}{2} - \phi\right)\cos\varphi \tag{3.114}$$

or

$$\cos\varphi = \frac{\cos\phi - \sin\xi\ sin\phi;}{\cos\xi\ \cos\phi} \tag{3.115}$$

where α has to satisfy the condition: $\alpha = \Phi + \varphi \geq \pi/p$.

Computation can be carried out to obtain the total number of satellites required for single coverage above a latitude ϕ, for given Φ, φ and s.

The above equations can be generalised for multiple coverage over a given point. Let j be the multiple level of coverage provided by satellites in a single plane; and let k be the multiple level of coverage provided by satellites in neighbouring planes. The total multiple level of coverage n can be factorised as $n = jk$. By making use of the Lagrange multipliers technique [ADA-87], it can be shown that:

$$\left.\begin{aligned} s &= \frac{2}{\sqrt{3}} j \frac{\pi}{\varphi} \\ p &= \frac{2}{3} k \frac{\pi}{\varphi} \\ N_{A-R} = ps &= \frac{4\sqrt{3}}{9} n \left(\frac{\pi}{\varphi} \right)^2 \end{aligned}\right\} \tag{3.116}$$

where j denotes the multiple coverage factor in the same orbital plane; k denotes the multiple coverage factor in different orbital planes; $n = jk$ denotes the multiple coverage factor of the constellation.

An analogous expression for optimum triple coverage using Adams and Rider's formula can be made by setting $n = 3$, $k = 1$.

3.4.3 Inclined Orbit Constellation

Optimisation techniques for inclined orbit constellations have been investigated by several researchers in the past [WAL-73, LEO-77, BAL-80]. Walker [WAL-73] has shown that world-wide single-satellite coverage can be accomplished by five satellites; while seven satellites can provide dual-coverage. Walker's work has been extended and generalised by Ballard [BAL-80]. The class of constellation considered by Ballard is characterised by circular, common-periods all having the same inclination with respect to an arbitrary reference plane. The orbits are uniformly distributed in a right ascension angle as they pass through the reference plane. The initial phase positions of satellites in each orbital plane is proportional to the right ascension of that plane. The optimisation parameter is the coverage angle (also called the central angle), φ, as indicated in Figure 3.14, from an observer anywhere on the Earth's surface to the nearest sub-satellite point. In inclined orbit constellations, all the orbital planes have the same inclination angle, i, with reference to the equatorial plane. Figure 3.14 shows the geometry adopted by Ballard for inclined orbit constellation optimisation. ψ_{ij} denotes the inter-satellite bearing angles. Ballard has named this type of constellation as *rosette constellation* since the orbital traces resemble the petals of flowers.

There are several steps to Ballard's optimisation. At first, a set of common altitude orbits which minimise the maximum value of φ is chosen, considering all possible observation points on Earth at all instants of time. The constellation altitude is then selected to obtain a guaranteed minimum elevation angle. The constellation is used to provide world-wide coverage such that all orbits are assumed to have the same altitude and period T. All the satellites move in a circular path following a celestial sphere with radius $R_e + h$. The position of the satellite is described by three constant orientation angles and a time-varying phase angle:

α_j	right ascension angle for the jth orbital plane
i_j	inclination angle
γ_j	initial phase angle of the jth satellite in its orbital plane at $t = 0$, measured from the point of the right ascension
χ_j	$2\pi t/T$ = time-varying phase angle for all satellites of the constellation

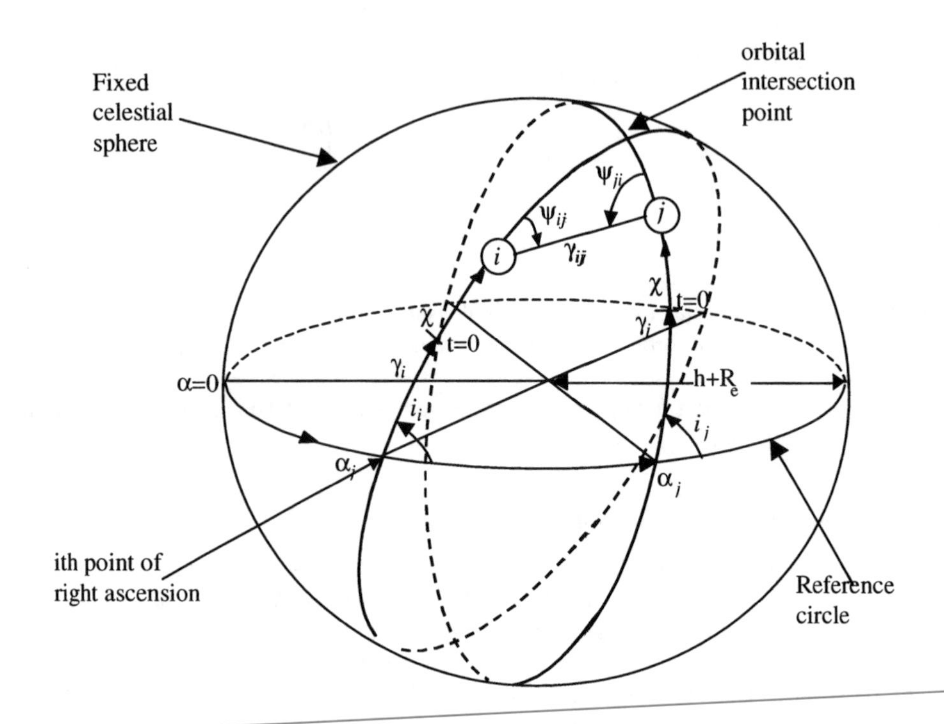

Figure 3.14 Geometry for inclined orbit constellation optimisation.

Thus, for a rosette constellation containing N satellites, p planes and s satellites in each plane, the orbital orientation angles have the following symmetric form:

$$\left.\begin{aligned} \alpha_j &= 2\pi j/p \\ i_j &= i \\ \gamma_j &= m\alpha_j = ms(2\pi j/N) \end{aligned}\right\} \tag{3.117}$$

Where m is the harmonic factor, which influences both the initial distribution of satellites over the sphere and the rate at which the constellation pattern progresses around the sphere.

The harmonic factor, m, can be an integer or an unreduced ratio of integers (fractional values). If m takes on integer values from 0 to $N-1$, i.e. $s=1$, widely different constellation patterns are generated, all of which contain a single satellite in each of the N separate orbital planes. For more generalised rosette constellations having s satellites in each of the p orbital planes, m takes the fractional values of (0 to $N-1)/s$. Hence, a rosette constellation is designated by the notation (N, p, m).

Referring to Figure 3.15, the angular range, $\Re_{jk}$ between any two satellites in the jth and kth planes, respectively, is expressed by the following formula:

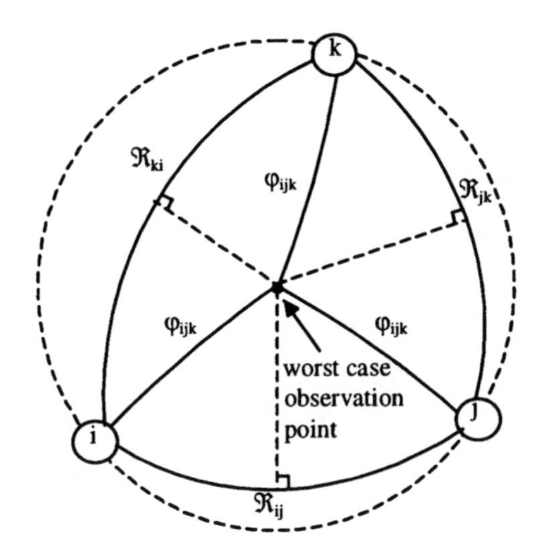

Figure 3.15 Triangle triad used by Ballard for optimisation of the arc range φ.

$$\begin{aligned}\sin^2(\Re_{jk}/2) = {} & \cos^4(i/2)\sin^2[(m+1)(k-j)(\pi/p)] \\ & +2\sin^2(i/2)\cos^2(i/2)\sin^2[m(k-j)(\pi/p)] \\ & +\sin^4(i/2)\sin^2[(m-1)(k-j)(\pi/p)] \\ & +2\sin^2(i/2)\cos^2(i/2)\sin^2[(k-j)(\pi/p)]\cos[2\chi+2m(j+k)(\pi/p)]\end{aligned} \tag{3.118}$$

Table 3.5 Best single visibility rosette constellations for $N = 5$–15 [BAL-80]

Constellation dimension			Optimum inclination	Minimax arc range	Lowest deployment of elevation ≥10°	
N	*P*	*m*	*i* (°)	φ (°)	*h*	*T*
5	5	1	43.66	69.15	4.232	16.90
6	6	4	53.13	66.42	3.194	12.13
7	7	5	55.69	60.26	1.916	7.03
8	8	6	61.86	56.52	1.472	5.49
9	9	7	70.54	54.81	1.314	4.97
10	10	7	47.93	51.53	1.066	4.19
11	11	4	53.79	47.62	0.838	3.52
12	3	1/4, 7/4	50.73	47.90	0.853	3.56
13	13	5	58.44	43.76	0.666	3.04
14	7	11/2	53.98	41.96	0.598	2.85
15	3	1/5, 4/5, 7/5, 13/5	53.51	42.13	0.604	2.87

Ballard proceeded to show that in order to provide world-wide single satellite coverage, the coverage angle, φ, has to exceed or be equal to φ_{ijk}, where φ_{ijk} is the equi-distance arc range from the mid-point of the spherical triangle. This is formed by joining the three sub-satellite points, to any of its three vertices of the spherical triangle as shown in Figure 3.15. If $\varphi < \varphi_{ijk}$, the mid-point of the spherical triangle will represent the worst observation point since none of the three satellite coverages would reach the mid-point. Conversely, if $\varphi \geq \varphi_{ijk}$, the number of satellites visible from the mid-point increases to three. φ_{ijk} can be obtained from the following formula:

$$\sin^2 \varphi_{ijk} = 4ABC/[(A + B + C)^2 - 2(A^2 + B^2 + C^2)] \qquad (3.119)$$

where $A = \sin^2(\Re_{ij}/2)$, $B = \sin^2(\Re_{jk}/2)$, $A = \sin^2(\Re_{ki}/2)$.

Equations (3.118) and (3.119) define the geometry of a satellite constellation for world-wide coverage. The coverage properties have to be analysed by examining the equi-distances φ_{ijk} of the spherical triangles at all instants of time over the orbital period to find the worst observation point. Ballard has shown that with a total number of N satellites, $(2N - 4)$ non-overlapping triangles are required to cover the whole sphere. The critical phase, χ, and the optimum inclination i, which produces the smallest equi-distance for the largest triangle are obtained by a trial and error process. Ballard has tabulated the results for the best single visibility rosette for N between 5 and 15 as shown in Table 3.5.

References

[ADA-87] W.S. Adams, L. Rider, "Circular Polar Constellations Providing Continuous Single or Multiple Coverage Above a Specified Latitude", *Journal of Astronautical Sciences*, 35(2), 1987; 155–192.

[AGR-86] B.N. Agrawal, *Design of Geosynchronous Spacecraft*, Prentice Hall, Englewood Cliffs, NJ, 1986.

[BAL-69] R.E. Balsam, B.M. Anzel, "A Simplified Approach for Correction of Perturbations on a Stationary Orbit", *Journal of Spacecraft and Rockets*, 6(7), July 1969; 805–811.

[BAL-80] A.H. Ballard, "Rosette Constellations of Earth Stations", *IEEE Transactions on Aerospace and Electronic Systems*, AES-16(5), September 1980; 656–673.

[BAT-71] R.R. Bate, D.D. Mueller, J.E. White, *Fundamentals of Astrodynamics*, Dover Publications, New York, 1971.

[BES-78] D.C. Beste, "Design of Satellite Constellation for Optimal Continuous Coverage", *IEEE Transactions on Aerospace and Electronic Systems*, AES-14(3), May 1978; 466–473.

[JEN-62] J.G. Jensen, J. Townsend, J. Kork, D. Kraft, *Design Guide to Orbital Flight*, McGraw-Hill, New York, 1962.

[KAL-63] F. Kalil, F. Martikan, "Derivation of Nodal Period of an Earth Satellite and Comparisons of Several First-Order Secular Oblateness Results", *AIAA Journal*, 1(9), September 1963; 2041–2046.

[LEO-77] C.T. Leondes, H.E. Emara, "Minimum Number of Satellites for Three Dimensional Continuous World-wide Coverage", *IEEE Transactions on Aerospace and Electronic Systems*, AES-13(2), March 1977; 108–111.

[LUD-61] R.D. Lüders, "Satellite Networks for Continuous Zonal Coverage", *ARS Journal*, 31, February 1961; 179–184.

[NAS-63] *Orbital Flight Handbook Part I – Basic Techniques and Data (NASA SP-33 Part 1)*, National Aeronautics and Space Administration, Washington, DC, 1963.

[PRA-86] T. Pratt, C.W. Bostian, *Satellite Communications*, Wiley, New York, 1986.

[PRI-93] W.L. Pritchard, H.G. Suyderhoud, R.A. Nelson, *Satellite Communication Systems Engineer*, 2nd edition, Prentice-Hall, Englewood Cliffs, NJ, 1993.

[WAL-73] J.G. Walker, "Continuous Whole Earth Coverage by Circular Orbit Satellites", *Proceedings of the IEE Satellite Systems for Mobile Communications Conference*, March 1973.

[WOL-61] R.W. Wolverton, *Flight Performance Handbook for Orbital Operations*, Wiley, New York, 1961.

4

Channel Characteristics

4.1 Introduction

This chapter considers the propagation environment in which a mobile-satellite system operates. The space between the transmitter and receiver is termed the channel. In a mobile-satellite network, there are two types of channel to be considered: the mobile channel, between the mobile terminal and the satellite; and the fixed channel, between the fixed Earth station or gateway and the satellite. These two channels have very different characteristics, which need to be taken into account during the system design phase. The more critical of the two links is the mobile channel, since transmitter power, receiver gain and satellite visibility are restricted in comparison to the fixed-link. The basic transmission chain is shown in Figure 4.1.

By definition, the mobile terminal operates in a dynamic, often hostile environment in which propagation conditions are constantly changing. In a mobile's case, the local operational environment has a significant impact on the achievable quality of service (QoS). The different categories of mobile terminal, be it land, aeronautical or maritime, also each have their own distinctive channel characteristics that need to be considered. On the contrary, the fixed Earth station or gateway can be optimally located to guarantee visibility to the satellite at all times, reducing the effect of the local environment to a minimum. In this case, for frequencies above 10 GHz, natural phenomena, in particular rain, govern propagation impairments. Here, it is the local climatic variations that need to be taken into account. These very different environments translate into how the respective target link availabilities are specified for each channel. In the mobile-link, a service availability of 80–99% is usually targeted, whereas for the fixed-link, availabilities of 99.9–99.99% for the worst-month case can be specified.

The following reviews the current status of channel modelling from a mobile and a fixed perspective.

4.2 Land Mobile Channel Characteristics

4.2.1 Local Environment

Spurred on by the needs of the mobile-satellite industry, the 10 years spanning the mid-1980s

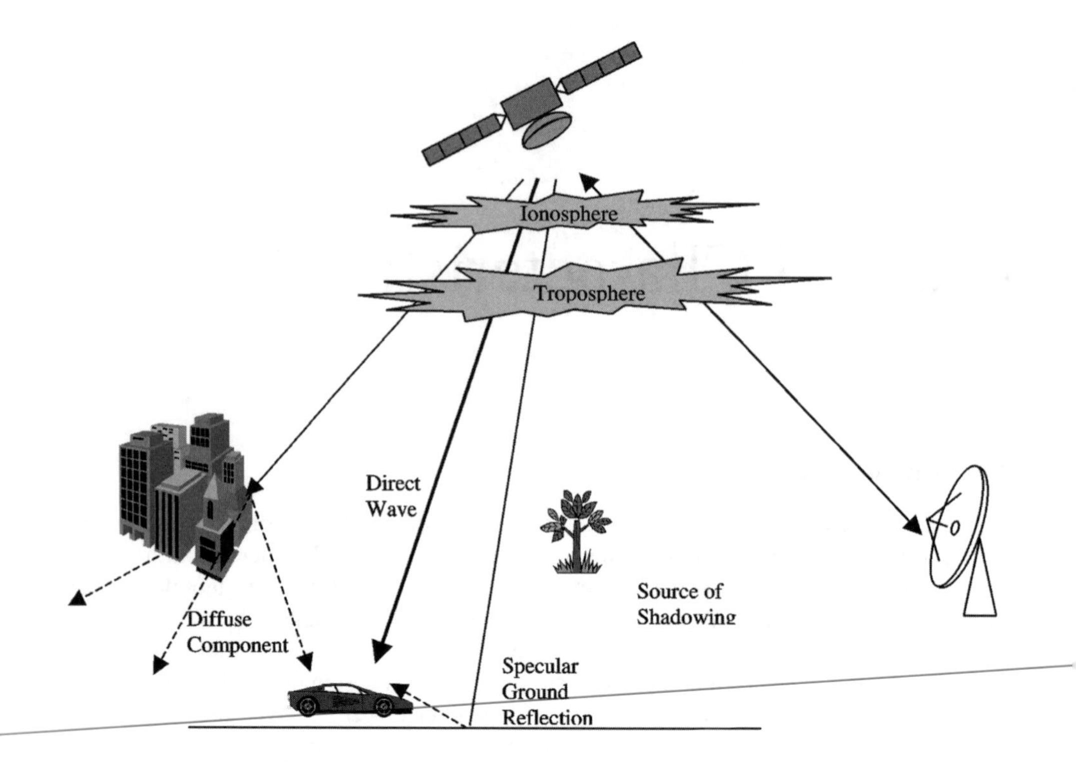

Figure 4.1 Mobile network propagation environment.

to the mid-1990s witnessed significant effort around the world in characterising the land mobile-satellite channel. The vast majority of these measurement campaigns were focused on the UHF and L-/S-bands, however, by the mid-1990s, with a number of mobile-satellite systems in operation, focus had switched to characterising the next phase in mobile-satellite development, that of broadband technology at the Ka-band and above.

The received land mobile-satellite signal consists of the combination of three components: the direct line-of-sight (LOS) wave, the diffuse wave and the specular ground reflection. The direct LOS wave arrives at the receiver without reflection from the surrounding environment. The only L-/S-band propagation impairments that significantly affect the direct component are free space loss (FSL) and shadowing. FSL is related to operating frequency and transmission distance. This will be discussed further in the following chapter. Tropospheric effects can be considered negligible at frequencies below 10 GHz, while impairments introduced by the ionosphere, in particular, *Faraday rotation* can be effectively counteracted by the selective use of transmission polarisation. Systems operating at above 10 GHz need to take into account tropospheric impairments and these will be considered further when discussing the fixed-link channel characteristics.

Shadowing occurs when an obstacle, such as a tree or a building, impedes visibility to the satellite. This results in the attenuation of the received signal to such an extent that transmissions meeting a certain QoS may not be possible.

The diffuse component comprises multipath reflected signals from the surrounding environment, such as buildings, trees and telegraph poles. Unlike terrestrial mobile networks, which rely on multipath propagation, multipath has only a minor effect on mobile-satellite links in most practical operating environments [VUC-92].

The specular ground component is a result of the reception of the reflected signal from the ground near to the mobile. Antennas of low gain, wide beamwidth operating via satellites with low elevation angle are particularly susceptible to this form of impairment. Such a scenario could include hand-held cellular like terminals operating via a non-geostationary satellite, for example.

The first step towards modelling the mobile-satellite channel is to identify and categorise typical transmission environments [VUC-92]. This is usually achieved by dividing the environment into three broad categories:

- Urban areas, characterised by almost complete obstruction of the direct wave.
- Open and rural areas, with no obstruction of the direct wave.
- Suburban and tree shadowed environments, where intermittent partial obstruction of the direct wave occurs.

As far as land mobile-satellite systems are concerned, it is the last two of the above environments that are of particular interest. In urban areas, visibility to the satellite is difficult to guarantee, resulting in the multipath component dominating reception. Thus, at the mobile, a signal of random amplitude and phase is received. This would be the case unless multi-satellite constellations are used with a high guaranteed minimum elevation angle. Here, satellite diversity techniques allowing optimum reception of one or more satellite signals could be used to counteract the effect of shadowing.

The fade margin specifies the additional transmit power that is needed in order to compensate for the effects of fading, such that the receiver is able to operate above the threshold or the minimum signal level that is required to satisfy the performance criteria of the link. The threshold value is determined from the link budget, which is discussed in the following chapter. The urban propagation environment places severe constraints on the mobile-satellite network. For example, in order to achieve a fade margin in the region of 6–10 dB in urban and rural environments, a continuous guaranteed minimum user-to-satellite elevation angle of at least 50° is required [JAH-00]. The compensation for such a fade margin should not be beyond the technical capabilities of a system and could be incorporated into the link design. However, to achieve such a high minimum elevation angle using a low Earth orbit constellation would require a constellation of upwards of 100 satellites. On the other hand, for a guaranteed minimum elevation angle of 20°, a fade margin in the region of 25–35 dB, would be required for the same grade of service, which is clearly unpractical. While these figures demonstrate the impracticalities of providing coverage in urban areas, in reality, for an integrated space/terrestrial environment, in an urban environment, terrestrial cellular coverage would take priority and this is indeed how systems like GLOBALSTAR operate.

In open and rural areas, where direct LOS to the satellite can be achieved with a fairly high degree of certainty, the multipath phenomenon is the most dominant link impairment. The multipath component can either add constructively (resulting in signal enhancement) or destructively (causing a fade) to the direct wave component. This results in the received mobile-satellite transmissions being subject to significant fluctuations in signal power.

In tree shadowed environments, in addition to the multipath effect, the presence of trees will result in the random attenuation of the strength of the direct path signal. The depth of the fade is dependent on a number of parameters including tree type, height, as well as season due to the leaf density on the trees. Whether a mobile is transmitting on the left or right hand side of the road could also have a bearing on the depth of the fade [GOL-89, PIN-95], due to the LOS path length variation through the tree canopy being different for each side of the road. Fades of up to 20 dB at the L-band have been reported due to shadowing caused by roadside trees in a suburban environment [GOL-92].

In suburban areas, the major contribution to signal degradation is caused by buildings and other man-made obstacles. These obstacles manifest as shadowing of the direct LOS signal, resulting in attenuation of the received signal. The motion of the mobile through suburban areas results in the continuous variation in the received signal strength and variation in the received phase.

The effect of moving up in frequency to the K-/Ka-bands imposes further constraints on the design of the link. Experimental measurement campaigns performed in Southern California by the jet propulsion laboratory as part of the Advanced Communications Technologies Satellite (ACTS) (not to be confused with the European ACTS R&D Programme) Mobile Programme reported results for three typical transmission environments [PIN-95]. For an environment in which infrequent, partial blocking of the LOS component occurred, fade depths of 8, 1 and 1 dB were measured at the K-band, corresponding to fade levels of 1, 3 and 5%, respectively. This implies that for 1% of the time, the faded received signal is greater than 8 dB below the reference pilot level and so on. In an environment in which occasional complete shadowing of the LOS component occurred, corresponding fade depths of 27, 17.5 and 12.5 dB were measured. Lastly, for an environment in which frequent complete blockage of the LOS occurred, fade depths of greater than 30 dB were obtained for as low as a 5% fade level. In all environments, fade depths at the K- and Ka-bands were found to be essentially the same. Similar degrees of fading were found during European measurement campaigns, such as under the European Union's ACTS programme SECOMS project. The results demonstrate the difficulty in providing reliable mobile communications to any environment in which the LOS to the satellite may be restricted.

Channel modelling is classified into two categories: narrowband and wideband. In the narrowband scenario, the influence of the propagation environment can be considered to be the same or similar for all frequencies within the band of interest. Consequently, the influence of the propagation medium can be characterised by a single, carrier frequency. In the wideband scenario, on the other hand, the influence of the propagation medium does not affect all components occupying the band in a similar way, thus causing distortion to selective spectral components.

4.2.2 Narrowband Channel Models

4.2.2.1 Overview

Narrowband channel characterisation is primarily aimed at establishing the amplitude variation of the signal transmitted through the channel. The vast majority of measurement campaigns have used the narrowband approach and, consequently, a number of narrowband models have been proposed. These models can be classified as being either (a) empirical with regression line fits to measured data, (b) statistical or (c) geometric-analytical.

Empirical models can be used to characterise the sensitivity of the results to critical parameters, e.g. elevation angle, frequency. Statistical models such as the Rayleigh, Rician and log-normal distributions or their combinations for use in different transmission environments, are especially useful for software simulation analysis; whilst geometric-analytical models provide an understanding of the transmission environment, through the modelling of the topography of the environment.

4.2.2.2 Empirical Regression Models

A number of measurement campaigns performed throughout the world have aimed to categorise the mobile-satellite channel. These measurement campaigns have attempted to emulate the satellite by using experimental air borne platforms, that is helicopters, aircraft, air balloons, or in some cases existing geostationary satellite systems have been used. The following briefly describes some of the most widely cited models.

Empirical Roadside Shadowing (ERS) Model This model is used to characterise the effect of fading predominantly due to roadside trees. The model is based upon measurements performed in rural and suburban environments in central Maryland, US, using helicopter-mobile and satellite-mobile links at the L-band [VOG-92]. Measurements were performed for elevation angles in the range 20–60°; the 20° measurements utilised a mobile-satellite link, and the remainder a helicopter. The subsequent empirical expression derived from the measurement campaign for a frequency of 1.5 GHz is given by

$$A_L(P, \theta, f_L) = -M(\theta)\ln P + N(\theta) \quad \text{dB for } f_L = 1.5 \text{ GHz} \tag{4.1}$$

$$M(\theta) = a + b\theta + c\theta^2 \tag{4.2}$$

$$N(\theta) = d\theta + e \tag{4.3}$$

$$a = 3.44,\ b = 0.0975,\ c = -0.002,\ d = -0.443 \text{ and } e = 34.76 \tag{4.4}$$

where $A_L(P, \theta, f_L)$ denotes the value of the fade exceeded in decibels, L denotes the L-band, f_L is the frequency at L-band in GHz and is equal to 1.5 GHz in the equation; P is the percentage of the distance travelled over which the fade is exceeded (in the range 1–20%) or the outage

Table 4.1 ERS model characteristics

Elevation angle (°)	$M(\theta)$	$N(\theta)$
20	4.59	25.90
25	4.63	23.69
30	4.57	21.47
35	4.40	19.26
40	4.14	17.04
45	3.78	14.83
50	3.32	12.61
55	2.75	10.40
60	2.09	8.18

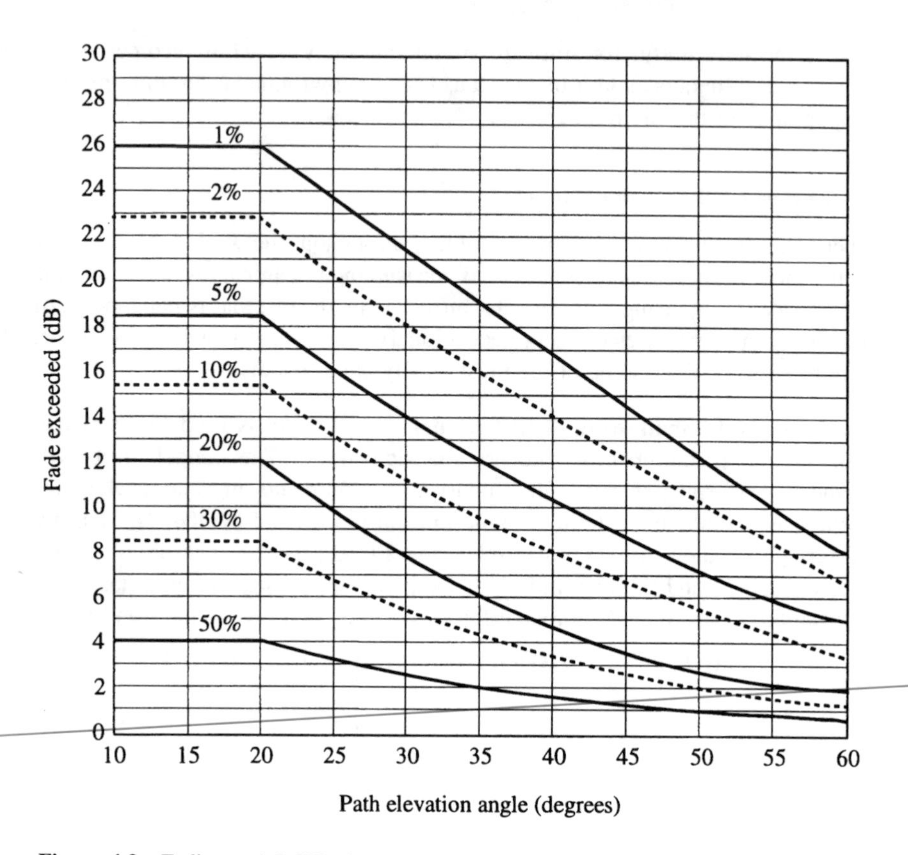

Figure 4.2 Fading at 1.5 GHz due to roadside shadowing versus path elevation angle.

probability in the range of 1–20% for a given fade margin. θ is the elevation angle. From the above, the following model characteristics can be derived (Table 4.1).

The following relationship between UHF and L-band has been derived for fade depth in tree shadowed areas for P in the range of 1–30% [VOG-88].

$$A_L(P, \theta, f_L) = A_{uhf}(P, \theta, f_{uhf})\sqrt{\frac{f_L}{f_{uhf}}} \text{ dB} \tag{4.5}$$

Similarly, it was found that when scaling from L-band (1.3 GHz) to S-band (2.6 GHz) the following relationship applies:

$$A_s(P, \theta, f_s) \approx 1.41 A_L(P, \theta, f_L) \text{ dB} \tag{4.6}$$

The ERS model can be found in Ref. [ITU-99a], which provides formulae that can be used to extend the operational range of the model. The conversion from L-band to K-band and *vice versa* in the frequency range 850 MHz to 20 GHz can be obtained from the formula, for an outage probability within the range $20\% \geq P \geq 1\%$ [ITU-99a]:

$$A_K(P, \theta, f_K) = A_L(P, \theta, f_L)\exp\left\{1.5\left[\frac{1}{\sqrt{f_L}} - \frac{1}{\sqrt{f_K}}\right]\right\} \text{ dB} \tag{4.7}$$

Elevation angles below 20° down to 7° are assumed to have the same value of fade as that at 20°. Figure 4.2 shows the results of application of the ERS model at 1.5 GHz for a range of elevation angles.

To increase the percentage of distance travelled (or outage probability) to the limits of $80\% \geq P \geq 20\%$, the ITU recommends the following formula:

$$A_{L-K}(P, \theta, f_{L-K}) = A_K(20\%, \theta, f_K)\frac{1}{\ln 4}\ln\left(\frac{80}{P}\right) \qquad (4.8)$$

where $A_{L-K}()$ denotes the attenuation for frequencies between 0.85 and 20 GHz.

An extension to the ERS model is also provided for elevation angles greater than 60° at frequencies of 1.6 and 2.6 GHz, respectively [ITU-99a]. This is achieved by applying the ERS model for an elevation angle of 60° and then linearly interpolating between the calculated 60° value and the fade values for an 80° elevation angle given in Table 4.2. Linear interpolation should also be performed between the figures given in Table 4.2 and 90° elevation, for which the fade exceeded is assumed to be 0 dB.

Table 4.2 Fades exceed (dB) at 80° elevation [ITU-99a]

P (%)	Tree-shadowed	
	1.6 GHz	2.6 GHz
1	4.1	9.0
5	2.0	5.2
10	1.5	3.8
15	1.4	3.2
20	1.3	2.8
30	1.2	2.5

As noted earlier, the season of operation effects the degree of attenuation experienced by the transmission. The following expression is used to take into account the effect of foliage on trees, at UHF, for P in the range 1–30%, indicating a 24% increase in attenuation due to the presence of leaves.

$$A(\text{full foliage}) = 1.24A(\text{no foliage})\ \text{dB} \qquad (4.9)$$

The Modified Empirical Roadside Shadowing (MERS) Model The European Space Agency modified the ERS model in order to increase the elevation angle range up to 80° and the percentage of optical shadowing up to 80%. One form of the MERS is given by

$$A(P, \theta)\ln(P) + B(\theta)\ \text{dB} \qquad (4.10)$$

where P and θ are as in the ERS model. $A(\theta)$ and $B(\theta)$ are defined by:

$$A(\theta) = a_1\theta^2 + a_2\theta + a_3$$

$$a_1 = 1.117 \times 10^{-4}, a_2 = -0.0701, a_3 = 6.1304$$

$$B(\theta) = b_1\theta^2 + b_2\theta + b_3$$

$$b_1 = 0.0032, b_2 = -0.6612, b_3 = 37.8581$$

Empirical Fading Model This model is based upon the measurements performed by the University of Surrey, UK, simultaneously using three bands (L, S and Ku) [BUT-92]. Elevation angles were within the range 60–80°. Its basic form is similar to the ERS model with the addition of a frequency-scaling factor. It is given by

$$M(P, \theta, f) = a(\theta, f)\ln(P) + c(\theta, f) \quad (4.11)$$

where

$$a(\theta, f) = 0.029\theta - 0.182f - 6.315$$

$$c(\theta, f) = -0.129\theta + 1.483f + 21.374$$

The model is valid for the following ranges: P, link outage probability, 1–20%; f, frequency, 1.5–10.5 GHz; θ, elevation angle, 60–80°.

The model has been extended by combining it with the ERS, resulting in the combined EFM (CEFM) model. This is valid for elevation angles in the range 20–80° with regression coefficients:

$$a(\theta, f) = 0.002\theta^2 - 0.15\theta - 0.2f - 0.7$$

$$c(\theta, f) = -0.33\theta + 1.5f + 27.2$$

4.2.2.3 Probability Distribution Models

Probability distribution models can be used to describe and characterise, with some degree of accuracy, the multipath and shadowing phenomena. This form of modelling allows the dynamic nature of the channel to be modelled. In turn, this enables the performance of the system to be evaluated for different environments. Essentially, a combination of three probability density functions (PDF) are used to characterise the channel: Rician (when a direct wave is present and is dominant over multipath reception), Rayleigh (when no direct wave is present and multipath reception is dominant) and log-normal (for shadowing of the direct wave when no significant multipath reception is present).

Complete Obstruction of the Direct Wave In an urban environment, the received signal is characterised by virtually a complete obstruction of the direct wave. In this case, the received signal will be dominated by multipath reception. The received signal comprises, therefore, of the summation of all diffuse components. This can be represented by two orthogonal independent voltage phasors, X and Y, which arrive with random phase and amplitude. The phase of the diffuse component can be characterised by a uniform probability density function within the range 0–2π, while the amplitude can be categorised by a Rayleigh distribution of the form

$$P_{\mathrm{Rayleigh}}(r) = \frac{r}{\sigma_{\mathrm{m}}^2} \exp\left(-\frac{r^2}{2\sigma_{\mathrm{m}}^2}\right) \tag{4.12}$$

where r is the signal envelope given by:

$$r = \sqrt{x^2 + y^2} \tag{4.13}$$

and σ_{m}^2 is the mean received scattered power of the diffuse component due to multipath propagation.

As noted in Chapter 1, for an unmodulated carrier, f_{c}, the Döppler shift, f_{d}, of a diffuse component arriving at an incident angle θ_{i} is given by:

$$f_{\mathrm{d}} = \frac{vf_{\mathrm{c}}}{\mathrm{c}} \cos\theta_{\mathrm{i}} \ \mathrm{Hz} \tag{4.14}$$

where θ_i is in the range 0–2π. This results in a maximum Döppler shift f_{m} of $\pm\, vf_{\mathrm{c}}/\mathrm{c}$, where c is the speed of light ($\approx 3 \times 10^8$ m/s).

Hence, at the receiver, a band of signals is received within the range $f_{\mathrm{c}} \pm f_{\mathrm{m}}$, where f_{m} is termed the fade rate. For uniform received power for all angles of arrival at the terminal, the resultant power spectral density is given by the expression:

$$S(f) = \frac{\sigma_{\mathrm{m}}^2}{\pi f_{\mathrm{m}}}\left[1 - \left(\frac{(f - f_{\mathrm{c}})}{f_{\mathrm{m}}}\right)^2\right]^{-\frac{1}{2}} \ (\mathrm{W/Hz}) \tag{4.15}$$

Unobstructed Direct Wave When in the presence of a direct source or wave of amplitude A, as in the mobile-satellite case in an open environment, the representation of the two-dimensional probability density function of the received voltage is given by [GOL-92]:

$$P_{XY}(x, y) = \frac{1}{2\pi\sigma_{\mathrm{m}}^2} \exp\left(-\frac{(x - A)^2 + y^2}{2\sigma_{\mathrm{m}}^2}\right) \tag{4.16}$$

Using the above expression, the p.d.f. of the random signal envelope follows a Rician distribution:

$$P_{\mathrm{Rice}}(r) = \frac{r}{\sigma_{\mathrm{m}}^2} \exp\left(-\frac{r^2 + A^2}{2\sigma_{\mathrm{m}}^2}\right) I_0\left(\frac{rA}{\sigma_{\mathrm{m}}^2}\right) \tag{4.17}$$

where $I_0(.)$ is the modified zero-order Bessel function of the first kind; $A^2/2$ is the mean received power of the direct wave component, r is the signal envelope and σ_{m}^2 is the mean received scattered power of the diffuse component due to multipath propagation.

It can be seen from the above equation that the Rayleigh distribution is a special case of the Rician distribution and arises when no LOS component is available, i.e. $A = 0$.

The power ratio of the direct wave to that of the diffuse component, $A^2/2\sigma_{\mathrm{m}}^2$, is known as the Rice-factor, which is usually expressed in dB.

Typical values of the Rice-factor, based on measurements in the US and Australia, are within the range 10–20 dB, with a fade rate of less than 200 Hz for a mobile travelling at less than 100 km/h [VUC-92]. Rician models can be used when an unobstructed LOS component is present along with coherent and incoherent multipath signals, such as occurs in open rural areas, for example.

Partial-Shadowing of the Direct Wave The log-normal density function is used to characterise the effect of shadowing of the direct wave, where no multipath component is present, here

$$P_{\text{Log-normal}}(r) = \frac{1}{\sigma_s r\sqrt{2\pi}} \exp\left(-\frac{(\ln r - \mu_s)^2}{2\sigma_s^2}\right) \tag{4.18}$$

where σ_s is the standard deviation of the shadowed component (ln r) and μ_s is the mean of the shadowed component (ln r).

The suburban environment is one in which random shadowing of the direct-wave occurs, due to the presence of trees, buildings, and so on. The log-normal distribution is used to model the effect of this environment on the direct wave component, however, a Rayleigh distribution needs also to be considered in order to take into account the multipath, diffuse component.

There is no set rule as to what parameters should be applied to the above models in order to generate a suitable representation of the environment of concern. How the above statistical models are combined to characterise the complete transmission environment is what identifies a particular model.

Two of the most widely referenced statistical models are those developed by Loo [LOO-85] and Lutz [LUT-91]. The modellers, however, differ in their approach. The Loo model is an example of how the constituents of the channel are combined into a single probability distribution with associated parameters. The Lutz approach, on the other hand, employs state-orientated statistical modelling, whereby each particular state of the channel is separately characterised by a probability distribution, with a specified probability of occurrence.

Joint Probability Distribution Modelling Loo's model is based upon a measurement campaign performed in Canada using a helicopter to mobile transmission link in rural environments. The model is valid for elevation angles up to 30°. Loo assumed: (a) received voltage due to diffusely scattered components is Rayleigh distributed; (b) voltage variations due to attenuation of the direct path signal are log-normally distributed. Further details can be found in Ref. [LOO-85].

Loo's PDF for a signal envelope r is given by:

$$p_{\text{Loo}}(r) = \frac{r}{\sigma_m^2\sqrt{2\pi\sigma_s^2}} \int_0^\infty \frac{1}{A} \exp\left(-\frac{(\ln A - \mu_s)^2}{2\sigma_s^2} - \frac{(r^2 + A^2)}{2\sigma_m^2}\right) I_0\left(\frac{rA}{\sigma_m^2}\right) \mathrm{d}A \tag{4.19}$$

where σ_m^2 is the mean received scattered power of the diffuse component due to multipath propagation, σ_s is the standard deviation of the shadowed component (ln A) and μ_s is the mean of the shadowed component (ln A).

The above expression can be simplified to either (4.18) when r is much greater than σ_m or (4.12) when r is much less than σ_m. Otherwise the above expression needs to be determined mathematically.

An alternative to Loo's approach for non-geostationary satellite constellations is presented in Ref. [COR-94], in which the direct and scattered components were both considered to be affected by shadowing. This is termed the Rice-log-normal model (RLM). A harmonisation of the RLM model with that of Loo's approach is presented in Ref. [VAT-95]. Further examples of statistical models can be found in Ref. [KAR-98].

Loo's model also provides an insight into the transient nature of the channel characteristics, which is useful when designing the radio parameters, in particular coding and interleaving techniques, the latter being used to disperse the effect of bursty errors over a transmitted frame or block. Radio interface aspects will be discussed further in the following chapter. Specifically, Loo derived expressions for the second-order statistics level crossing rate (LCR) and average fade duration (AFD). The LCR is defined as the rate at which a signal envelope transcends a threshold level, R, with a positive slope. The AFD is the mean duration for which a signal falls below a given value, R.

The LCR, normalised with respect to the maximum Döppler shift f_{m} in order to make it independent of vehicular velocity, is given by the formula:

$$\mathrm{LCR} = \sqrt{2\pi}\sqrt{1-\rho^2\sigma_{\mathrm{m}}^2}\,\frac{\sqrt{\left(\sigma_{\mathrm{m}}^2 + 2\rho\sqrt{\sigma_{\mathrm{m}}\sigma_{\mathrm{s}}} + \sigma_{\mathrm{s}}^2\right)}p_{\mathrm{Loo}}(r)}{\sigma_{\mathrm{m}}^2(1-\rho^2) + 4\rho\sigma_{\mathrm{s}}\sigma_{\mathrm{m}}} \tag{4.20}$$

where ρ is the correlation coefficient between multipath and shadowing. A figure of between 0.5 and 0.9 gives a good indication of the LCR [LOO-98].

The AFD is given by:

$$\mathrm{AFD} = \frac{1}{\mathrm{LCR}}\int_0^R \rho(r)\mathrm{d}r \tag{4.21}$$

An alternative solution to model the fade duration is presented by the ITU [ITU-99a]. Based on experiments performed in the US and Australia, the following expression has been derived to express the probability of fade duration in terms of travelled distance:

$$P(FD > dd|A > A_{\mathrm{q}}) = \frac{1}{2}\left(1 - erf\left[\frac{\ln dd - \ln\alpha}{\sqrt{2}\sigma}\right]\right) \tag{4.22}$$

where $P(FD > dd|A > A_{\mathrm{q}})$ represents the probability that a random fade duration, FD, is exceeded for a distance dd, given the condition that the attenuation A exceeds A_{q}. σ is the standard deviation of $\ln(dd)$ and $\ln(\alpha)$ is the mean value of $\ln(dd)$.

***N*-State Markov Modelling** The finite state Markov model can be used to represent the different environments in which a mobile operates. While, in the long-term, statistical properties of a mobile channel are dynamic, within a particular environment, the statistical properties that characterise it can be considered to be static and predictable. In order to apply the Markov model, all possible static environments are identified, which are then statistically categorised. For M identified static environments (or states in statistical term), the component, w_j, of the $1 \times M$ statistical matrix, W, of the form shown below, defines the probability of existence of the jth identified state:

$$[W] = [w_1, w_2, \ldots w_M] \tag{4.23}$$

The probability of existence of a given state depends only on the previous state.

The switching between static environments is then defined by a transition matrix, which

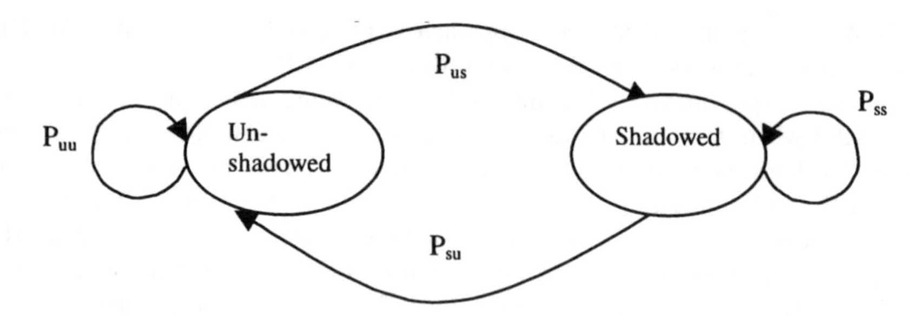

Figure 4.3 Two-state Markov process indicating shadowed and un-shadowed operation.

comprises transition probabilities between states and is of the form

$$[P] = \begin{bmatrix} P_{11} & P_{12} & \cdots & P_{1M} \\ P_{21} & P_{22} & \cdots & P_{2M} \\ \cdots & \cdots & \cdots & \cdots \\ P_{M1} & P_{M2} & \cdots & P_{MM} \end{bmatrix} \tag{4.24}$$

The statistical matrix, W, and the transition matrix, P, satisfy the following [VUC-92]:

$$[W][P] = [W] \text{ and } [W][E] = [I] \tag{4.25}$$

where E is a column matrix with all entities equal to 1 and I is the identity matrix.

Lutz's model, which employs a two-state Markov model of the form shown in Figure 4.3, was developed following an extensive measurement campaign across Europe using the geostationary MARECS satellite for satellite elevation angles in the range 13–43°. Various environments were characterised for different satellite elevation angles and antenna types. Lutz assumed that the propagation link has two distinct states: shadowed and un-shadowed. In the un-shadowed or "good" state, the received signal, comprising the direct component and multipath reflections, is assumed to be Rician distributed. In the shadowed or "bad" state, the received signal is characterised by a Rayleigh distribution, with a short-term time-varying mean received power S_0, for which a log-normal distribution is assumed. The resultant probability density function of S:

$$P_{\text{Lutz}}(s) = (1 - A)P_{\text{Rice}}(s) + A\int_o^\infty P_{\text{Rayleigh}}(s\,|\,s_0)P_{\text{Log-normal}}(s_0)\mathrm{d}s_0 \tag{4.26}$$

An important factor in the above expression is the parameter A, which is the proportion of time spent in each state, i.e. "good" and "bad".

An example of a four-state Markov model can be found in Ref. [VUC-92], which was derived from measurements performed in Australia using the Japanese ETS-V satellite. This model comprised two states modelled using Rician distributions with Rice-factors of 14 and 18 dB, respectively and two linearly combined Rayleigh/log-normal states.

4.2.2.4 Geometric Analytical Models

Geometric analytical models provide useful information on the mechanism of fading. Moreover, this method has certain advantages over the other methods of modelling in that it does not require the purchase of expensive test equipment or the performance of lengthy measurement campaigns. A powerful, flexible model will enable a channel to be characterised for different transmission environments using different satellite constellations.

To achieve this requires the modeller to have a representation of the transmission environment. Over recent years, there has been significant advancement in the digital mapping of our environment. This has been largely driven by cellular and wireless network operators who need as detailed a map as possible of their operating environment in order to ensure the optimum placement of base stations. Interpretation of satellite images, along with detailed landscape characterisation campaigns by such organisations as the UK's National Remote Sensing Centre (NRSC), has resulted in a large database of topographical and geographical information. These models can be used to study the dependence of signal variations on parameters such as antenna pattern, azimuth, elevation, etc. In addition to the topography of the propagation environment, it is also required to simulate the dynamics of the satellite orbit. Such an approach has been performed by [DOT-98] to derive a wideband channel model.

In order to attempt to understand the effects of local geography on satellite visibility and hence the channel, optical, photogrammetric measurement techniques have recently been developed. This method involves processing fisheye lens photographs of the environment, which results in a three-state description of the environment, as clear, shadowed or blocked. This technique has been used to determine the effectiveness of satellite diversity to increase mobile-satellite service availability in an urban environment [AKT-97].

Geometric analytical models are not as prevalent or as widely used as the other modelling techniques. However, with increasing computer power and as greater modelling information on the environment becomes available, there is an opportunity to develop sophisticated tools along similar lines to cellular planning tools to facilitate understanding of the transmission environment.

4.2.3 Wideband Channel Models

The overwhelming majority of measurement campaigns have concentrated on characterising the narrowband channel. In recent years, however, with the advent of satellite-UMTS and CDMA technologies, a number of wideband measurement campaigns have been undertaken [JAH-95a].

As noted previously, as a consequence of the multipath phenomenon, a transmitted signal will arrive at the receiver via many different paths, each of which will be delayed by a different amount of time. If, for example, an impulse signal is transmitted, at the receiver the signal would take the form of a pulse, distributed over a time period known as the *delay spread*. Related to the delay spread is the *coherence bandwidth*, which is the bandwidth over which two signals will exhibit a high degree of similarity, in terms of amplitude or phase. A signal occupying a bandwidth greater than the coherence bandwidth will exhibit *frequency selective fading*. The coherence bandwidth is defined for two fading amplitudes as in Ref. [LEE-89]:

$$B_c = \frac{1}{2\pi\Delta} \tag{4.27}$$

where B_c is the channel bandwidth and Δ is the delay spread.

The wideband channel model is usually presented as a tap-delayed filter structure. This is used to represent the reception of multiple signal echoes due to reflections from the surrounding environment [JAH-95b].

Measurements performed by DLR, Germany, have concluded that an impulse response with N echoes can be divided into three regions:

- The direct path. The amplitude of the direct signal is determined by the propagation environment, for example Rician distributed in an open environment;
- Near echoes, where the majority of echoes reside, with delays of 0 to τ_e where $\tau_e \approx 600$ ns. The number of near echoes varies according to the transmission environment. The power of the new echoes is governed by a Rayleigh distribution with the delay decreasing exponentially.
- Far echoes, $\tau_e > 600$ ns. Far echoes are characterised by a uniform distribution of delay and a log-normally-distributed power. The number of far echoes decreases exponentially.

Propagation measurements presented in Ref. [JAH-95a] for various environments suggest that: (a) for provincial roads, echoes of any significance are limited to less than 500 ns, attenuated by 10–20 dB; (b) in mountainous regions, echoes up to 2 μs were measured, attenuated by approximately 30 dB; (c) on highways, echoes were limited to less than 500 ns for a 65° elevation angle. In general, echoes were found to be mainly in the range 500 ns to 2 μs. Parameters for the wideband channel model are presented in Refs. [JAH-95b, JAH-00].

4.3 Aeronautical Link

The ability of communications to aircraft via satellite is becoming increasingly important. Due to safety regulations, communication channels to an aircraft need to be specified to a high degree of reliability.

A typical flight on board an aircraft comprises the following phases:

- Embarking of passengers and crew onto the aircraft and taxiing to a position on the runway ready for take-off.
- Take-off and ascension to cruise altitude.
- Cruising at constant height, which is normally above the cloud layer for a jet aircraft.
- Descent from cruise altitude to landing on the runway.
- Taxiing from the runway to a place of disembarkment for passengers and crew.

Each of the above phases can be considered to have particular channel characteristics. For example, while the aircraft is in the airport, a land mobile channel type environment could be envisaged, where the channel is subject to sporadic shadowing due to buildings, other aircraft and other obstacles. The aeronautical channel is further complicated by the manoeuvres performed by an aircraft during the course of a flight, which could result in the aircraft's structure blocking the LOS to the satellite. The body of the aircraft is also a source of multipath reflections, which also need to be considered. Moreover, the speed of an aircraft introduces large Döppler spreads.

The effect of multipath reflections from the sea for circularly polarised L-band transmissions can be found in Ref. [ITU-92]. This recommendation includes a methodology to derive the mean multipath power as a function of elevation angle and antenna gain. By applying this method for an aircraft positioned 10 km above the sea, and for a minimum elevation angle of 10°, the relative multipath power will be in the range of approximately −10 to −17 dB, for antenna gains varying from 0 to 18 dBi, respectively.

Presently, aeronautical mobile-satellite systems are served by the L-band. Due the bandwidth restrictions at this frequency, services are limited to voice and low data rate applications. The need to provide broadband multimedia services, akin to those envisaged by satellite-UMTS/IMT-2000 will require the move up in frequency band to the next suitable bandwidth, the K-/Ka-bands. At these frequencies, tropospheric effects will have an impact on link availability during the time when the aircraft is below the cloud layer. The channel characteristics for a K-band system have been investigated in Europe and the US. In Europe, the SECOMS/ABATE project, funded under the EU's ACTS programme performed a series of trials to investigate the characteristics of the aeronautical channel and to demonstrate the feasibility of transmitting multimedia type applications to the ground through a satellite link [HOL-99]. Here a Rice-factor of 34 dB is reported for LOS operation, while shadowing introduced by the aircraft's wing during a turning manoeuvre resulted in a fade of up to 15 dB.

4.4 Maritime Link

Commercial mobile-satellite services began with the introduction of communications to the maritime sector. Clearly, the maritime sector offers a very different propagation scenario to that of the land mobile case. Whereas in the land mobile channel, the modelling of the specular ground reflection is largely ignored, in the maritime case reflections from the surface of the sea provide the major propagation impairment. Such impairments are especially severe when using antennas of wide beamwidth, when operating at a low elevation angle to the satellite. Such a scenario is not untypical in the maritime environment.

In comparison to the modelling of the land mobile channel, results for the maritime channel are relatively few. However, [ITU-99b] provides a methodology similar to that for the aeronautical link, for determining the multipath power resulting from specular reflection from the sea.

4.5 Fixed Link

4.5.1 Tropospheric Effects

4.5.1.1 Atmospheric Gasses

Water vapour and oxygen (dry air) are the main contributors to gaseous absorption. The total attenuation due to atmospheric gases can therefore be determined by summing the respective contributions of these two components. At frequencies below 10 GHz, the influence of attenuation due to atmospheric gases can be effectively ignored. Moreover, at frequencies above 150 GHz, water vapour dominates the total attenuation. Since gaseous absorption is highly related to temperature, pressure and humidity, which are variables and

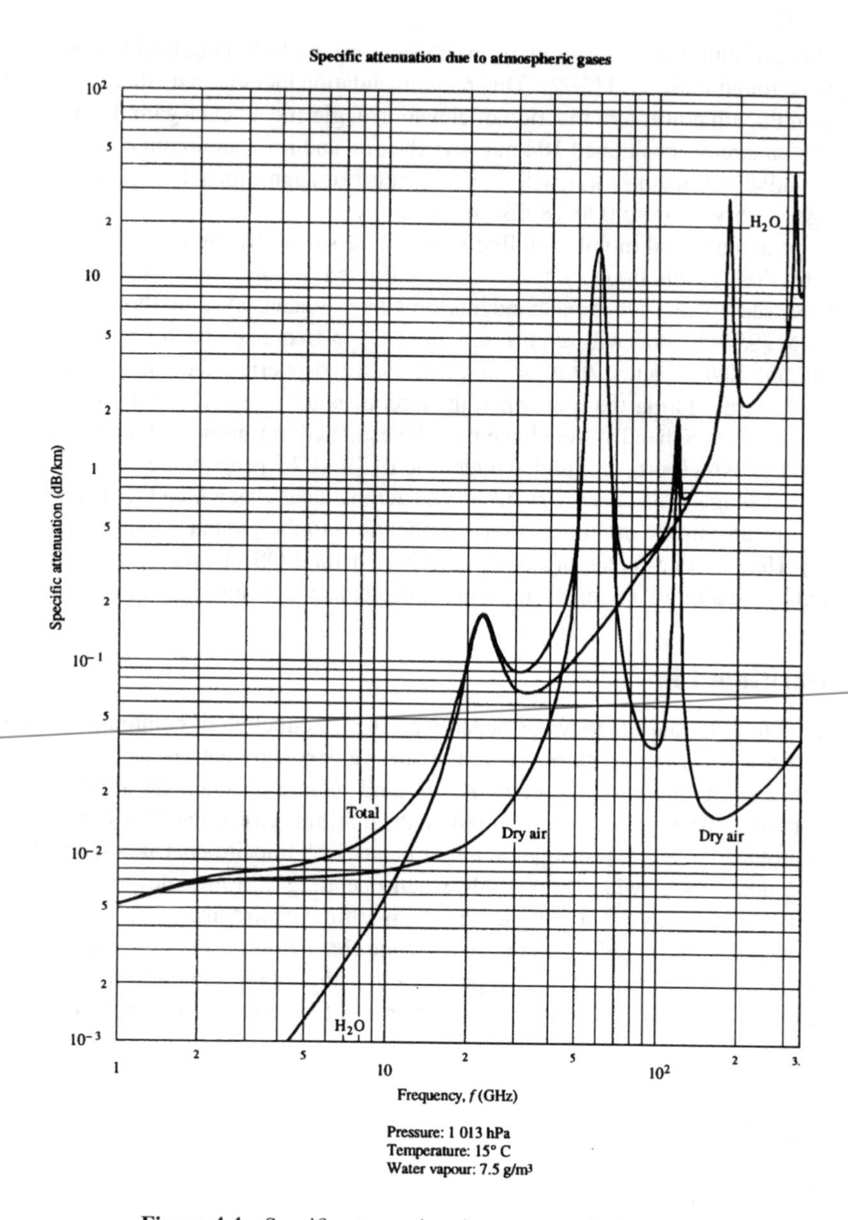

Figure 4.4 Specific attenuation due to atmospheric gasses.

are functions of height, it is usual to calculate gaseous absorption with reference to a standard exponential atmosphere.

Oxygen Attenuation Models Oxygen has a number of resonant lines between about 50 and 70 GHz and an isolated line at 118.75 GHz. At frequencies higher than 150 GHz, the attenuation due to oxygen is negligible relative to that of water vapour. A number of models are available for the calculation of specific attenuation due to oxygen absorption. A relatively straightforward formulae for the calculation of oxygen absorption at sea level can be found in Ref. [ITU-99c] (pressure = 1013 hPa, temperature = 15°C, water vapour density = 7.5 g/m^3). In this case, the attenuation due to oxygen is obtained by multiplying the specific attenuation due to dry air, γ_0 (dB/km), with an equivalent height h_0 (km). This is valid for elevation angles up to 10°. For elevation angles greater than 10° the path attenuation is obtained using the Cosecant Law (see below). Specific attenuation due to gases, as provided by the ITU-R, is shown in Figure 4.4. The application of the ITU-R procedures results in zenith attenuation accuracy of $\pm 10\%$ from sea level up to an altitude of about 2 km.

More accurate predictions can be obtained if an Earth station's height, h_s, above sea level is known. In this case, at an elevation angle, θ, greater than 10°, the path attenuation due to oxygen is obtained using the expression:

$$A_o = \frac{h_0 \gamma_0 e^{-\frac{h_s}{h_0}}}{\sin\theta} \text{ dB} \tag{4.28}$$

The use of the ITU-R model in the frequency range 50–70 GHz, may introduce an error of up to 15% due to the approximation of the equivalent height. In this region, a line-by-line calculation is recommended when greater accuracy is required.

An alternative model for specific attenuation due to oxygen can be found in Ref. [ULA-82], which is expressed as a function of frequency, atmospheric pressure and temperature. The model also takes into account the rotational quantum number of molecular oxygen. Another specific attenuation model for oxygen absorption is that by Liebe [LIE-88]. Input variables to this model are barometric pressure, ambient temperature, relative humidity, hygroscopic aerosol concentration, suspended water droplet concentration and rain rate.

Salonen et al. have derived methods to take the temperature dependence into account for slant path oxygen attenuation [SAL-92]. The model relates the effective height of oxygen with the outage probability. This enables the calculation of oxygen attenuation with respect to a specific link availability. The model has been compared with Liebe's model and good correlation has been obtained for five European sites.

Water Vapour Attenuation Models Water vapour absorption lines occur at 22.2, 67.8, 120, 183.3 and 325 GHz plus numerous lines in the sub-millimetre wave and infrared bands. As with the case of oxygen absorption, the ITU-R provides simple formulae for the calculation of water vapour specific attenuation at sea level for a pressure of 1013 hPa and at a temperature of 15°C. The method is based on the calculation of the specific attenuation due to water vapour at sea level, γ_w, and on the use of the equivalent height of water vapour h_w to describe the path length to zenith. The ITU formula for oxygen attenuation of zenith, A_w, is as follows:

$$A_w = \gamma_w h_w \tag{4.29}$$

Total path Attenuation due to Gaseous Absorption Combining oxygen attenuation with water vapour attenuation, the total path attenuation at zenith at sea level is given by:

$$A_G = \gamma_0 h_0 + \gamma_w h_w \text{ dB} \tag{4.30}$$

Figure 4.5 shows the total attenuation at sea level for dry air and water vapour derived by the ITU-R.

For elevation angles above 10°, the cosecant law is applied of the form:

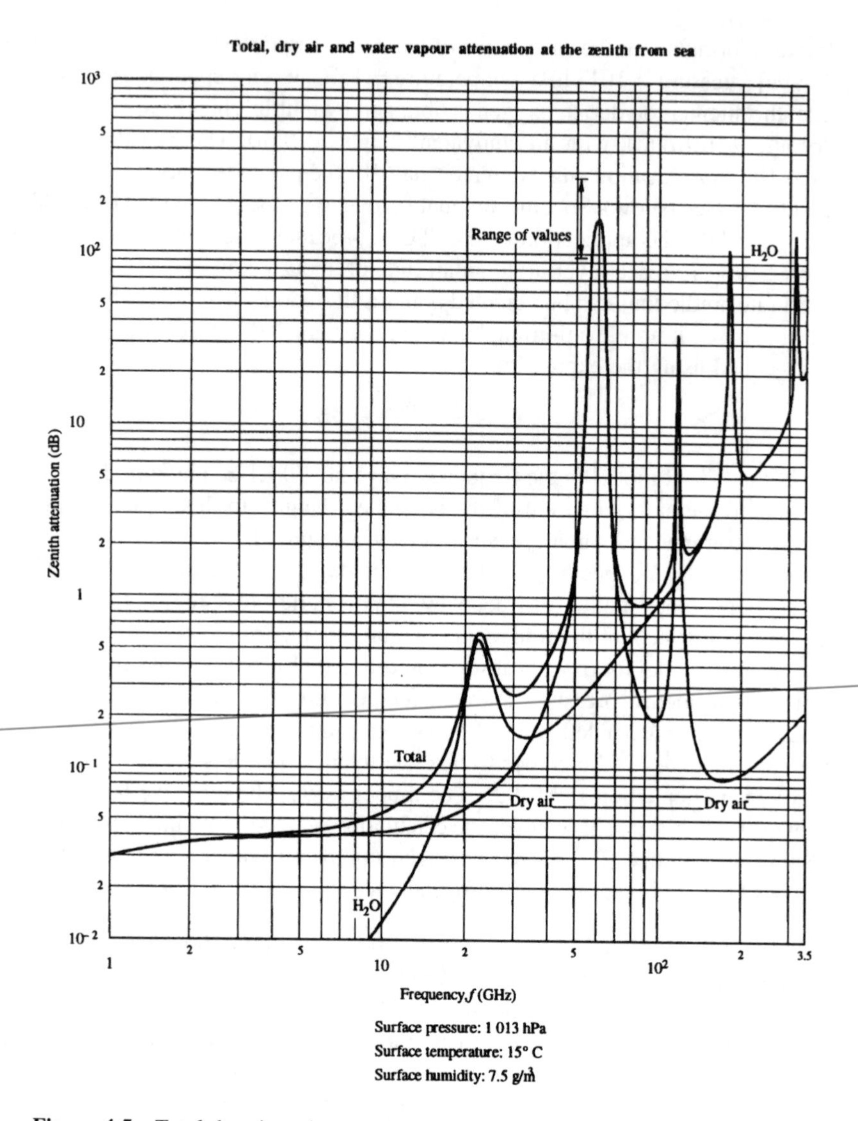

Figure 4.5 Total dry air and water vapour attenuation at the zenith from sea level.

$$A_G = \frac{h_0 \gamma_0 + h_w \gamma_w}{\sin\theta} \text{ dB} \tag{4.31}$$

where θ is the elevation angle.

For an Earth station operating above sea level, at a height of h_s km, the above expression is modified to:

$$A_G = \frac{\gamma_0 h_0 e^{-\frac{h_s}{h_0}} + \gamma_w h_w}{\sin\theta} \text{ dB} \tag{4.32}$$

Salonen et al. have derived a method for the statistical distribution of water vapour

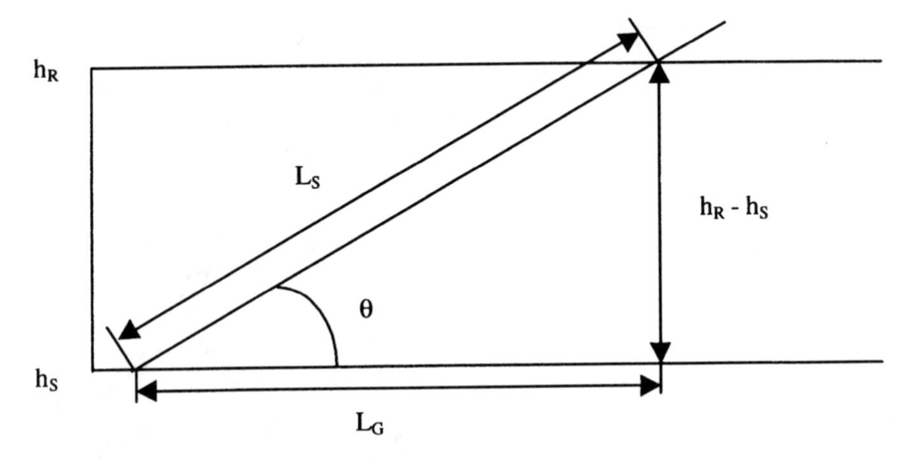

Figure 4.6 Relationship between slant path length and rain height.

attenuation at a given frequency [SAL-92]. The model is based on Liebe's specific attenuation model, which takes into account 30 water vapour absorption lines in the frequencies below 1000 GHz, at a specified reference atmospheric condition.

4.5.1.2 Rain Attenuation

The most significant contribution to atmospheric attenuation is that due to rain. Numerous prediction models for rain attenuation exist. Generally, these prediction methods employ two stages of modelling. The first stage involves the prediction of rain fall rate probability distributions and the second stage relates the path attenuation with the specific attenuation, γ_R (dB/km). γ_R is expressed in terms of the rain fall rate R (mm/h) and the slant path length, L_s, and rain height, h_R, as shown in Figure 4.6.

A simple step-by-step method for the derivation of attenuation due to rain is provided by the ITU [ITU-99d]. Initially, the slant path, L_s, is determined from the rain height, h_R. Here, the mean rain height, h_R, can be approximated by h_0, the mean 0° isotherm height [ITU-99e].

For elevation angles above 5°, the slant path length below the rain height is given by the formula:

$$L_s + \frac{(h_R - h_s)}{\sin\theta} \text{ km} \tag{4.33}$$

or for elevation angles less than 5°.

$$L_s = \frac{2(h_R - h_s)}{\left(\sin^2\theta + \frac{2(h_R - h_s)}{R_e}\right)^{\frac{1}{2}} + \sin\theta} \tag{4.34}$$

where R_e is the effective radius of the Earth (8500 km).

Using simple trigonometry, the horizontal projection of the slant path length, L_G, is then given by:

$$L_G = L_s \cos\theta \text{ km} \tag{4.35}$$

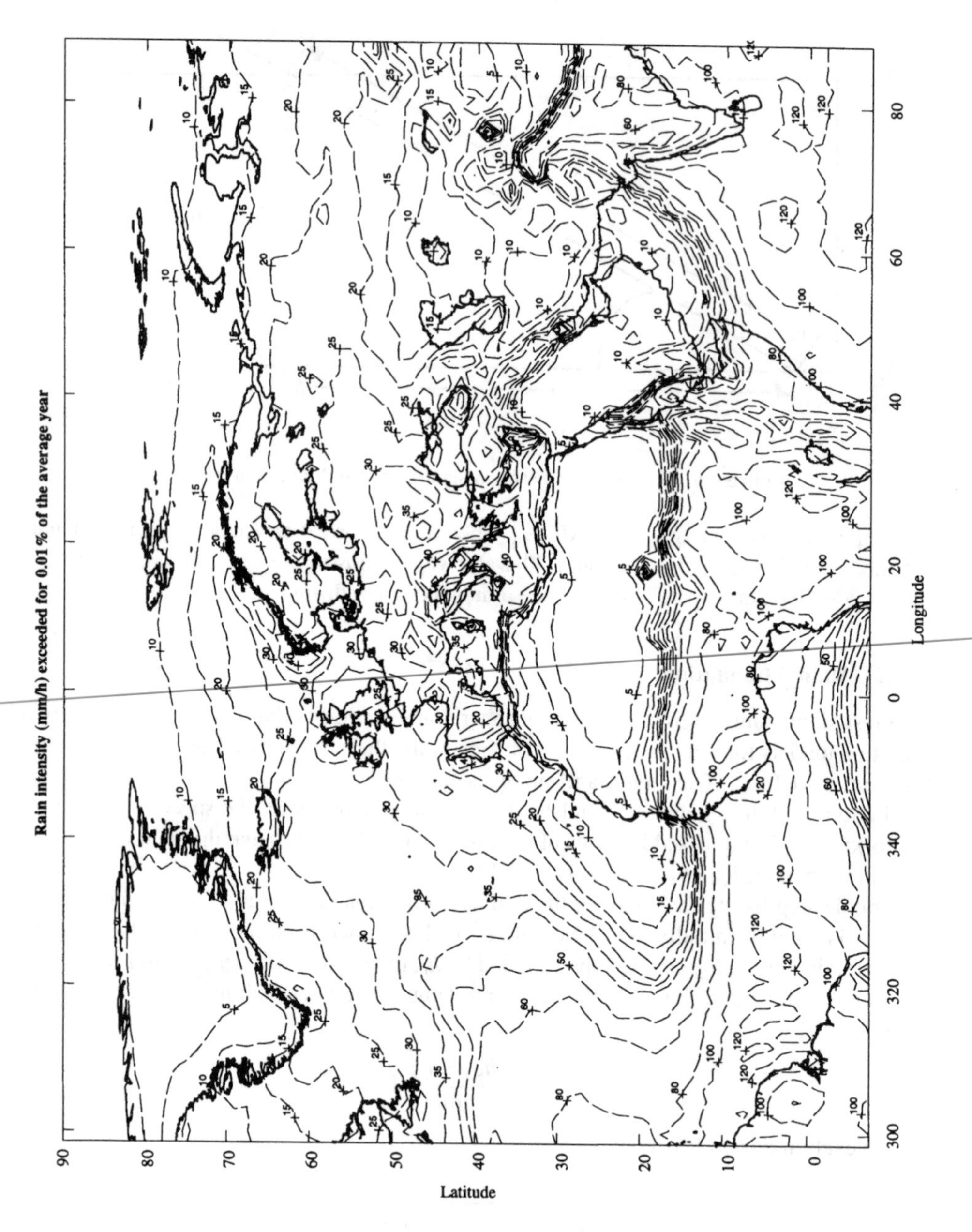

Figure 4.7 Rain intensity (mm/h) exceeded for 0.01% of an average year.

The next step is to obtain the rain fall rate exceeded for 0.01% of the average year, $R_{0.01}$. A procedure for the derivation of rain fall rate for a given latitude and longitude, anywhere in the world, is recommended by the ITU [ITU-99f]. This enables the rain fall rate exceeded for p% of the average year to be obtained. This recommendation should be used to obtain the rain fall rate in the absence of experimental data. Figure 4.7 shows a map of the rainfall rate in

Table 4.3 ITU-R regression coefficients for estimating attenuation

Frequency (GHz)	k_h	k_v	α_h	α_v
1	0.0000387	0.0000352	0.912	0.880
2	0.000154	0.000138	0.963	0.923
4	0.000650	0.000591	1.121	1.075
6	0.00175	0.00155	1.308	1.265
7	0.00301	0.00265	1.332	1.312
8	0.00454	0.00395	1.327	1.310
10	0.101	0.00887	1.276	1.264
12	0.0188	0.0168	1.217	1.200
15	0.0367	0.0335	1.154	1.128
20	0.0751	0.0691	1.099	1.065
25	0.124	0.113	1.061	1.030
30	0.187	0.167	1.021	1.000
35	0.263	0.233	0.979	0.963
40	0.350	0.310	0.939	0.929
45	0.442	0.393	0.903	0.897
50	0.536	0.479	0.873	0.868
60	0.707	0.642	0.826	0.824

mm/h for 0.01% of the average year. From the rain fall rate, the specific attenuation due to rain, γ_R, can be derived using the expression:

$$\gamma_R = k(R_{0.01})^{\alpha} \text{ dB/km} \tag{4.36}$$

where k and α are frequency dependent coefficients, the values for linear polarisation's vertical and horizontal paths are shown in Table 4.3, as defined in Ref. [ITU-99g].

The ITU provides the following formulae to derive k and α using the values in Table 4.3, for linear and circular polarisation, for all path geometries.

$$k = \left[k_h + k_v + (k_h - k_v)\cos^2\theta \cos 2\tau\right]/2 \tag{4.37a}$$

$$\alpha = [k_h \alpha_h + k_v k_v + (k_h \alpha_h - k_v \alpha_v)\cos^2\theta \cos 2\tau]/2k \tag{4.37b}$$

where θ = path elevation angle and τ = polarisation tilt angle relative to the horizontal (45° for circular polarisation).

Values of k and α for frequencies not given in Table 4.3 can be obtained by interpolation using logarithmic scales for frequency and k, and a linear scale for α. The ITU report that its method provides sufficient accuracy up to 55 GHz.

The next step is to calculate the horizontal and vertical reduction factors, $r_{0.01}$ and $v_{0.01}$, where

$$r_{0.01} = \frac{1}{1 + 0.78\sqrt{\frac{L_G \gamma_R}{f}} - 0.38(1 - e^{-2L_G})} \tag{4.38}$$

To evaluate $v_{0.01}$, consider the following:

$$\varsigma = \tan^{-1}\left(\frac{h_R - h_s}{L_G r_{0.01}}\right) \tag{4.39}$$

For $\varsigma > \theta$

$$L_R = \frac{L_G r_{0.01}}{\cos\theta} \text{ km} \tag{4.40a}$$

Otherwise

$$L_R = \frac{L_G r_{0.01}}{\sin\theta} \text{ km} \tag{4.40b}$$

If $|\varphi| < 36°$, $\chi = 36 - |\varphi|$ degrees.
Else

$$\chi = 0°$$

$$v_{0.01} = \frac{1}{1 + \sqrt{\sin\theta}\left(31\left(1 - e^{-\left(\frac{\theta}{1+\chi}\right)}\right)\frac{\sqrt{\gamma_R L_R}}{f^2} - 0.45\right)} \tag{4.41}$$

From which the effective path length, L_E, is derived:

$$L_E = L_R v_{0.01} \text{ km} \tag{4.42}$$

The attenuation due to rain exceeded for 0.01% of an average year, $A_{0.01}$, is then given by

$$A_{0.01} = \gamma_R L_E \text{ dB} \tag{4.43}$$

Estimated attenuation exceeded for other percentages of the average year, in the range 0.001–5% can then be obtained using the expression:

$$A_p = A_{0.01}\left(\frac{p}{0.01}\right)^{-(0.655+0.033\ln p - 0.045\ln A_{0.01} - \beta(1-p)\sin\varphi)} \text{ dB} \tag{4.44}$$

where $\beta = 0$ if $p \geq 1\%$ and $|\varphi| \geq 36°$; $\beta = -0.005(|\varphi| - 36°)$ if $p < 1\%$ and $|\varphi| < 36°$ and $\theta \geq 25°$; $\beta = -0.005(|\varphi| - 36°) + 1.8 - 4.25\sin\theta$ otherwise.

In the above, φ is the latitude location and θ is the elevation angle to the satellite.

Occasionally, it is important to know the dimension of the worst-month statistics. The ITU's expression for converting the annual time percentage of excess, p, to the worst-month time percentage of excess, p_W, on a global basis is as follows [ITU-99h]:

$$p(\%) = 0.3 p_w(\%)^{1.15} \tag{4.45}$$

for

$$1.9 \times 10^{-14} < p_w(\%) < 7.8$$

More accurate estimations for specific regions of the world are also provided in Ref. [ITU-99h].

Apart from the ITU model, the two most commonly used models derived from measurement data are the Crane 2-component model [CRA-89] and the Leitao–Watson European Model [LEI-86, WAT-87]. Both are based on radar studies of rainstorm structure and are able to give attenuation predictions at 11 GHz for high availabilities with typical rms error in dB of <25% for the US and for Europe. For high availability (>99.9%), showery rain is the dominant factor, whereas for availabilities less than or near to 99.9% of the average year, it is expected that many sites' contributions will be made from both widespread and showery rain. Leitao–Watson's model applied the excess zenith attenuation concept to the showery rain model to include a region of melting particles to be modelled as rain. This allows the rain height to be defined as the 0°C isotherm height plus seasonal climatic adjustments.

Due to the significantly different frequency dependence of specific attenuation in the rain and melting regions, Watson–Hu adopted a more physically based approach, modelling the rain region and melting zone entirely separately [WAT-94]. In Watson–Hu's model, the rain height was taken as the 0°C isotherm height minus the thickness of the melting zone and a correction was made for the variation of specific attenuation with height (an empirically determined factor of 0.95 from Leitao's radar study [LEI-85]). The thickness of the melting zone was taken from the Russchenberg–Lighthart melting layer model [RUS-93]. When the rain height is separated from the thickness of the melting layer, the melting zone attenuation needs to be derived. In Watson–Hu's model, the melting layer model of Russchenberg–Lighthart was used to determine the melting layer thickness, as well as to calculate melting zone attenuation for a specific ground rainfall intensity. From Watson–Hu, the rain process contains at least three components: the rain component, the melting layer and the water vapour component, which is due to a saturated water vapour column up to the rain height. Thus, for a given point rain rate R, the attenuation A due to either showery rain or widespread rain together with the melting layer and a saturated water vapour column is given by

$$A(p) = CA_{\text{rain}}(H_0 - H_{\text{m}}, R) + A_{\text{bb}}(H_{\text{m}}, R) + A_{\text{watervapour}}(H_0 - H_{\text{m}})\ \text{dB} \tag{4.46}$$

where H_0 is the rain height as defined by Leitao–Watson, H_{m} is the thickness of the melting layer, C ($\approx$0.95) is the correction factor for the non-uniform vertical profile of rain-specific attenuation in the rain region, $A_{\text{rain}}()$ is the attenuation due to rain; $A_{\text{bb}}()$ is the attenuation due to the bright band (or melting layer) and $A_{\text{water vapour}}()$ is the attenuation due to a saturated water vapour column.

For frequencies of 30 GHz and above, a correction for cloud attenuation, concurrent with rain is performed, using the semi-empirical model of Altshuler–Marr [ALT-89].

Figure 4.8 demonstrates the accuracy of the Watson–Hu model when compared with measured data obtained from an ITU database.

4.5.1.3 Fog and Clouds

For systems operating at frequencies below 30 GHz, the effect of cloud and fog is not as significant as other impairments and can be largely neglected. On the other hand, for low

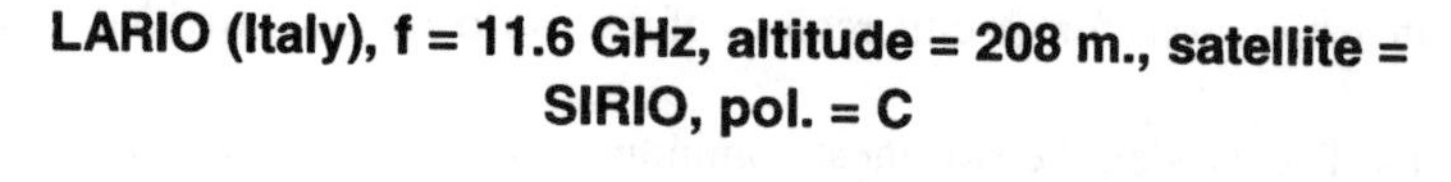

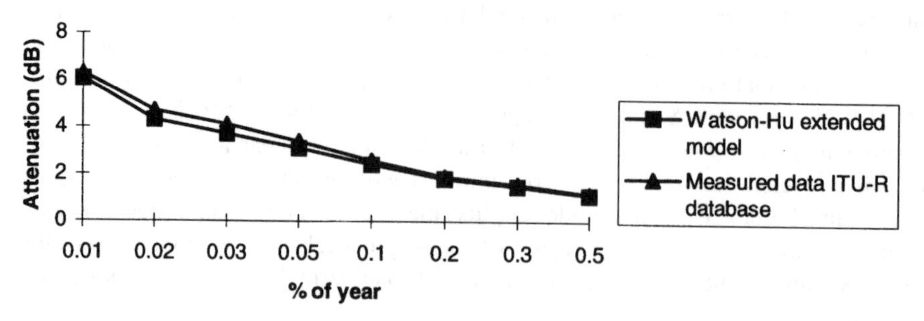

Figure 4.8 Predicted (Watson–Hu model) and measured attenuation at Lario station (Italy).

availability systems operating above 30 GHz, such as in a transportable terminal environment, the effect of cloud attenuation may become significant.

The specific attenuation of clouds and fog depends on frequency, liquid water density (g/m^3) and temperature. Empirical models for prediction of cloud and fog attenuation are in the relatively early stages of development, however, a model for the prediction of attenuation due to clouds and fog is recommended by the ITU, which is valid for frequencies up to 200 GHz [ITU-99i].

In order to determine the attenuation due to clouds, the statistics of total columnar liquid content (kg/m^2) or millimeters of precipitatable water for a site needs to be known. In the absence of local measurements, ITU-R Recommendation 840 can be used, which provides global maps for the total columnar content of cloud liquid water for four different yearly exceedance probabilities (20%, 10%, 5%, 1%). Using one of these maps, for an annual exceedance probability of 1% over Europe, the attenuation due to clouds can be seen to vary from less than 0.5 dB at 20 GHz to in the region of 9 dB at 100 GHz for a 30° elevation angle to the satellite. This clearly illustrates the significance of cloud attenuation at higher frequencies.

Further examples of empirical models for the derivation of cloud attenuation can be found in Refs. [ALT-89, DIN-89, SAL-92].

4.5.1.4 Scintillation

At elevation angles of less than 10°, signals operating in the Ku-band and above are subject to generally small, moderately rapid fluctuations attributed to irregularities in the tropospheric refractive index. These fluctuations can result in both enhancements and fades to the signal, as in the reception of multipath signals in the mobile environment. Systems operating below 10 GHz and at elevation angles above 10° are unlikely to be affected by tropospheric scintillation.

Most scintillation measurements have been made at 11–12 GHz, due to the fact that beacon signals at these frequencies have been widely available. A model developed by INTELSAT [ROG-87] gives a worst season scintillation fading distribution for a 3-m diameter reference antenna, at 11 GHz, for an elevation angle 5° and an assumed antenna efficiency 60%. Scaling factors are provided for other values of frequencies, elevation angles and antenna sizes. Karasawa et al. derived an empirical model for the rms value of scintillation based on measurements at 11.5 and 14.2 GHz, with elevation angles of 4°, 6.5° and 9° and antenna diameter of 7.6 m in Yamaguchi, Japan [KAR-88]. The cumulative distribution due to scintillation is then expressed as a function of the rms value.

The ITU-R model is based upon measurements within a frequency range of 7–14 GHz and is applicable to elevation angles of greater than 4° [ITU-99d]. This model requires site information on:

- the average surface ambient temperature for a period of 1 month or longer (from which the saturation water vapour pressure is determined [ITU-99j]); and
- the average surface relative humidity for a period of 1 month or longer (which is used with the average surface ambient temperature and the calculated saturation water vapour pressure to calculate the wet term of the radio refractivity, N_{wet} [ITU-99j]).

The other parameters used by the ITU-R model are the operating frequency, within the limits of 4–20 GHz, the antenna diameter and its efficiency, and the path elevation angle.

Application of the model results in the calculation of the fade depth, for a specified time percentage within the range 0.01–50.

4.5.1.5 Total Attenuation

The ITU recommends that all of the possible tropospheric contributions to signal attenuation are combined as follows:

$$A_T(p) = A_G + \sqrt{(A_R(p) + A_c(p))^2 + A_s(p)^2} \tag{4.47}$$

where $A_T(p)$ is the total attenuation for a given probability; $A_R(p)$ is the attenuation due to rain for a given probability; $A_C(p)$ is the attenuation due to clouds for a given probability; A_G is the gaseous attenuation due to water vapour and oxygen and $A_S(p)$ is the attenuation due to scintillation for a given probability.

Various simplifications to the above formula can be made depending on the operating environment. For example, the scintillation parameter can be discarded when operating with an elevation angle above 10°.

Two alternative approaches to the ITU model are those presented by Salonen et al. and Watson–Hu. Salonen et al. combined the fading processes on an equi-probability basis, which represents the worst case situation. In the Watson–Hu model, two main parts are involved in the fading model: the rain-fading component and the non-rain component. These two components are combined as mutually exclusive sets. Within the rain process, contributions from rain, melting zone, rain-cloud and water vapour are considered as being highly correlated. Within the non-rain process, non-precipitating cloud, water vapour and scintillation effects are added on an equi-probability basis.

4.5.1.6 Depolarisation

The electromagnetic wave transmitted by an antenna comprises two components: the electric-field vector and the magnetic-field vector. These two components travel in the direction of the

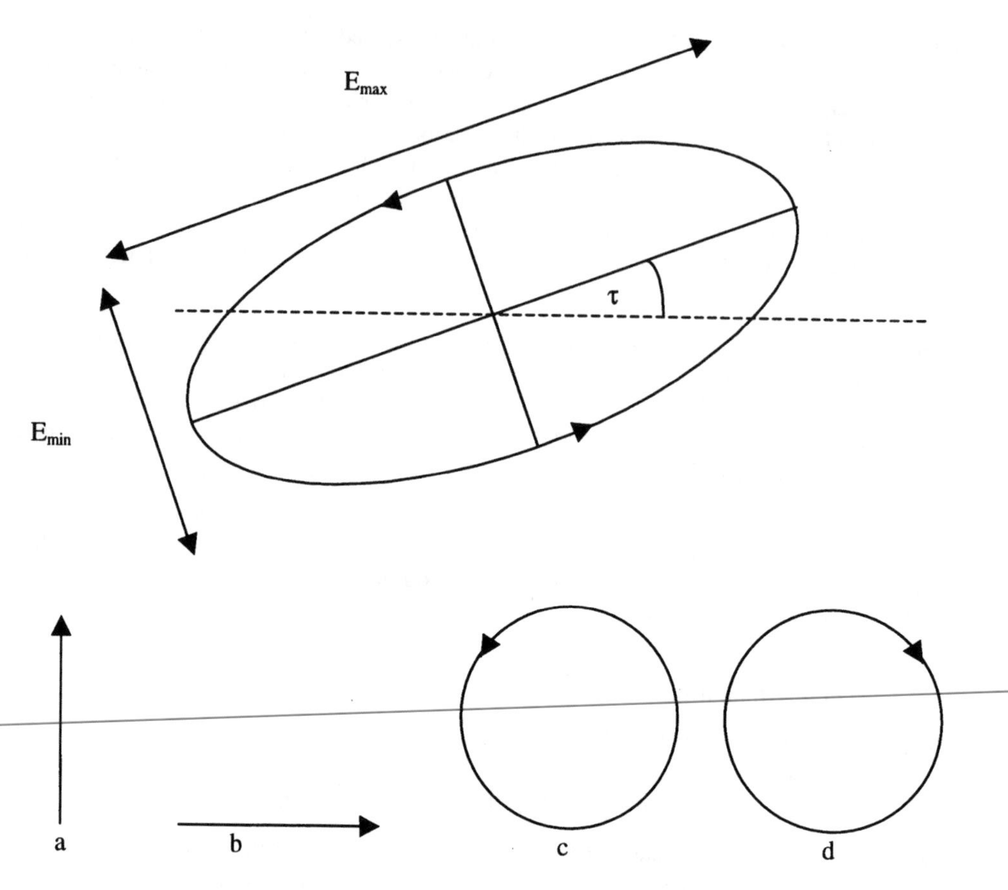

Figure 4.9 Generalised elliptical waveform also illustrating: (a) vertical polarisation, (b) horizontal polarisation, (c) left hand circular polarisation, (d) right hand circular polarisation. The direction of travel is into the paper.

transmission and are orthogonal. By convention, the orientation of the electric-field vector defines the polarisation of the transmitted wave. The general form of a polarised wave, when viewed perpendicular to the direction of travel, is elliptical in shape. This is shown in Figure 4.9. Two important parameters in this figure are the *axial ratio*, and the angle τ, the inclination angle with respect to the reference axis.

The major and minor axes of the ellipse, shown in Figure 4.9 as E_{max} and E_{min}, define the *axial ratio*, A_{XR},by using the following expression:

$$A_{XR} = 20\log\left(\frac{E_{max}}{E_{min}}\right) \text{ dB} \tag{4.48}$$

In satellite communications, four types of polarisation are employed in transmissions: vertical linear polarisation; horizontal linear polarisation; left hand circular polarisation (LHCP); and right hand circular polarisation (RHCP). Horizontal and vertical polarisations are defined with respect to the horizon, while left hand or right hand circular is determined by the anti-clockwise or clockwise rotation when viewed from the antenna in the direction of travel. For circular polarisation, the axial ratio is equal 0 dB, while for linear polarisation, a theoretical value of infinity would be obtained.

Left and right hand circular polarisation's are orthogonal, as are vertical and horizontal. Thus, an antenna receiving a horizontally polarised wave cannot receive a vertically polarised wave and *vice versa*. Similarly, an antenna receiving a LHCP wave cannot receive a RHCP wave and *vice versa*. This property lends itself to frequency re-use, the other means of achieving this being spatial separation of beams transmitting in the same frequency band.

In practice, a polarised wave will comprise both the wanted polarisation plus some energy transmitted on the orthogonal polarisation. The degree of the coupling of energy between polarisation's is termed the *cross polar discrimination* (XPD). This is given by:

$$\mathrm{XPD} = 20\log\left|\frac{E_{\mathrm{CP}}}{E_{\mathrm{XP}}}\right| \mathrm{dB} \tag{4.49}$$

where E_{CP} is the received co-polarised electric field strength and E_{XP} is the received cross-polarised electric field strength.

For the case of co-existing circular polarised waves, the XPD is determined from the axial ratio, using the expression:

$$\mathrm{XPD} = 20\log\frac{A_{\mathrm{XR}} + 1}{A_{\mathrm{XR}} - 1} \mathrm{dB} \tag{4.50}$$

The ITU provides a step-by-step method for the calculation of hydrometer induced cross polarisation, which is valid for frequencies within the range 8 GHz $< f <$ 35 GHz and for elevation angles less than 60° [ITU-99d]. In addition to operating frequency and elevation angle, the attenuation due to rain exceeded for the required percentage of time p, and the polarisation tilt angle, τ, with respect to the horizontal also needs to be known. The XPD due to rain is given by the following equation:

$$\mathrm{XPD}_{\mathrm{rain}} = C_f - C_A + C_\tau + C_\theta + C_\sigma \ \mathrm{dB} \tag{4.51}$$

where the following terms are derived:

frequency dependent term: $C_f = 30\ \log f\ f$ in GHz

rain dependent term: $C_A = V(f)\log A_p$

$$V(f) = \begin{cases} 12.8f^{0.19} & 8 \le f \le 20 \text{ GHz} \\ 22.6 & 20 \le f \le 35 \text{ GHz} \end{cases}$$

polarisation improvement factor: $C_\tau - 10\log[1 - 0.484(1 + \cos 4\tau)]$

elevation angle dependent term: $C_\theta = -40\log(\cos\theta)$

Canting angle term: $C_\sigma = 0.0052\sigma$

In the above, the canting angle refers to the angle that a falling raindrop arrives at the Earth with respect to the local horizon. τ, θ and σ are expressed in degrees. σ has values of 0°, 5°, 10° and 15° for 1, 0.1, 0.01 and 0.001% of the time, respectively.

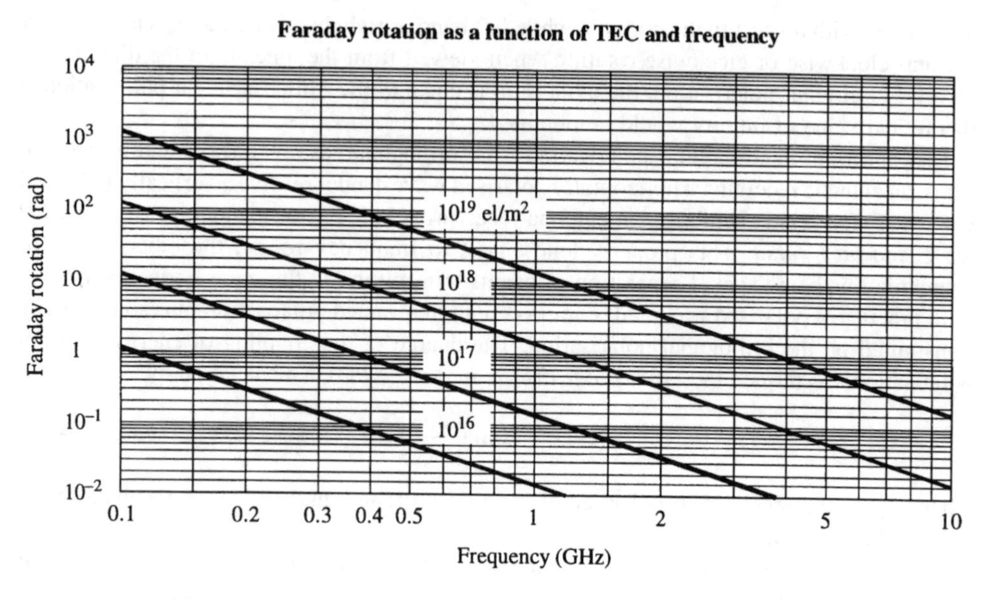

Figure 4.10 Faraday rotation as a function of TEC and frequency.

Taking ice into account, the XPD not exceeded for $p\%$ of time is given by:

$$\mathrm{XPD}_p = \mathrm{XPD}_{\mathrm{rain}} - C_{\mathrm{ice}} \text{ dB} \tag{4.52}$$

where $C_{\mathrm{ice}} = \mathrm{XPD}_{\mathrm{rain}} \times (0.3 + 0.1 \log p)/2$.

4.5.2 Ionospheric Effects

4.5.2.1 Faraday rotation and Group Delay

Ionospheric effects are significant for frequencies up to about 10 GHz and are particularly significant for non-geostationary satellites operating below 3 GHz.

The total electron content (TEC) accumulated through the trans-ionospheric transmission path results in the rotation of the linear polarisation of the carrier and a time delay in addition to the anticipated propagation path delay. The delay is known as the *group delay*, while the rotation of the linear polarisation of the carrier is known as *Faraday rotation*.

The TEC, denoted by N_{T}, is given by the formula [ITU-99k].

$$N_{\mathrm{T}} = \int_S n_{\mathrm{e}}(s)\mathrm{d}s \tag{4.53}$$

where: n_{e} is the electron concentration (el/m^3) and s is the propagation path length through the ionosphere (m).

Typically, N_{T} has a value in the range 10^{16}–10^{18} el/m^2.

The Faraday rotation, ϕ, is then given by:

$$\phi = 2.36 \times 10^2 B_{\mathrm{av}} N_{\mathrm{T}} f^2 \text{ Radian} \tag{4.54}$$

where B_{av} is the average Earth magnetic field (Wb/m^2) and f is frequency (GHz).

The occurrence of Faraday rotation is well understood and can be predicted with a high degree of accuracy and compensated for by adjusting the polarisation tilt angle at the Earth station. Figure 4.10 shows the characteristics of Faraday rotation as recommended by the ITU-R. Faraday rotation does not affect transmissions employing circular polarisation, hence the use in mobile-satellite systems.

Group delay, denoted by T is obtained using the formula:

$$T = 1.34 N_T / f^2 \times 10^{-7} \text{ s} \tag{4.55}$$

where f is frequency (Hz).

4.5.2.2 Ionospheric Scintillation

Ionospheric scintillation is caused by irregularities in the electron density of the ionosphere. Ionospheric scintillation is particularly significant at frequencies below 1 GHz, and is dependent on location, season, solar activity (sunspots) and local time.

High latitude areas near the Arctic polar region and the Equatorial region bounded between $\pm 20°$ are susceptible to intense scintillation activity. Scintillation activity is at a maximum during night-time, lasting from 30 minutes to a number of hours in duration.

Further information can be found in Ref. [ITU-99k].

References

[AKT-97] R. Akturan, K. Penwarden, "Satellite Diversity as a Propagation Impairment Mitigation Technique for Non-GSO MSS Systems", *Proceedings of Fifth International Mobile Satellite Conference*, Pasadena, 16–18 June 1997; 189–194.

[ALT-89] E.E. Altshuler, R.A. Marr, "Cloud Attenuation at Millimetre Wavelengths" *IEEE Transactions on Antennas and Propagation*, 37(11), November 1989; 1473–1479.

[BUT-92] G. Butt, B.G. Evans, "Narrowband Channel Statistics from Multiband Propagation Measurements Applicable to High Elevation Angle Land-Mobile Satellite Systems", *IEEE Journal on Selected Areas in Communications*, 10(8), October 1992; 1219–1226.

[COR-94] G.E. Corazza, F. Vatalaro, "A Statistical Model for Land Mobile Satellite Channels and its Applications to Non-Geostationary Orbit Systems", *IEEE Transactions on Vehicular Technology*, 43(2), August 1994; 738–742.

[CRA-89] R.K. Crane, H.C. Shieh, "A Two-Component Rain Model for the Prediction of Site Diversity Performance", *Radio Science*, 24(6), September – October 1989; 641–665.

[DIN-89] F. Dintelmann, G. Ortgies, "Semi-empirical Model for Cloud Attenuation Prediction", *Electronic Letters*, 25(22), 26 October 1989; 1487–1488.

[DOT-98] M. Döttling, H. Ernst, W. Wesbeck, "A New Wideband Model for the Land Mobile Satellite Propagation Channel", *Proceedings of IEEE 1998 International Conference on Universal Personal Communications*, 1, 5–9 October 1998, Florence, 647–651.

[GOL-89] J. Goldhirsh, W.J. Vogel, "Mobile Satellite System Fade Statistics for Shadowing and Multipath from Roadside Trees at UHF and L-band", *IEEE Transactions on Antennas and Propagation*, 37(4), April 1989; 489–498.

[GOL-92] J. Goldhirsh, W.J. Vogel, "Propagation Effects for Land Mobile Satellite Systems: Overview of Experimental and Modelling Results", *NASA Reference Publication 1274*, February 1992.

[HOL-99] M. Holzbock, C. Senninger, "An Aeronautical Multimedia Service Demonstration at High Frequencies", *IEEE Multimedia*, 6(4), October – December 1999; 20–29.

[HOW-92] R.G. Howell, J.W. Harris, M. Mehler, "Satellite Copolar Measurements at BT Laboratories" *BT Technical Journal*, 10(4), October 1992; 34–51.

[ITU-92] *ITU-R Rec. P.682-1*, "Propagation Data Required for the Design of Earth-Space Aeronautical Mobile Telecommunication Systems", 1992.

[ITU-99a] *ITU-R Rec. P.681-4*, "Propagation Data Required for the Design of Earth-Space Land Mobile Telecommunication Systems", 1999.

[ITU-99b] *ITU-R Rec. P.680-3*, "Propagation Data Required for the Design of Earth-Space Maritime Mobile Telecommunication Systems", 1999.

[ITU-99c] *ITU-R Rec. P.676-4*, "Attenuation by Atmospheric Gases", 1999.

[ITU-99d] *ITU Rec. P 618-6*, "Propagation Data and Prediction Methods Required for the Design of Earth-Space Telecommunication Systems", 1999.

[ITU-99e] *ITU-R Rec. P.839-2*, "Rain Height Model for Prediction Methods", 1999.

[ITU-99f] *ITU-R Rec. P.837-2*, "Characteristics of Precipitation for Propagation Modelling", 1999.

[ITU-99g] *ITU-R Rec. P.838-1*, "Specific Attenuation Model for Rain for use in Prediction Methods", 1999.

[ITU-99h] *ITU-R Rec. P.841-1*, "Conversion of Annual Statistics to Worst-Month Statistics", 1999.

[ITU-99i] *ITU-R Rec. P.840-3*, "Attenuation Due to Clouds and Fog", 1999.

[ITU-99j] *ITU-R Rec. P.453-7*, "The Radio Refractive Index: Its Formula and Refractivity Data", 1999.

[ITU-99k] *ITU-R Rec. P.531-5*, "Ionospheric Propagation Data and Prediction Methods Required for the Design of Satellite Services and Systems", 1999.

[JAH-95a] A. Jahn, S. Buonomo, M. Sforza, E. Lutz, "Narrow- and Wide-band Channel Characterization for Land Mobile Satellite Systems: Experimental Results at L-Band", *Proceedings of Fourth International Mobile Satellite Conference*, Ottawa, 4–6 June 1995; 115–121.

[JAH-95b] A. Jahn, S. Buonomo, M. Sforza, E. Lutz, "A Wideband Channel Model for Land Mobile Satellite Systems", *Proceedings of Fourth International Mobile Satellite Conference*, Ottawa, 4–6 June 1995; 122–127.

[JAH-00] A. Jahn, "Propagation Characterisation for Mobile Multimedia Satellite Systems from L-Band to EHF-Band", *Proceedings of AP2000 Millennium Conference on Antennas and Propagation*, Davos, 9–14 April 2000.

[KAR-88] Y. Karasawa, M. Yamada, J.E. Allnut, "A New Prediction Method for Tropospheric Scintillation on Earth-Space paths" *IEEE Transactions on Antennas and Propagation*, 36(11), November 1988; 1608–1614.

[KAR-98] M.S. Karaliopoulos, F.-N. Pavlidou, "Modelling of the Land Mobile Satellite Channel: A Review", *Electronics& Communications Engineering Journal*, 11(5), October 1998; 235–248.

[LEE-89] W.C.Y. Lee, *Mobile Cellular Telecommunications Systems*, McGraw-Hill, 1989.

[LEI-85] M.J.M. Leitao, "Propagation factors affecting the design of Satellite Communication Systems" *PhD Thesis*, University of Bradford, 1985.

[LIE-88] H.J. Liebe, "Atmospheric Attenuation and Delay Rates between 1 GHz and 1 THz" *Proceedings of Conference on Microwave Propagation in the Marine Boundary Layer*, Monterey, CA, 1988.

[LEI-86] M.J. Leitao, P.A. Watson, "Method for Prediction of Attenuation on Earth-Space Links based on Radar Measurements of the Physical Structure of Rainfall", *IEE Proceedings – Part F*, 133(4), July 1986; 429–440.

[LOO-85] C. Loo, "A Statistical Model for a Land Mobile Satellite Link", *IEEE Transactions on Vehicular Technology*, VT-34(3), August 1985; 122–127.

[LOO-98] C. Loo, J.S. Butterworth, "Land Mobile Satellite Channel Measurements and Modelling", *Proceedings of IEEE*, 8(7), July 1998; 1442–1463.

[LUT-91] E. Lutz, D. Cygan, M. Dippold, F. Doliansky, W. Papke, "The Land Mobile Satellite Communication Channel - Recording, Statistics and Channel Model", *IEEE Transactions on Vehicular Technology*, 40(2), May 1991; 375–386.

[PIN-95] D.S. Pinck, M. Rice, "Ka-Band Channel Characterization for Mobile Satellite Systems", *ACTS Mobile Program*, Jet Propulsion Laboratory, April 1995; 29–36.

[ROG-87] D.V. Rogers, J.E. Allnut, "A Practical Tropospheric Scintillation Model for Low Elevation Angle Satellite Systems", *Proceedings of Fifth International Conference on Antennas and Propagation*, ICAP 87, Part 2: Propagation. IEE, London, 1987; 273–276.

[RUS-93] H.W.J. Russchenberg, L.P. Lighthart, "Backscattering by and Propagation through the Melting Layer of Precipitation", *ESA/ESTEC Contract, Final Report PO122859*, 1993.

[SAL-92] E. Salonen, S. Karhu, P. Jokela, W. Zhang, S. Uppala, H. Aulamo, S. Sarkkula, J.P.V.P. Baptista,

"Modelling and Calculation of Atmospheric Attenuation for Low Fade Margin Satellite Communications", *ESA Journal*, 16(3), 1992; 299–317.

[ULA-82] F.T. Ulaby, R.K. Moore, A.K. Fung, *Microwave Remote Sensing. Active and Passive. Vol. I: Microwave Remote Sensing Fundamentals and Radiometry*, Addison-Wesley, Reading, MA.

[VAT-95] F. Vatalaro, "Generalised Rice-Lognormal Channel Model for Wireless Communications", *Electronic Letters*, 31(22), 26 October 1995; 1899–1900.

[VOG-88] W.J. Vogel, U.-S. Hong, "Measurement and Modelling of Land Mobile Satellite Propagation at UHF and L-band", *IEEE Transactions on Antennas and Propagation*, 36(5), May 1988; 707–719.

[VOG-92] W.J. Vogel, J. Goldhirsh, "Mobile Satellite Systems Propagation Measurements at L-band Using MARECS-B2", *IEEE Transactions on Antennas and Propagation*, 38(2), February 1990; 259–264.

[VUC-92] B. Vucetic, J. Du, "Channel Modelling and Simulation in Satellite Mobile Communication Systems", *IEEE Journal on Selected Areas in Communications*, 10(8), October 1992; 1209–1218.

[WAT-87] P.A. Watson, M.J. Leitao, V. Sathiaseelan, M. Gunes, J.P.V.P. Baptista, B.A. Potter, N. Sengupta, O. Turney, G. Brussaard, "Prediction of Attenuation of Satellite-Earth Links in the European Region", *IEE Proceedings – Part F*, 134(5), October 1987; 583–596.

[WAT-94] P.A. Watson, Y.F. Hu, "Prediction of Attenuation on Satellite-Earth Links for Systems operating with Low Fade Margins", *IEE Proceedings Microwave Antennas and Propagation*, 141(6), December 1994; 467–472.

5

Radio Link Design

5.1 Introduction

Unlike terrestrial cellular networks, in a mobile-satellite network, transmissions are constrained by available power. As illustrated in the previous chapter, the mobile-satellite channel provides a challenging environment in which to operate. Consequently, efficient coding and modulation techniques need to be employed in order to achieve a system margin above the minimum needed to guarantee a particular Quality of Service (QoS).

The transmission chain for a satellite communication system is shown in Figure 5.1.

In Figure 5.1, the transmit (Tx)/receiver (Rx) hardware includes the application of the multiple access scheme. Of course, not all of the above need be applied to a particular system, although there is an obvious need for certain components, such as the modulator/demodulator, for example. The selection of particular elements of the chain is driven by the needs of the system design. This chapter initially considers the approach to developing a link budget analysis. Here, the influence of the satellite payload characteristics, as well as other operational characteristics such as frequency, transmit power, and so on, on the overall link design

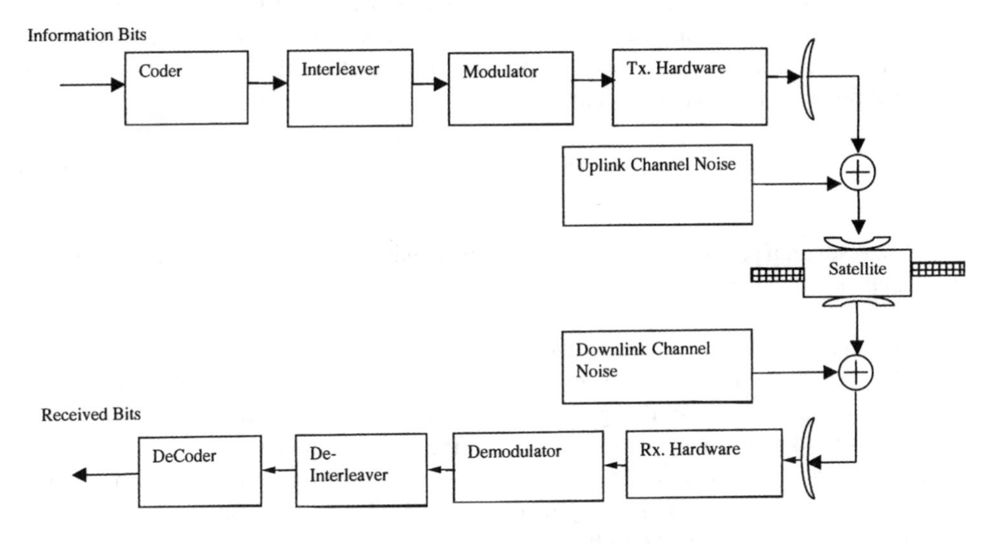

Figure 5.1 Simplified transmission chain.

are considered. This is followed by a description of the modulation schemes and coding techniques that are employed on the link. This chapter concludes with a presentation on the multiple access schemes that are applicable to a mobile-satellite system, followed by an assessment of the current status of the standardisation of the multiple access scheme for S-UMTS/IMT-2000.

5.2 Link Budget Analysis

5.2.1 *Purpose*

A link budget analysis forms the cornerstone of the system design. Link budgets are performed in order to analyse the critical factors in the transmission chain and to optimise the performance characteristics, such as transmission power, bit rate and so on, in order to ensure that a given target quality of service can be achieved.

5.2.2 *Transmission and Reception*

The strength of the received signal power is a function of the transmitted power, the distance between transmitter and receiver, the transmission frequency, and the gain characteristics of the transmitter and receiver antennas.

An ideal isotropic antenna radiates power of uniform strength in all directions from a point source. The power flux density (PFD) on the surface of a sphere of radius R, which has at its centre an isotropic antenna radiating in free space a power P_t (Watts), is given by:

$$\mathrm{PFD} = \frac{P_t}{4\pi R^2}\ \mathrm{Wm}^{-2} \tag{5.1}$$

In practice, antennas with directional gain are used to focus the transmitted power towards a particular, wanted direction. Here, an antenna's gain in direction (θ, ϕ), that is $G(\theta, \phi)$, is defined as the ratio of the power radiated per unit solid angle in the direction (θ, ϕ) to the same total power, P_T, radiated per unit solid angle from an isotropic source:

$$G(\theta, \phi) = \frac{P(\theta, \phi)}{\dfrac{P_T}{4\pi}} \tag{5.2}$$

Antenna radiation patterns are three-dimensional in nature, however, it is usual to represent an antenna radiation pattern from the point of view of a single-axis plot. Such a plot is shown in Figure 5.2.

An antenna's gain is normally calculated with reference to the boresight, the direction at which the maximum antenna gain occurs. In this case $\theta, \phi = 0°$. Gain is usually expressed in dBi, where i refers to the fact that gain is relative to the isotropic gain. An important parameter that is used in an antenna's specification is the 3-dB beamwidth, which represents the angular separation at which the power reduces to 3-dB, or half-power, below that of boresight. For a parabolic antenna, the simplified relationship between the antenna diameter and 3-dB beamwidth, θ_{3db}, as shown in Figure 5.2, is given by:

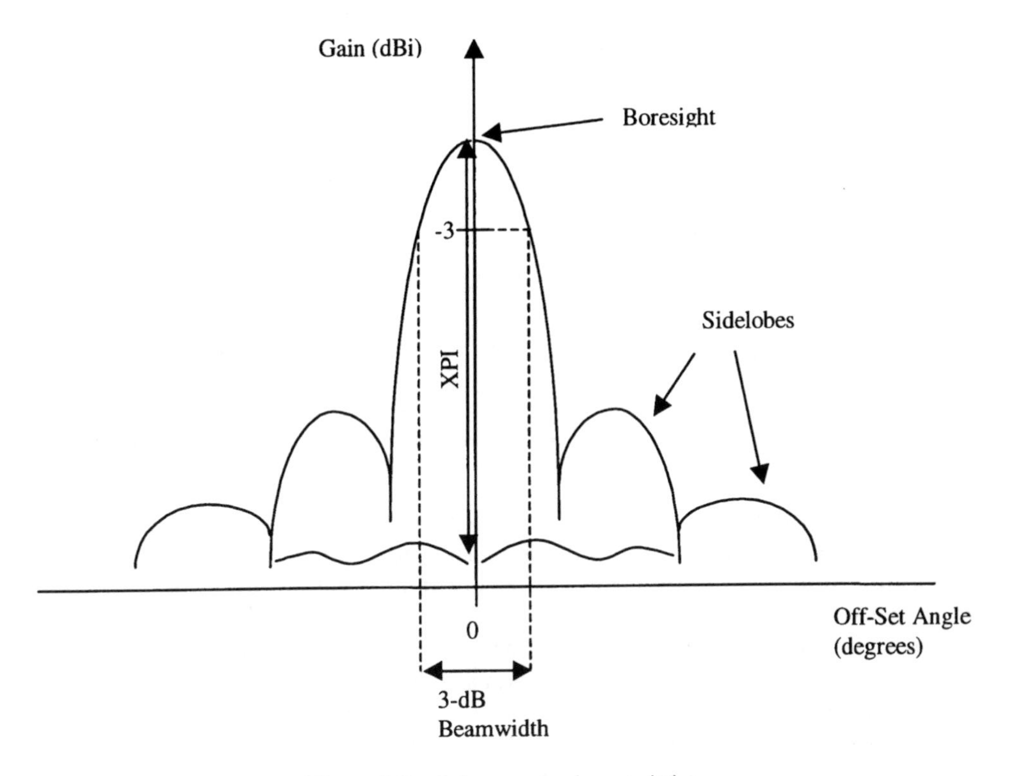

Figure 5.2 Antenna gain characteristics.

$$\theta_{3\text{dB}} \approx \frac{65\lambda}{D} \text{ degrees} \tag{5.3}$$

where λ is the transmission wavelength (m); D is the antenna diameter (m).

Here, it can be seen that the half-power beamwidth is inversely proportional to the operating frequency and the diameter of the antenna. For example, a 1 m receiver antenna operating in the C-band (4 GHz) has a 3-dB beamwidth of roughly 4.9°. The same antenna operating in the Ku-band (11 GHz) has a 3-dB beamwidth of approximately 1.8°.

The level of the antenna pattern's sidelobes is also important, as this tends to represent gain in unwanted directions. For a transmitting gain this leads to the transmission of unwanted power, resulting in interference to other systems, or in the case of a receiving antenna, the reception of unwanted signals or noise. The ITU-R recommend several reference radiation patterns, with respect to the antenna's sidelobe characteristics [ITU-93, ITU-94a], depending on the application and the antenna characteristics. For example, for a reference earth station:

$$\begin{aligned} G &= 32 - 25\log\phi \text{ dBi}, \text{ for } \varphi_{\min} \le \varphi \le 48^\circ \\ &= -10 \text{ dBi for } 48^\circ \le \varphi \le 180^\circ \end{aligned}$$

where $\varphi_{\min}$ is the greater of 1° or $100\lambda/D$.

Figure 5.3 is the recommended radiation pattern for a vehicular-mounted near-omni-directional antenna operating within the 1–3 GHz band. Here, the gain of the antenna is restricted to less than or equal to 5 dBi for elevation angles in the range −20 to 90°.

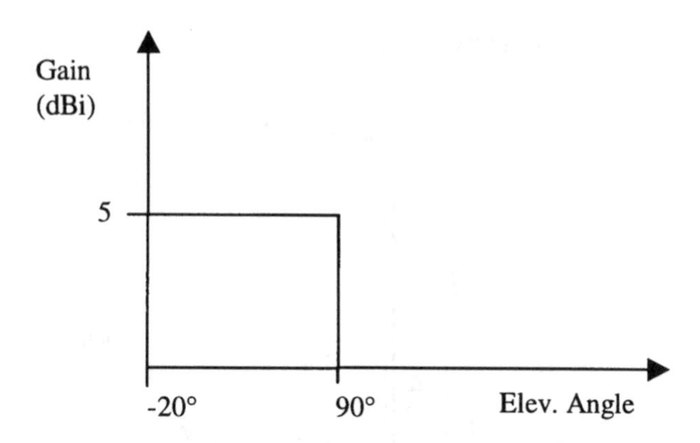

Figure 5.3 Reference radiation pattern for vehicle mounted antennas operating in the 1–3 GHz band.

As was discussed in Chapter 4, antennas have co- and cross-polar gains, where the reception of unwanted, orthogonally polarised cross-polar signals will add as interference to the co-polar signal. As was noted in Chapter 4, the ability of an antenna to discriminate between a wanted polarised waveform and its unwanted orthogonal component is termed its *cross-polar discrimination* (XPD). When dual polarisation is employed, an antenna's ability to differentiate between the wanted polarised waveform and the unwanted signal of the same polarisation, introduced by the orthogonally polarised wave, is termed the *cross-polar isolation* (XPI). Typically, an antenna would have an XPI > 30 dB.

If an antenna of gain G_t is transmitting power in the direction of a receiver located on the boresight of the antenna, then the power flux density at the receiver at a distance R from the receiver, is given by:

$$\text{PFD} = \frac{P_t G_t}{4\pi R^2} \quad \text{Wm}^{-2} \tag{5.4}$$

The product $P_t G_t$ is termed the effective isotropic radiated power (EIRP).

For an ideal receiver antenna of aperture area A, the total received power at the receiver is given by:

$$P_r = \frac{P_t G_t A}{4\pi R^2} \quad \text{W} \tag{5.5}$$

In reality, not all of the transmitted power will be delivered, due to antenna reflections, shadowing due to the feed, manufacturer imperfections, etc. Antenna efficiency is taken into account by the term *effective collecting area*, A_e, which is given by:

$$A_e = \eta A \tag{5.6}$$

where η, the antenna efficiency factor, is generally assumed to be in the region of 50–70%.

Therefore, the actual received power is given by:

$$P_r = \frac{P_t G_t A_e}{4\pi R^2} \quad \text{W} \tag{5.7}$$

An antenna of maximum gain G_r is related to its effective area by the following equation:

$$G_r = \eta \frac{4\pi A}{\lambda^2} \tag{5.8}$$

where λ is the wavelength of the received signal.

For a parabolic antenna of diameter D, this equation can be re-written as:

$$G_r = \eta \frac{\pi^2 D^2}{\lambda^2} \tag{5.9}$$

Using equation (5.9), the variation in antenna gain for a range of transmission frequencies that are employed in satellite communications is shown in Figure 5.4, assuming an efficiency of 60%.

Rearranging equation (5.8) and substituting in (5.7) gives:

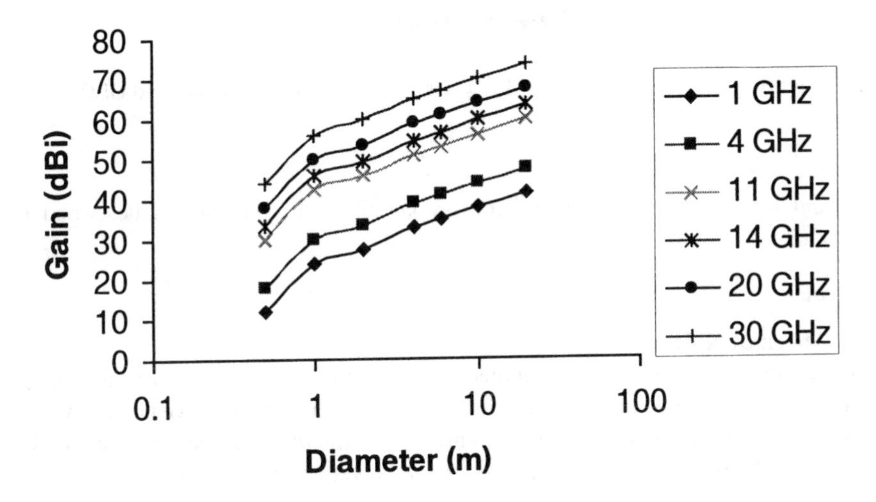

Figure 5.4 Variation in antenna gain with frequency.

$$P_r = \frac{P_t G_t G_r \lambda^2}{(4\pi R)^2} \quad \text{W} \tag{5.10}$$

The term $(\lambda/4\pi R)^2$ is known as the *free space loss (FSL)*. The variation in free space loss against frequency for LEO, MEO and GEO satellites is illustrated in Figure 5.5.

Usually, it is more convenient to express the parameters of the link in terms of dB ratio. For power ratios, parameters are expressed in terms of dBW or dBm. Here, the term dBW refers to the ratio, expressed in dB, of the parameter power to 1 W. Similarly, dBm refers to the ratio of parameter power to 1 mW. So, for example, 20 W is equal to 13 dBW or 43 dBm.

Expressing the above equation in terms of dB results in:

$$P_r = \text{EIRP} + \text{FSL} + G_r + A_p \quad \text{dBW} \tag{5.11}$$

In the above expression, an additional parameter, A_p, has been added to the equation to take

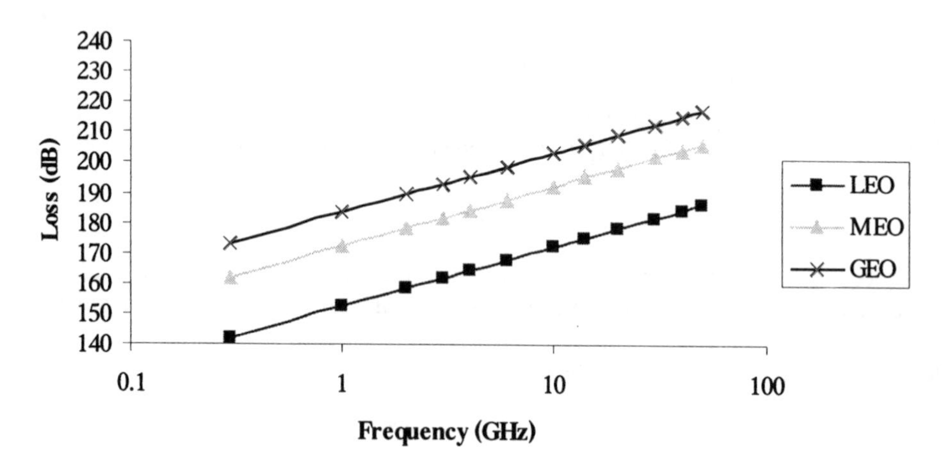

Figure 5.5 Free space loss of: LEO (1000 km); MEO (10000 km); and GEO.

into account the losses introduced by the propagation environment, as described in the previous chapter.

5.2.3 Noise

5.2.3.1 Thermal Noise

Generally, receiver antennas are specified in terms of G/T_e, where G is the antenna power gain, and T_e is the effective noise temperature of the receiver. The effective noise temperature, T_e, comprises the equivalent noise temperature of the antenna and feed plus the total noise temperature of the receiver equipment. A typical receiver architecture is illustrated in Figure 5.6.

In this example, the receiver comprises of five blocks: the antenna; the lossy feeder link; the first stage low noise amplifier (LNA); the first stage local oscillator (LO); and the intermediate frequency (IF) amplifier. Each of these devices contributes to the overall noise temperature of the receiver. To attain the overall system noise temperature, T_s, a specific point in the receiver chain from which every other noise temperature is referenced is assumed. Usually, this is at the input to the first amplifier of the receiver chain, although sometimes it is referred to at the input to the feeder link.

The thermal noise power generated by a particular device is given by the expression:

$$N = kTB \quad \text{Watts} \tag{5.12}$$

where k is the Boltzmann's constant ($1.38 \times 10 - {}^{23}$ J/K or alternatively -228.6 dBW/K/Hz); T is the noise temperature of the device, K; B is the equivalent noise bandwidth (Hz).

From equation (5.12), it can be seen that the output noise power of the above receiver chain is given by the expression:

$$P_o = k(T_{in} + T_1)G_1G_2G_3B + kT_2G_2G_3B + kT_3G_3B \quad \text{Watts} \tag{5.13}$$

where T_{in} represents the equivalent noise temperature of the antenna and the lossy feed.

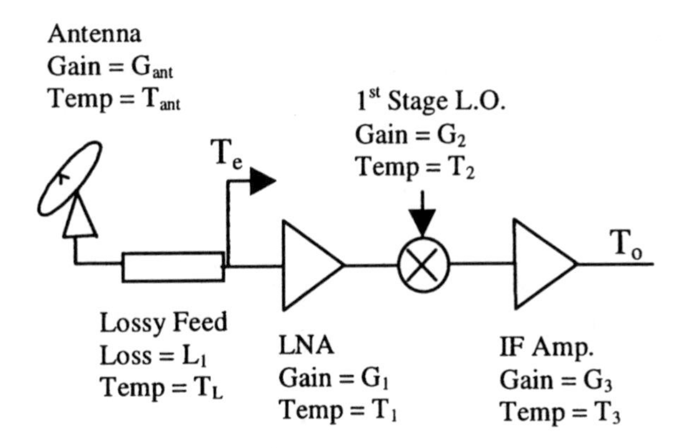

Figure 5.6 Typical receiver chain.

When referred back to the input to the first stage LNA, the above expression becomes:

$$P_o = kB\left((T_{in} + T_1) + \frac{T_2}{G_1} + \frac{T_3}{G_1 G_2}\right)G_1 G_2 G_3 \text{ Watts} \tag{5.14}$$

From which the equivalent noise temperature of the receiver is given by:

$$T_e = \left((T_{in} + T_1) + \frac{T_2}{G_1} + \frac{T_3}{G_1 G_2}\right) \quad \text{K} \tag{5.15}$$

It can be seen that to optimise the receiver chain in order to reduce the equivalent noise temperature, it is important that the first stage device has a large gain and a low noise temperature. As can be seen from the above equations, the contribution of a device to the overall performance of the link rapidly decreases the further the device is down the receiver chain.

From (5.12), the total noise power of the receiver chain, N, is then:

$$N = kT_e B \text{ Watts} \tag{5.16}$$

5.2.3.2 Background Noise

In the above expression, the noise contributions due to the antenna and the lossy feed were simplified into a single parameter, T_{in}.

For a lossy network, of gain L dB, the equivalent noise temperature is given by the equation:

$$T_e = T_0\left(1 - \frac{1}{L}\right) \quad \text{K} \tag{5.17}$$

where L is the lossy gain, given by the ratio of the input to the output powers, P_i/P_o; T_0 is the ambient temperature, usually assumed to be 290 K.

Referring to the above equation, and by performing a similar analysis as in (5.13–5.15) T_{in} can now be expressed as:

$$T_{in} = T_a/L + 290(1 - 1/L) \quad \text{K} \tag{5.18}$$

The antenna noise temperature, T_a, is due to the reception of unwanted noise sources from the sky and the ground within proximity of the antenna. Such unwanted noise sources are usually expressed in terms of brightness temperature, T_b. The antenna noise temperature, T_a, is given by the convolution of the antenna gain and the brightness temperature:

$$T_a = \frac{1}{4\pi} \int_0^{2\pi} \int_0^{\pi} G(\theta, \phi) T_b(\theta, \phi) \mathrm{d}\Omega \qquad \mathrm{K} \tag{5.19}$$

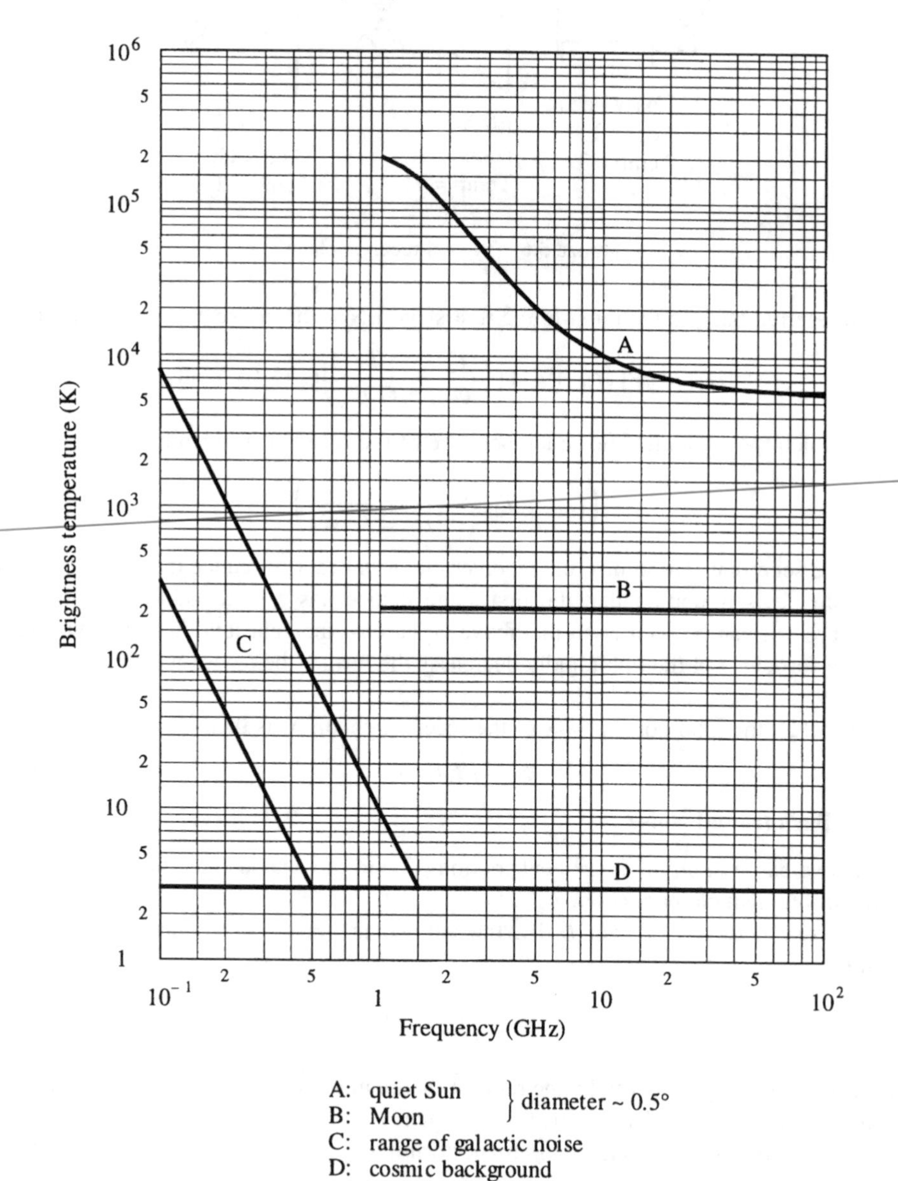

Figure 5.7 Brightness temperature variation with frequency for extra terrestrial noise sources.

where $T_b(\theta, \phi)$ is the brightness temperature (K) of a radiating body located in a direction (θ, ϕ). $G(\theta, \phi)$ represents the gain of the antenna at elevation angle θ and azimuth angle ϕ. $d\Omega$ is the elementary solid angle in the direction Ω.

An Earth station's antenna noise temperature is a result of the combination of two types of noise source, namely cosmic sources, denoted by T_{sky}, and noise due to the reception of unwanted signals from the ground in the proximity of the antenna, denoted by T_{ground}. This results in the expression:

$$T_a = T_{sky} + T_{ground} \quad \text{K} \tag{5.20}$$

Possible sources of sky noise are the Sun, Moon, oxygen and water vapour absorption and rain. The Sun has a brightness temperature of in excess of 10 000 K at frequencies below 10 GHz, and for this reason, Earth stations avoid pointing in the direction of the Sun. Similar considerations apply to the Moon, which has a brightness temperature of on average 200 K. General cosmic background noise has a value of about 3 K and is independent of frequency. For all intents and purposes, cosmic background noise can be neglected. Variation of the brightness temperature with frequency for extra terrestrial noise sources is shown in Figure 5.7 [ITU-94a].

The major sources of sky noise are atmospheric absorption gases and rain. From the discussion in the previous chapter, it can be deduced that the noise temperature is related to the operating frequency and the elevation angle. When operating in clear sky conditions, a noise temperature of about 15–30 K occurs for frequencies in the range 4–11 GHz at an elevation angle of 10°. Noise from the ground is due to the reception of unwanted signals via the antenna sidelobes and to a lesser extent, the main beam of the antenna. This requires consideration when antennas are operating at low elevation angles to the satellite, say less than 10°. As an antenna's elevation angle increases, the influence of ground temperature on the overall antenna noise temperature reduces significantly.

For systems operating above 10 GHz, rain not only attenuates the wanted signal, as discussed in the previous chapter, it also increases the antenna noise temperature. From equation (5.18), by substituting the attenuation due to rain for the lossy gain, L, it can be seen that the effect of rain attenuation, A_{rain}, on the noise temperature is as follows:

$$T_a = \frac{T_{sky}}{A_{rain}} + T_o\left(1 - \frac{1}{A_{rain}}\right) + T_{ground} \quad \text{K} \tag{5.21}$$

where $T_o = 290$ K.

The antenna noise temperature of a satellite is influenced by the satellite's location, its operating frequency, and the area covered by the satellite's antenna. Coverage over land areas has a higher noise temperature than over oceanic regions. The effect of geostationary satellite location and frequency on the brightness temperature is illustrated in Ref. [ITU-94b]. For example, when positioned over the Pacific Ocean a brightness temperature of 110 K at 1 GHz, rising to near 250 K at 51 GHz is reported. Similarly, when located over Africa, a brightness temperature of 180 K at 1 GHz, rising to nearly 260 K at 51 GHz is reported.

5.2.3.3 Noise Figure

A convenient means of specifying the noise performance of a device is by its noise figure, F,

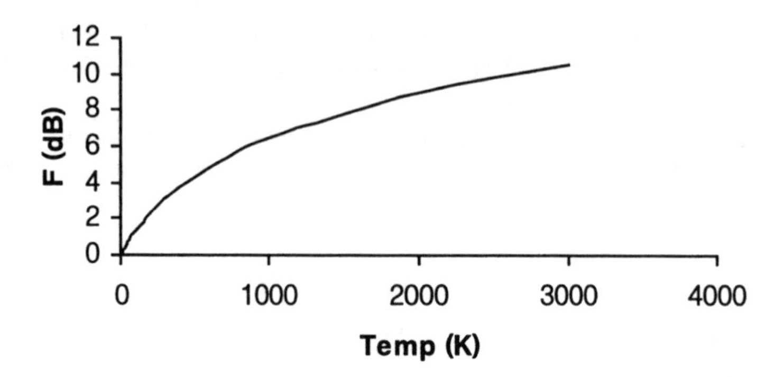

Figure 5.8 Noise figure variation.

which is defined as the ratio of the signal to noise ratio at the input to the device to that at the output of the device.

$$F = \frac{S_i}{N_i} / \frac{S_o}{N_o} \tag{5.22}$$

This can be shown to be equal to:

$$F = 10\log\left(1 + \frac{T_e}{T_o}\right) \text{ dB} \tag{5.23}$$

where T_e is the effective noise temperature of the device (K); T_o is the ambient temperature (usually assumed to be 290 K).

The variation in noise figure with temperature is shown in Figure 5.8.

For a series of devices in cascade, such as that shown in Figure 5.6, the overall noise figure can be determined using the expression:

$$F = F_1 + \frac{F_2 - 1}{G_1} + \frac{F_3 - 1}{G_1 G_2} + \ldots + \frac{F_{n-1}}{G_1 G_2 \ldots G_{n-1}} \tag{5.24}$$

Example:

The receiver chain shown in Figure 5.6, comprises components with the following values: $G_{ant} = 48.5$ dBi, $T_{ant} = 20$ K; $L_1 = 1.5$, $T_L = 290$ K; $G_1 = 30$ dB, $T_1 = 150$ K; $G_2 = 10$ dB, $T_2 =$ 600 K; $G_3 = 20$ dB, $T_3 = 1000$ K. Calculate the equivalent noise temperature of the receiver, T_e, and hence derive the receiver's Figure of Merit (G/T_e).

Taking the input to the first stage amplifier as the reference point and using equations (5.15) and (5.18).

$$T_e = \frac{20}{1.5} + 290\left(1 - \frac{1}{1.5}\right) + 150 + \frac{600}{1000} + \frac{1000}{10.1} \text{ K} = 260.7 \text{ K}$$

$$\text{Figure of Merit} = \frac{G}{T_e} = 48.5 - 10\log(260.7) = 24.3 \text{ dBK}^{-1}$$

5.2.3.4 Carrier to Noise Ratio

By combining the above expressions for received power (5.10) and total noise (5.16), the received carrier-to-noise ratio is given by:

$$\frac{C}{N} = P_t G_t \frac{G_r}{T} \left(\frac{\lambda}{4\pi R} \right)^2 \frac{1}{kB} \frac{1}{A_p} \tag{5.25}$$

or expressing the above equation in dB:

$$\frac{C}{N_0} = 10\log(P_t G_t) - 20\log\left(\frac{4\pi R}{\lambda} \right) + 10\log\left(\frac{G}{T} \right) - 10\log A_p - 10\log kB \qquad \text{dBWHz} \tag{5.26}$$

where N_o, the one-sided noise power spectral density (dBWHz^{-1}), is equal to: $N - 10\log(B)$ and $-10\log A_p$ represents the atmospheric attenuation in dB.

The above expression is valid for either the uplink to the satellite or the downlink to the Earth station.

In the uplink case, EIRP refers to the transmission of the mobile terminal or Earth station and *G/T* refers to the satellite antenna. Similarly, in the downlink, EIRP refers to the satellite transmission and *G/T* refers to Earth station or mobile terminal. An example link budget is shown in Table 5.1. Here, parameters shown in **bold** are those derived from the known link parameters.

Table 5.1 Example link budget

Earth station		
	Transmit power (W)	**10**
	Antenna diameter (m)	2
	Antenna efficiency (%)	55
	3-dB beamwidth (°)	**1.9**
	Transmit gain (dBi)	**28.6**
	Transmit EIRP (dBW)	**38.6**
Up path losses		
	Transmit frequency (GHz)	6.0
	Transmission distance (km)	38000
	Free space loss (dB)	**−199.6**
	Atmospheric attenuation (dB)	0.3
Satellite		
	Received power flux density (dB Wm^{-2})	**−124**
	G/T (dB/K)	−1.0
	Bandwidth (kHz)	150
Link parameters		
	***C/N* (dB)**	**14.2**
	Target *C/N* (dB)	8
	Link margin (dB)	**6.2**

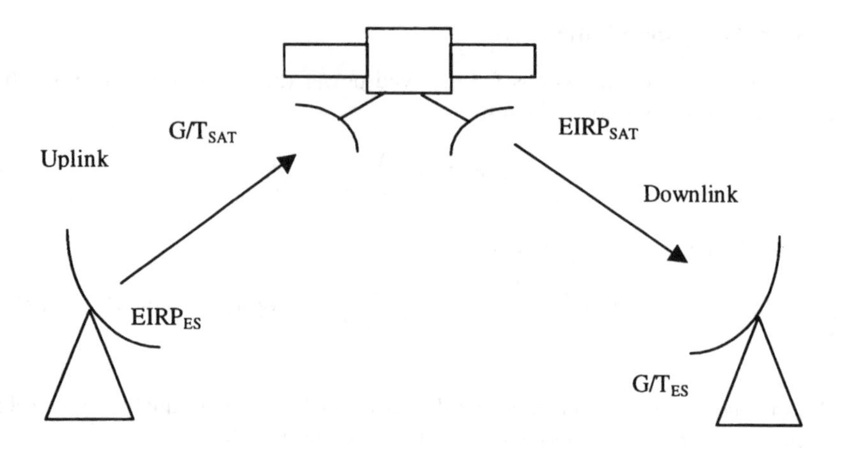

Figure 5.9 Composite transmission chain.

5.2.4 Satellite Transponder

5.2.4.1 Role

When considering the overall performance of the transmission link between the transmitter and the receiver, the influence of the satellite payload needs to be taken into account. The basic form of such a link is illustrated in Figure 5.9. Here, the link is shown between two fixed Earth stations but applies equally to the mobile terminal scenario.

Based on the type of coverage and level of complexity provided, satellite payload technology can be broadly classified under one of the following categories.

5.2.4.2 Simple Wide-Beam Coverage over a Region

Such a scenario is employed by the INMARSAT-2 satellites, for example. This is the simplest means of implementing a mobile-satellite system, the satellite merely relays transmissions from one area to another.

The satellite transponder, in this case, that is used to receive/transmit a single carrier is shown in Figure 5.10.

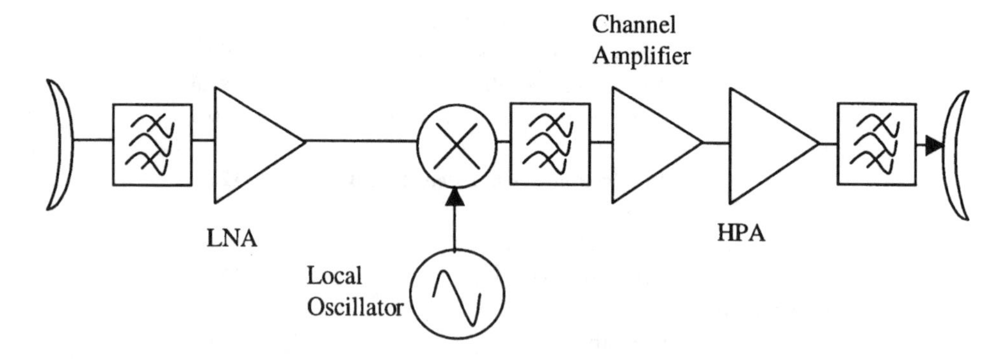

Figure 5.10 Simple transparent transponder.

Initially, the received signal is bandpass filtered prior to low noise amplification. The output of the low noise amplifier (LNA) device is then fed into a local oscillator, which performs the required frequency translation from uplink to downlink. The output of the local oscillator is then filtered to remove the unwanted image frequency, prior to undergoing two stages of amplification. The first stage, known as *channel amplification* is used to ensure that the input power level to the second stage high power amplification remains at a constant level. The output of the high power amplifier (HPA) is then bandpass filtered prior to transmission. In order to attain the required gain of the high power amplifier, a travelling wave tube amplifier (TWTA) is employed, the characteristics of which are shown in Figure 5.11.

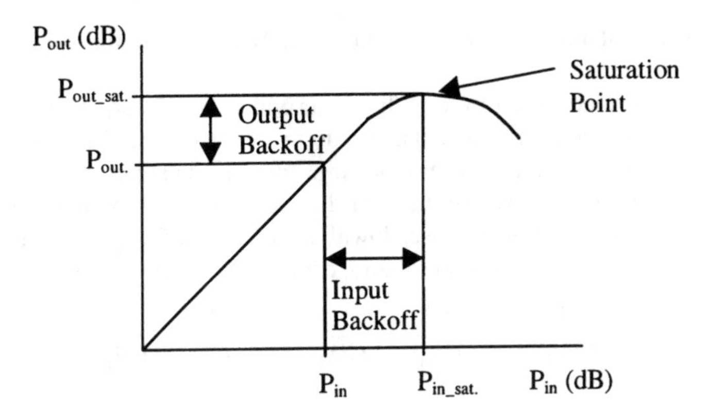

Figure 5.11 Satellite TWTA operational characteristics.

In the case of a linear transponder, where the received signal at the satellite is merely frequency translated and amplified without introducing any signal distortion, the overall carrier-to-noise ratio between a transmitter and receiver located on the ground, via a satellite, in an interference-free environment is given by:

$$\frac{1}{C/N_{\mathrm{Tot}}} = \frac{1}{C/N_{\mathrm{Up}}} + \frac{1}{C/N_{\mathrm{Down}}} \tag{5.27}$$

where carrier-noise-ratios are expressed numerically, not in dB.

As can be seen from Figure 5.11, TWTAs are non-linear in the region of the saturation point. When operating using two or more carriers within the non-linear region of the amplifier, the signals will interact resulting in the generation of unwanted harmonic frequencies. These spurious signals are collectively termed *intermodulation* products.

If two carriers at frequencies ω_1 and ω_2 are applied to the non-linear region of the transponder, it can be shown that after filtering, the major unwanted signals of concern are at the output frequencies $2\omega_1 - \omega_2$ and $2\omega_2 - \omega_1$. These are termed third-order intermodulation products and can appear within the wanted signal bandwidth. Any harmonics above third-order will be at such a low power level that they can be effectively ignored.

In order to avoid the non-linear region, an input signal's power is reduced to below a level which would cause saturation. The degree of reduction is termed the *input backoff*. The reduction in the input power will correspond to a reduction in the output power with respect to the saturation level. This is termed the *output backoff*. When employinig multicarrier operation, the maximum input PFD_{sat}, corresponding to the satellite's saturation level, is shared between all carriers simultaneously transmitting through the transponder. For the case, where n carriers are all transmitting with the same PFD, a carrier's uplink PFD_{up} is given by the expression:

$$PFD_{up} = PFD_{sat} - 10 \log n - BO_{ip} \qquad dBWm^{-2} \tag{5.28a}$$

where BO_{ip} is the input backoff (dB) and n is the number of carriers sharing the transponder. Correspondingly, the downlink $EIRP_d$ is given by the expression:

$$EIRP_d = EIRP_{sat} - 10 \log n - BO_{op} \qquad dBW \tag{5.28b}$$

Where $EIRP_{sat}$ is the satellite EIRP when at saturation, BO_{op} is the output transponder backoff (dB).

The carrier-to-intermodulation ratio, C/Im, determined at the output of the TWTA, is a further detrimental effect on the link that needs to be accounted for in the link budget. Similarly, sources of interference from other unwanted signals can also be considered as an additional noise source. Therefore, the total link noise is given by the summation of all noise sources on the link, i.e. uplink noise, downlink noise, interference and intermodulation.

In this case, the overall carrier-to-noise ratio, expressed linearly, is given by the equation:

$$\frac{1}{C/N_{Tot}} = \frac{1}{C/N_{Up}} + \frac{1}{C/N_{Down}} + \frac{1}{C/I} + \frac{1}{C/I_m} \tag{5.29}$$

5.2.4.3 Spot-Beam Coverage Employing Static Switching Between Spot-Beams

In this case, the transmission path between a particular uplink spot-beam and the corresponding downlink spot-beam is fixed. No processing of the received signal is performed by the satellite prior to transmission.

Spot-beam coverage provides the possibility of frequency re-use, thus increasing the traffic handling capacity of the satellite. However, the uniform allocation of spot-beams within a satellite's coverage area fails to take into account the variation in traffic density. To overcome this problem, beam forming networks (BFN) may be used to change the orientation of the spot-beams in response to traffic variations.

In a satellite system employing multi-carrier transmission, it is generally infeasible to drive a single HPA due to the intermodulation between carriers. To alleviate this problem, carriers are separated into single paths, or channels, such that each is amplified by a separate HPA. Such a scenario is illustrated in Figure 5.12.

The separation of carriers into individual paths is performed after the first stage low noise amplifier. This is achieved by what is known as a *demultiplexer*, comprising of a bank of bandpass filters. After frequency translation, each individual carrier is then amplified. The individual channels are then re-grouped through the satellite's *multiplexer*, again comprising of a bank of bandpass filters, prior to feeding into the beam forming network for transmission.

In addition to frequency re-use capabilities, spot-beam transmission enables the transmission power constraints to be relaxed in comparison to wide-beam coverage. In this instance, the HPA can be replaced by solid state power amplifiers (SSPA), which not only weigh less but also have an improved linear response [EVA-99].

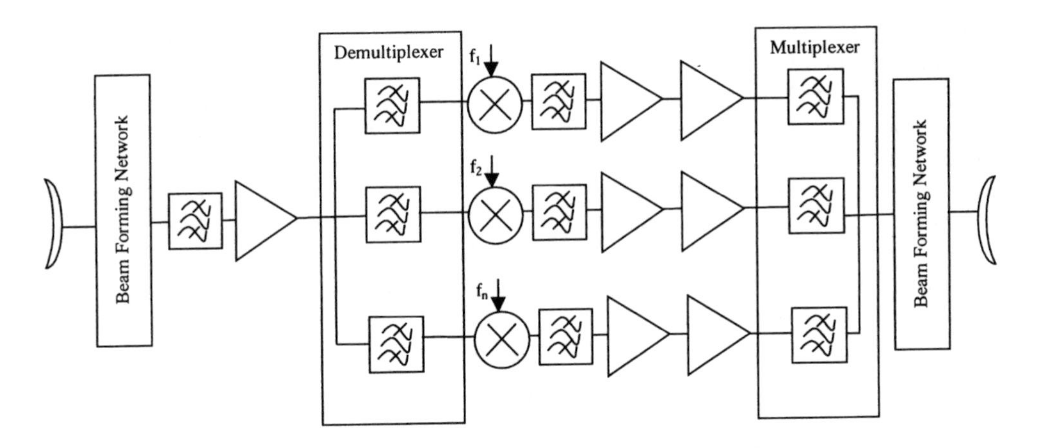

Figure 5.12 Multi-carrier payload configuration.

In terms of deriving the link performance, the calculations performed for the transparent payload may be applied.

5.2.4.4 Spot-Beam Coverage Employing Satellite Switched Time Division Multiple Access (SS-TDMA)

SS-TDMA employs high-speed dynamic switch matrices on-board the satellite, the state of which changes automatically according to a pre-assigned switching sequence which is repeated every TDMA frame. Switching between uplink and downlink spot-beams can either take place at microwave or baseband. As a result, each uplink spot-beam can access each downlink spot-beam for a specific duration of time during each TDMA frame. The payload for the microwave switched SS-TDMA is similar to that of the fixed switch, the difference being the time multiplexed microwave switch. This is shown in Figure 5.13. For simplicity, an on-board reference clock is used for timing purposes. Alternatively, a timing reference may be transmitted from the network controller, however, in this instance, some form of demodulation would be required on the satellite, thus increasing on-board complexity.

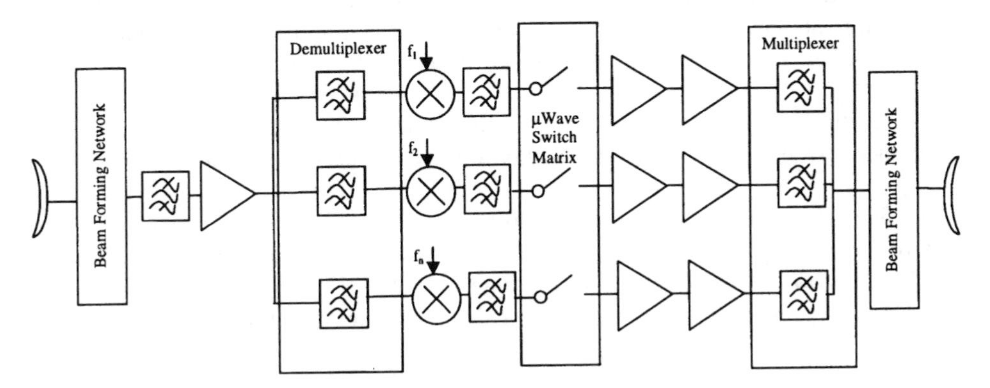

Figure 5.13 SS-TDMA payload employing microwave switching.

Where no on-board processing of the signal is performed, the overall carrier to noise ratio of a link is given by equation (5.29).

For cases where baseband switching of the signal is performed, the uplink and downlink transmission paths are de-correlated, hence the overall performance is determined on a link-by-link basis. For digital communications, the performance of the link is generally categorised by the relationship between the bit error rate (BER) and the ratio, E_b/N_0; where E_b is the energy per information bit, and N_0 is the one-sided noise power spectral density. In this instance, uplink and downlink bit error rates, not noise levels, are added together when determining the overall performance of the link. In this instance, the overall performance of the link is given by:

$$\frac{E_b}{N_{0\_Total}} = \frac{E_b}{N_{0\_Up}} + \frac{E_b}{N_{0\_Down}} \tag{5.30}$$

E_b is related to the carrier power and the information bit rate, R_b, by the following expression:

$$E_b = \frac{C}{R_b} \text{ J/bit} \tag{5.31}$$

The overall performance of a transmission is made of the sum of the bit error rates on each path. The relationship between the above expression and modulation and coding techniques is discussed in the following sections.

One major advantage in using baseband switching is that the uplink and downlink modulation and access schemes can be optimised for the particular transmission environment. Furthermore, since the uplink and downlink noise levels are uncorrelated, the transmission power requirements can be reduced on both links.

5.2.4.5 Spot-Beam Coverage Incorporating Path Routing On-board the Satellite

In this scenario, routing may be via beams of the same satellite or between satellites using inter-satellite link technology.

The "digital exchange satellite" configuration represents the maximum level of complexity on-board the satellite. In this configuration, all call control functionalities, such as routing, are performed by the satellite, as opposed to the terrestrial network management station, that is required in all previous configurations. This configuration allows direct mobile-to-mobile calls without the need of a double-hop, provided that:

1. Both mobiles are within the coverage area of the same satellite;
2. Or alternatively, satellites have the ability to perform inter-satellite link routing.

This implies that the satellite must perform the functions normally associated with the terrestrial network functionality. The payload configuration follows on from the baseband SS-TDMA payload. The fact that traffic routing is to be performed by the satellite implies that after demodulation the signal must be analysed for the required service, destination address and so on.

5.3 Modulation

5.3.1 Overview

This section is restricted to discussion on digital methods of transmission, as the use of analogue transmission techniques is becoming less and less relevant to the mobile-satellite industry. There are several excellent sources which provide in-depth treatment of digital transmission techniques to a level which is beyond the scope of this book, and the interested reader is referred to [PRO-95, SKL-88] as two such examples.

Digital signals can be used to modulate the amplitude, frequency or phase of the carrier. Amplitude modulation is known as amplitude shift keying (ASK), also known as on-off keying (OOK), while modulation of the frequency of the carrier is termed frequency shift keying (FSK).

In terms of performance, ASK and FSK require twice as much power to attain the same bit error rate performance as phase shift keying (PSK) (Figure 5.14). Consequently, the vast majority of mobile-satellite systems employ a method of phase modulation, known as PSK.

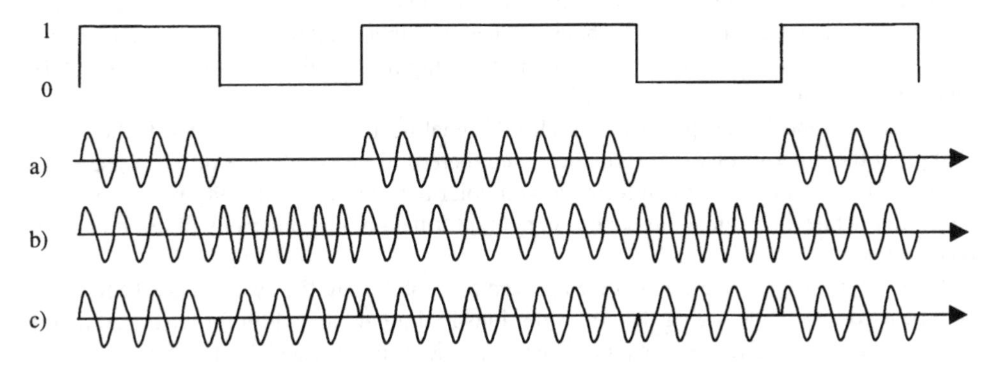

Figure 5.14 Comparison of: (a) ASK; (b) FSK; and (c) PSK.

5.3.2 Phase Shift Keying

5.3.2.1 M-PSK

PSK involves changing the phase of the carrier in accordance with the information content. The general form of a PSK waveform is given by the expression:

$$S_{\mathrm{m}}(t) = A(t)\cos\left[\omega_{\mathrm{c}}t + \frac{2\pi}{M}(m-1)\right] \tag{5.32}$$

where *A(t)* is the signal pulse shape; *M* is the number of states; and *m* is an integer in the range $1 \leq m \leq M$.

A PSK waveform can be generated in *M*-states, the two most widely used techniques being binary PSK (BPSK) ($M = 2$) and quadrature PSK (QPSK) ($M = 4$). The simplest form of PSK is BPSK, which entails the changing of the phase of the carrier between two states (0° and 180°), representing the binary states "1" and "0".

QPSK introduces four states, each phase representing a symbol comprising of 2-bits of

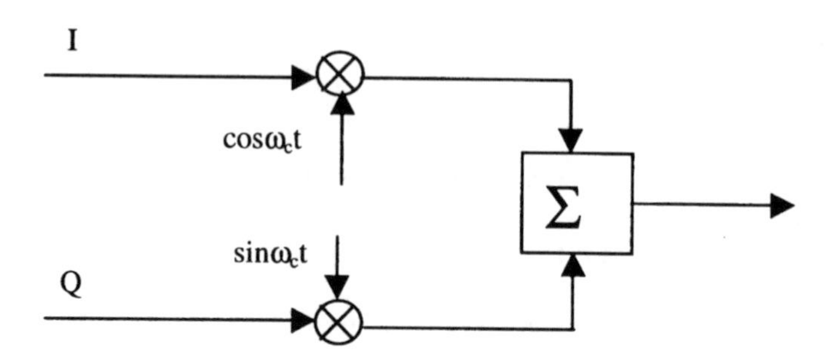

Figure 5.15 QPSK modulation.

information (i.e. "00", "01", "10", "11"). A QPSK signal can be thought of as the combination of two uncorrelated, orthogonal BPSK signals. By convention, a QPSK signal is said to comprise an in-phase and quadrature components, termed the I-channel and Q-channel, respectively. Conversion of serial input bit streams to the parallel I- and Q-channel formats can be achieved by alternative bit sampling. Thus, if an information stream arrives at the QPSK modulator at a rate of $1/T$ bits/s, then even bits in the sequence would be directed to the I-channel and the odd bits to the Q-Channel. These bits then modulate the carrier components at a rate of $1/2T$ bits/s, prior to combination, resulting in a symbol transmission rate of $1/T$ bits/s. This is illustrated in Figure 5.15.

A change in the carrier phase of the QPSK signal occurs every $1/2T$ bit/s. If there is no change in the polarity of the information bits, there will be no change in the phase of the carrier. A change in one of the information bit polarities, will result in a 90° phase change of the carrier, while a simultaneous change in both information bits' polarity will result in a 180° phase change.

The other possible technique to be employed for mobile-satellite systems is 8-PSK, which is to be used to provide higher data transmissions in limited bandwidth for terrestrial mobile systems, such as EDGE (see Chapter 1). Higher order schemes, such as 16-PSK, in general, are only used when power margins are sufficient to ensure correct reception, which is unlikely in a mobile-satellite system, and where bandwidth is at a premium.

Figure 5.16 provides signal space diagrams for the three major PSK methods of modulating the carrier.

Optimum detection of PSK is achieved by the use of a coherent demodulator, which

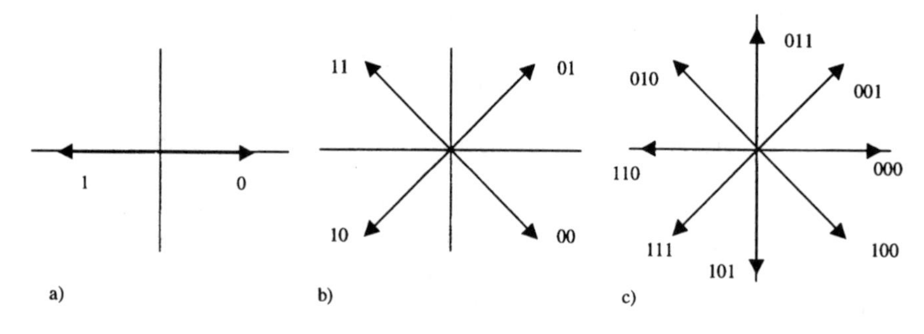

Figure 5.16 PSK phasor diagrams: (a) BPSK; (b) QPSK; (c) 8-PSK.

multiplies the incoming carrier with that of a locally generated reference carrier, the product of which is then passed into matched filters or product integrators prior to threshold detection. The local carrier is generated either by making use of a transmitted pilot, or derived from the received carrier by some form of carrier recovery circuit, such as the Costas Loop or the signal squaring circuit. QPSK reduces by half the required spectral occupancy for the same bit rate as BPSK. However, the receiver circuitry is more complicated, mainly due to the increased complexity of the carrier recovery circuitry. Here, the received phase angle can be derived by obtaining the arctangent of the ratio of quadrature to in-phase components, from which a phase angle is obtained. This is then compared with the anticipated reference values, from which a decision is made. This is illustrated in Figure 5.17.

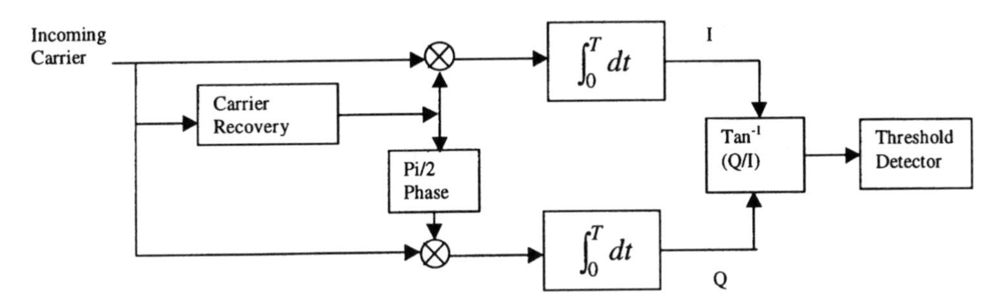

Figure 5.17 PSK demodulator.

When using coherent detection for BPSK, the relationship between the probability of bit error, P_b, and E_b/N_0 is given by the following expression:

$$\mathrm{P_b} = Q\left(\sqrt{\frac{2E_b}{N_0}}\right) \tag{5.33}$$

The value of $Q(x)$, also known as the *complementary error function*, can be obtained from standard tables or alternatively can be approximated when $x > 3$ using the equation:

$$Q(x) = \frac{1}{x\sqrt{2\pi}}\exp\left(\frac{-x^2}{2}\right) \tag{5.34}$$

The relationship between equivalent bit error probability P_b to that of symbol error probability P_s for M-ary PSK can be obtained using the expression:

$$P_b \approx \frac{P_s}{\log_2 M} \tag{5.35}$$

Hence, it can be seen from the above that for BPSK, the symbol probability is equal to that of the bit error probability. Whereas, for QPSK, the probability of symbol error is twice that of bit error. As noted previously, BPSK and QPSK exhibit the same bit error probability.

For voice-type services, where the effect of bit errors on the perceived quality by the user is not so pronounced, a target bit error rate (BER) of 10^{-3} is usually adequate. Data services tend to have a target BER in the range 10^{-5}–10^{-10}, depending on the application and the required integrity of the data.

5.3.2.2 Differential PSK

At the receiver, after performing coherent demodulation, there is generally a 180° ambiguity in the received signal phase, which cannot be resolved unless some known reference signal is transmitted and a comparison made. In the worst case, if such a situation is left unresolved, the received signal could end up being the complement of that transmitted.

Differentially encoded PSK can be used to remove sign ambiguity at the receiver. Differentially encoding data prior to modulation occurs when a binary "1" is used to indicate that the current message bit and the prior code bit are of the same polarity; and "0" to represent the condition when the two pulses are of opposite polarity (Table 5.2).

Table 5.2 Example of D-PSK coding table

	Transmitter			Receiver	
Tx_bit$_{n-1}$	Message_bit$_n$	Transmit_bit	Rx_pulse$_{n-1}$	Rx_pulse$_n$	Received bit
0	0	1	0	1	0
1	1	1	1	1	1
1	0	0	1	0	0
0	0	1	0	1	0
1	1	1	1	1	1
1	0	0	1	0	0
0	0	1	0	1	0

As shown in Figure 5.18, a complex carrier recovery circuit is no longer required at the receiver; the demodulator operates by comparing the carrier's current phase to its previous state. Hence, a positive value is detected as a transmitted message value of "1", and negative as "0".

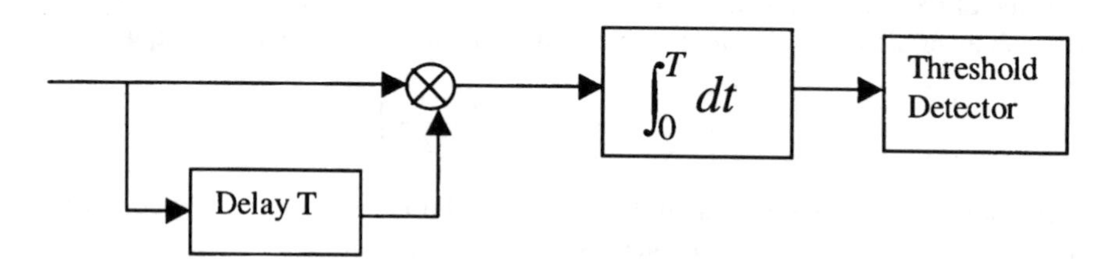

Figure 5.18 D-PSK demodulator.

Non-coherent detection results in the degradation in the bit error performance with respect to coherent detection. In this case, the bit error rate probability P_b for binary D-PSK is given by:

$$P_b = \frac{1}{2}\exp\left(\frac{-E_b}{N_0}\right) \quad (5.36)$$

Hence, when considering employing D-PSK, a trade-off between a simplified receiver complexity against a reduced performance in the presence of noise, in particular when employing higher order modulation techniques, needs to be made.

Figure 5.19 illustrates the relationship between symbol error rate and energy per noise density ratio, comparing BPSK with D-PSK.

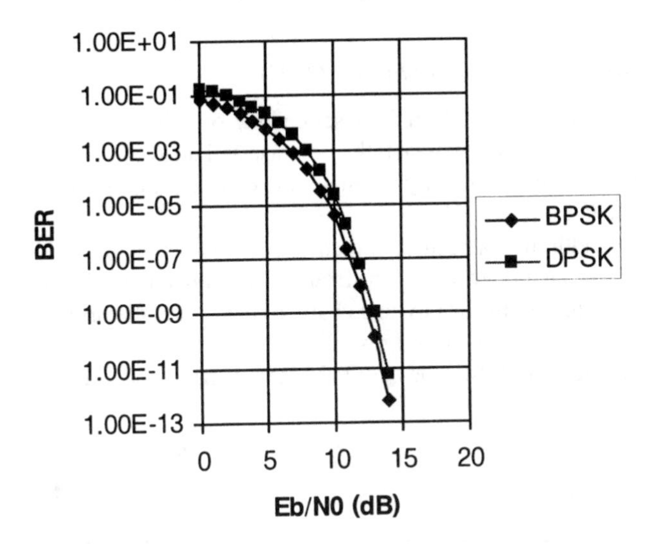

Figure 5.19 Relationship between P_s and E_b/N_0.

5.3.2.3 Offset-QPSK

In QPSK, a simultaneous change in the I- and Q-channel polarities of the information bits can result in a ±180° phase change in the transmitted signal. During such instances, the signal can momentarily transcend the zero level. This can result in a satellite transponder inadvertently amplifying the signal's harmonic components, resulting in the transmission of unwanted, out-of-band interference levels.

Offset-QPSK (OQPSK) can be used to alleviate this problem. This technique is implemented by the delaying of the quadrature component information stream with respect to the in-phase by a half-bit period, $T/2$. This implies that a simultaneous change in the polarity of the I- and Q-channel information pulses can no longer occur, thus ensuring that the maximum phase change of the carrier is limited to ±90°, since the signal can no longer reduce to zero.

OQPSK has the same theoretical bit error rate performance as BPSK and QPSK.

5.3.3 Minimum Shift Keying

Minimum shift keying (MSK) is a binary form of continuous phase frequency shift keying (CPFSK), where the frequency deviation, Δf, from the carrier is set at half the reciprocal data rate, $1/2T$. MSK may also be viewed as a special form of offset-QPSK, consisting of two sinusoidal envelope carriers, employing modulation at half the bit rate. For this reason, the MSK demodulator is usually a coherent quadrature detector, similar to that for QPSK. In this case, the error rate performance is the same as that of BPSK and QPSK. Similarly, differentially encoded data has the same error performance as D-PSK. MSK can also be received as an FSK signal using coherent or non-coherent methods, however, this will degrade the performance of the link.

The sidelobes of MSK are usually suppressed using a Gaussian filter and the modulation method is then referred to as GMSK, which is the modulation scheme adopted by GSM.

5.3.4 *Quadrature Amplitude Modulation (QAM)*

QAM is a combination of amplitude and phase modulation. Modulation can be achieved in a similar manner to that of QPSK, by which the in-phase and quadrature carrier components are independently amplitude modulated by the incoming data streams. Signals are detected at the receiver using matched filters.

In terms of bandwidth, it is a highly efficient method for transmitting data. However, the sensitivity of the QAM method to variations in amplitude, in practice limits its applicability to satellite systems, where non-linear payload characteristics may distort the waveform, resulting in the reception of erroneous messages.

Figure 5.20 illustrates the MSK and 16-QAM signal space diagrams.

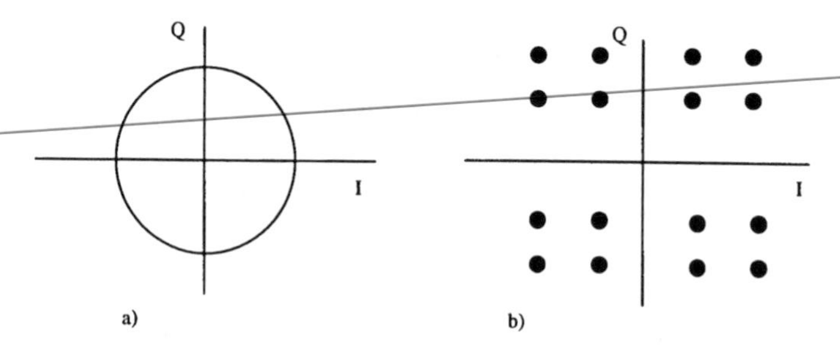

Figure 5.20 Signal space diagrams: (a) MSK; (b) 16-QAM.

5.4 Channel Coding

5.4.1 *Background*

For an additive white Gaussian noise (AWGN) channel, the Shannon–Hartley law states that:

$$C = B\log_2(1 + S/N) \text{ bit/s} \tag{5.37}$$

where C is the capacity of the channel (bit/s); B is the channel bandwidth (Hz); S/N is the signal-to-noise ratio at the receiver.

According to Shannon, if information is provided at a rate R, which is less than the capacity of the channel, then a means of coding can be applied such that the probability of error of the received signal is arbitrarily small. If the rate, R, is greater than the channel capacity, then it is not possible to improve the link quality through means of coding. Indeed, its application could have a detrimental effect on the link.

Re-arranging the above equation in terms of energy-per-bit and information rate, where the information rate is equal to the channel capacity, results in the following:

$$C/B = \log_2(1 + E_b C/N_0 B) \tag{5.38}$$

This expression can be used to derive the Shannon limit, the minimum value of E_b/N_0 below which there can be no error free transmission of information. As C/B tends to zero, this can be shown to be equal to -1.59 dB ($1/\log_2 e$).

As was noted earlier, satellite communication systems are generally limited by available power and bandwidth. It is therefore of interest if the signal power can be reduced while maintaining the same grade of service (BER). This can be achieved by adding extra or redundant bits to the information content, using a channel coder. The two main classes of channel coder that are most widely used for satellite communications are: block encoders and convolutional encoders. At the receiver, the additional bits are used to detect any errors introduced by the channel. There are two techniques employed in satellite communications to achieve this:

- Forward error correction (FEC), where errors are detected and corrected for at the receiver;
- Automatic repeat request (ARQ), where a high degree of integrity of the data is required, and latency is not a significant factor.

It is also possible to combine FEC and ARQ in the form of a hybrid scheme.

The effectiveness of a coding technique is expressed by the term *coding gain*, defined as the difference in dB between the E_b/N_0 for a given BER in the case of ideal signalling and that of the particular coding scheme.

5.4.2 Block Codes

5.4.2.1 Code Generation – Linear Codes

Binary linear block codes are expressed in the form (n, k), where k is the number of message bits that are converted into n code word bits. The difference between n and k accounts for the number of redundancy check bits, r, that are added by the coder. The code rate or code efficiency is given by the ratio of k/n. Mapping between message sequences and code words can be achieved using look-up tables, although as the size of the code block increases, such an approach becomes impractical. This is not such a problem, however, since linear code words can be generated using some form of linear transformation of the message sequence. A code sequence, $\mathbf{c}$, comprising of the row vector elements $[c_1, c_2, \ldots c_n]$, is generated from a message sequence, $\mathbf{m}$, comprising of the row vector elements $[m_1, m_2, \ldots m_k]$ by a linear operation of the form:

$$\mathbf{c} = \mathbf{m}G \tag{5.39}$$

where G is known as the generator *matrix*.

In general, all c code bits are generated from linear combinations of the k message bits.

A special category of code, known as a *systematic code*, occurs when the first k digits of the code are the same as the first k message bits. The remaining $n - k$ code bits are then generated from the k message bits using a form of linear combination. These bits are termed the *parity data bits*. The generator matrix for systematic code generation is of the form:

$$G = \begin{bmatrix} 1 & 0 & 0 & \dots & 0 & P_{10} & \dots & P_{1r} \\ 0 & 1 & \dots & \dots & .. & P_{20} & \dots & P_{2r} \\ \dots & \dots & \dots & \dots & \dots & \dots & \dots & \dots \\ \dots & \dots & \dots & \dots & \dots & \dots & \dots & \dots \\ 0 & 0 & 0 & \dots & 1 & P_{k0} & \dots & P_{kr} \end{bmatrix}$$

I_m[k x k] P[kx r]

Here the matrix can be seen to consist of the identity matrix, I_m of dimension $[k \times k]$ and the parity generating matrix, P, of dimension $[k \times r]$.

In practice, code words are conveniently generated using a series of simple shift registers and modulo-2 adders.

5.4.2.2 Decoding

The *Hamming distance* is defined by the number of bit positions by which two code words in an (n, k) block code differ. This can be found by simply performing a modulo-2 addition of the two code words, for example:

Let $s1 = 110100$, $s2 = 101101$
$s1 \oplus s2 = 011001$, in which case the distance is equal to 3.

The concepts of modulo-2 arithmetic are shown in Table 5.3.

Table 5.3 Modulo-2 arithmetic

A	B	$A \oplus B$	$A \otimes B$
0	0	0	0
1	0	1	0
0	1	1	0
1	1	0	1

For a set of code words, the capability of a decoder to detect and correct errors is defined by the *minimum distance*, d_{min}, between code words, the smallest value of the Hamming distance.

From the *minimum distance*, the number of errors in the received code word that can be detected by the receiver, e, is given by:

$$e = (d_{min} - 1) \text{ bits} \tag{5.40}$$

and the corresponding number of errors that can be subsequently corrected for t is given by:

$$t = (d_{\min}/2 - 1) \quad \text{for even } d_{\min} \tag{5.41a}$$

$$= 1/2(d_{\min} - 1) \quad \text{for odd } d_{\min} \tag{5.41b}$$

Hence, in the example above which has a minimum distance of 3, two error bits can be detected and one can be corrected. At the receiver, block codes are decoded either by referring to an identical look-up table as applied at the transmitter or by applying the inverse of the known linear transformation using what is known as the *parity check matrix*, H, which is of the form:

$$H = \left[P^{\mathrm{T}} I_{\mathrm{m}}\right] \tag{5.42}$$

Under error free conditions, at the receiver, $rH^{\mathrm{T}}=0$, for each row vector of the received code matrix, **r**. In practice, however, a received codeword will be made up of the wanted codeword, **c**, plus any errors introduced by the channel, **e**. In other words, the ith row of the received code matrix can be represented by:

$$r_j = c_j \oplus e_j$$

In the case, where errors are introduced at the receiver, the vector product, rH^{T}, will be equal to a non-zero row vector, and this is termed *syndrome*, s. Here, it can be seen that the syndrome at row i is equal to $e_i H^{\mathrm{T}}$.

For a block code containing r redundancy bits, the maximum number of syndromes is given by 2^r. Each syndrome is not necessarily unique to a specific code error, hence at the receiver a pre-defined set of correctable code errors for each syndrome is stored in a look-up table and a form of maximum likelihood decision making is used to select the most probable error vector. This is then added to the received code word in order to nullify the error. Once this has been performed, the receiver can then determine the linear combination of message bits that would be required to generate the corresponding parity bit sequence.

Popular linear block codes include Hamming codes, Hadamard codes and extended Golay codes.

Hamming codes, which have a minimum distance of 3, are characterised by the expression:

$$(n,\ k) = (2^m - 1,\ 2^m - 1 - m) \tag{5.43}$$

where m=2, 3, 4,...

Typical examples of Hamming codes are (7, 4), (15, 11) and (31, 26).

The probability of error for coherently demodulated BPSK coded symbols over an AWGN channel can be expressed in a similar manner to that shown in (5.33), that is:

$$P_{\mathrm{e}} = Q\left(\sqrt{\frac{2E_{\mathrm{c}}}{N_0}}\right) \tag{5.44}$$

where E_c/N_0 is the ratio of the energy per code symbol to the one-sided noise spectral density.

E_c/N_0 is related to E_b/N_0 by the expression [SKL-88]

$$\frac{E_{\mathrm{c}}}{N_0} = \left(\frac{k}{n}\right)\frac{E_{\mathrm{b}}}{N_0} \tag{5.45}$$

which for Hamming codes, from (5.43), can be expressed as:

$$\frac{E_{\mathrm{c}}}{N_0} = \frac{2^m - 1 - m}{2^m - 1}\frac{E_{\mathrm{b}}}{N_0} \tag{5.46}$$

The extended Golay code offers superior performance to Hamming codes but at a cost of increased receiver complexity. The code, which has a minimum distance of eight, is of the form (24, 12). The extended Golay code refers to the fact that the code is a derivative of the perfect Golay (23, 12) code, which has a minimum distance of 7, by the addition of a parity bit.

5.4.2.3 Code Generation – Cyclic Codes

Cyclic codes are a subclass of linear codes. In this case, a code word is generated simply by performing a cyclic shift on its predecessor. In other words, each bit in a code sequence generation is shifted by one place to the right and the end bit is fed back to the start of the sequence. Hence the term cyclic. The linear Hamming and extended Golay codes discussed above, have equivalent cyclic code generators.

Non-systematic cyclic codes are generated using a *unique generator polynomial*, of the form:

$$g(p) = p^{n-k} + g_{n-k}p^{k-1} + \ldots + g_1 p + 1$$

where the generator polynomial is a factor of p^{n+1}.

Defining the message polynomial by:

$$m(p) = m_{k-1}p^{k-1} + m_{k-2}p^{k-2} + \ldots + m_1 p + m_0$$

where the values of $[m_{k-1}\ldots m_0]$ represent the information bits. When this is multiplied by the generator polynomial, it results in the generation of a code word of the form:

$$c(p) = (m_{k-1}p^{k-1} + m_{k-2}p^{k-2} + \ldots + m_1 p + m_0)g(p)$$

An alternative to this approach is to generate systematic codes. The concept of systematic codes was introduced in the previous section, whereby message information is incorporated into the code word sequence.

Systematic cyclic codes can be generated in three stages, involving the use of feedback shift registers:

- Step 1: the message polynomial is multiplied by p^{n-k} which is equivalent to shifting the message sequence by $n-k$ bits. This is necessary to make space for the insertion of the parity bits.
- Step 2: the product of step 1, $p^{n-k}m(p)$ is divided by the generator polynomial, $g(p)$.
- Step 3: the remainder from step 2 is the parity bit sequence, which is then added to the message sequence prior to transmission.

BCH and Reed–Solomon Of the cyclic codes, Bose–Chadhuri–Hocquenghem (BCH) and Read–Solomon (RS) are the most widely used, the former being the most powerful of all of the cyclic codes, while the latter is especially suited for correcting the effect of burst errors. The latter consideration is particularly important in the context of the mobile-satellite channel and hence RS codes are usually incorporated into the system design. Note: RS codes are a subset of the BCH codes.

Reed–Solomon codes are specified using the convention RS(n, k), where n is the symbol

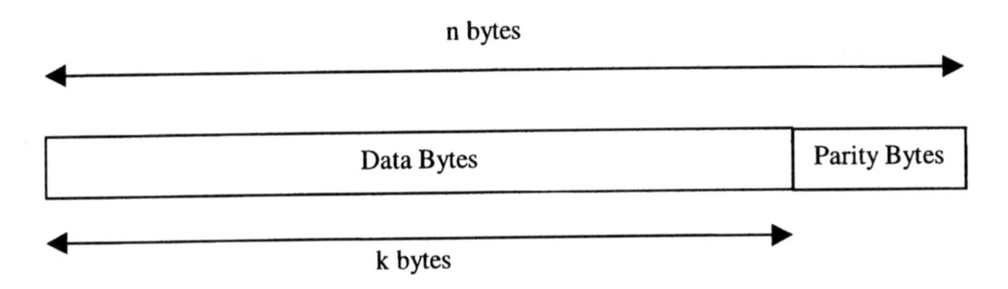

Figure 5.21 RS code structure.

code word length, k is the number of data symbols of s-bits, and the difference between n and k is the number of parity symbols added to the data. The minimum distance of an RS code is given by:

$$d_{\min} = (n - k) + 1 \tag{5.47}$$

Hence, from (5.41), the RS code is capable of correcting up to $n-k$ symbols in a code word.

For a given symbol size in terms of bits, s, the maximum length of an RS code in bytes is given by 2^s-1. So, for example, if an 8-bit symbol representation is employed, the maximum length of the RS code would be 255 bytes. In fact, the RS(255, 223) code is one of the most commonly used of all RS codes in satellite communications. This equates to each code word comprising of 223 bytes of data and 32 bytes are used for parity. Hence, this code is capable of correcting up to 16 bytes received in error in a code word. The structure of an RS code word is shown in Figure 5.21.

A symbol can be received in error as a result of a single bit error within a symbol or, in the worst case, due to all bits within a symbol being in error. Irrespective of the two cases, as was noted in the previous paragraph, the RS code is capable of correcting errors in symbols, not specific bits, hence in the RS(255, 223) example, a minimum of 16 bits, up to a maximum of 128 (16×8) bits can be corrected for in a corrupted code word. This illustrates why RS codes are particularly useful for counteracting the effect of burst errors.

Shortened RS code words, whereby a sequence of all 0s are added to the data symbols, prior to coding, then removed before transmission, and then re-inserted at the receiver, are also used. For example, the shortened RS(204, 188) is generated from the RS(255, 239) by adding 51 zero-bytes to the 188 bytes of data, prior to adding the parity bytes. The zero bytes

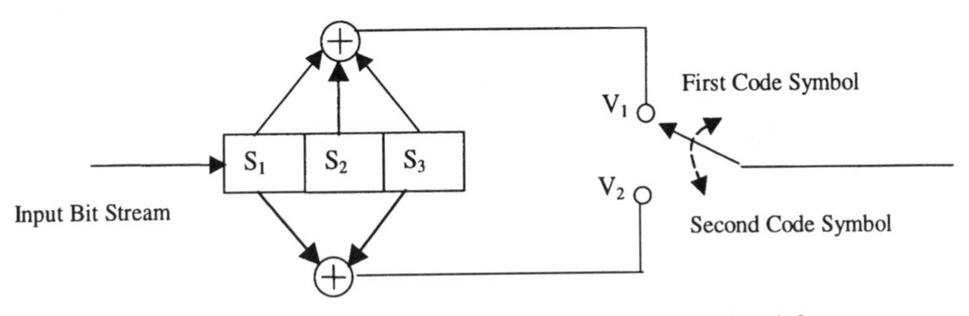

Figure 5.22 Half-rate convolutional encoder of constraint length 3.

are then removed before transmission and then re-inserted at the receiver. This, in effect, reduces the number of bits transmitted, while retaining the error correcting properties of the code.

5.4.3 *Convolutional Codes*

5.4.3.1 Code Generation

Convolutional codes are generated by a tapped-shift register and two or more modulo-2 adders, which are connected to particular stages of the register. The number of bits stored in the shift register is termed the *constraint length*, K. Bits within the register are shifted by k input bits. Each new input generates n output bits, which are obtained by sampling the outputs of the modulo-2 adders. The ratio of k to n is known as the code rate. Convolutional code types are usually classified according to the following convention: (n, k, K), so for example (2, 1, 7) refers to an half-rate encoder of constraint length 7.

It is of interest to know what sequence of output code bits will be generated for a particular input stream. There are several techniques available to assist with this understanding, the most popular being connection pictorial, state diagram, tree diagram and trellis diagram. To illustrate how these methods are applied, consider the following example, shown in Figure 5.22.

Figure 5.22 illustrates using connection pictorial representation, an encoder of constraint length 3, which generates two coded bits for every new input bit. To illustrate how the encoder operates, consider an input stream comprising of the bit sequence 10110, where the most significant bit represents the first input into the encoder. Initially, the encoder is considered to be flushed with all 0s. When the first input bit, 1, is entered into the encoder, the content of S_1 changes to a 1; S_2 and S_3 contain 0s, as before. This results in the outputs of the modulo-2 adders, V_1 and V_2 being equal to 1 and 1, respectively, which is the coded output. The next input bit, 0, results in the previous inputs being shifted to the right, that is the status of S_2 now becomes 1, S_1 and S_3 are now 0. The sampled coder output now becomes 10. Proceeding along similar lines, it can be shown that the subsequent encoded output stream for the input sequence 10110 is given by the following sequence of digits 11 10 00 01 01 11 00 00.

Here, it should be noted that the last bit in the message stream, 0, needs to be passed all of the way through the encoder, this is achieved by following the message bits with a further three zeroes in order to flush the coder. This results in there being generated $(m+K)$ code words for m input bits.

The configuration of the encoder, in terms of connections to the modulo-2 adders, can be defined using generator polynomials. The polynomial for the nth generating arm of the encoder is of the form:

$$g_n(p) = g_0 p^0 + g_1 p^1 + \dots g_n p^n$$

where the value of g_i takes on the value of 0 or 1. A "1" is used to indicate that there is a connection between a particular element of the shift register and the modulo-2 adder.

In the above example, the encoder can be represented by the generator polynomials:

$$g_1(p) = 1 + p + p^2$$

$$g_1(p) = 1 + p^2$$

In addition to providing a simple representation of the encoder, generator polynomials are used to predict the output coded message sequence for a given input sequence. For example, the input sequence 10110 can be represented by the polynomial:

$$m(p) = 1 + p^2 + p^3$$

Combining this with the respective generator polynomials and using the rules of modulo-2 arithmetic results in the following:

$$m(p)g_1(p) = (1 + p^2 + p^3)(1 + p + p^2)$$
$$= 1 + p + p^5$$

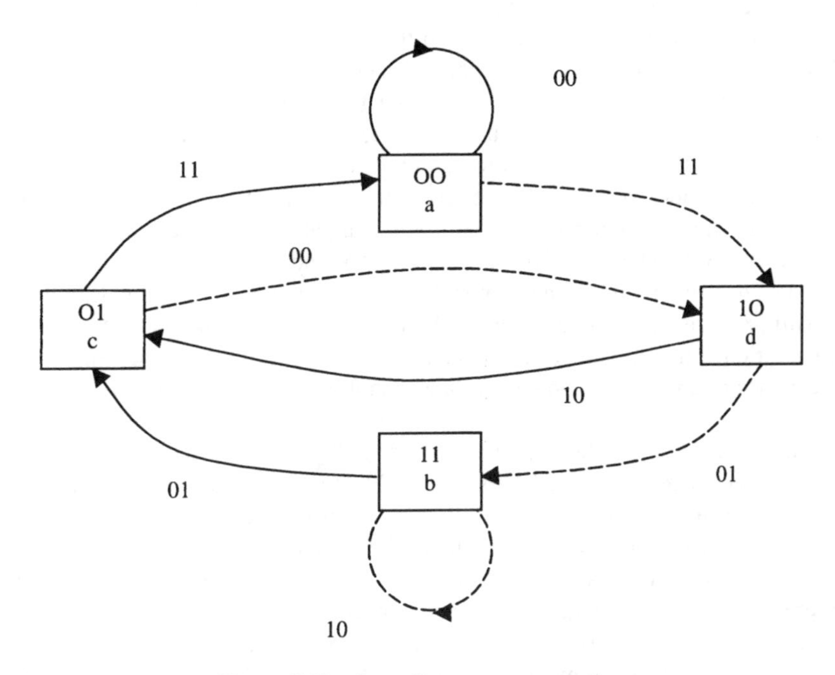

Figure 5.23 State diagram representation.

$$m(p)g_2(p) = (1 + p^2 + p^3)(1 + p^2)$$
$$= 1 + p^3 + p^4 + p^5$$

The output code sequence, $c(p)$, is then obtained by interleaving the above two products, that is:

$$c(p) = [1, 1]p^0 + [1, 0]p^1 + [0, 0]p^2 + [0, 1]\, p^3 + [0, 1]p^4 + [1, 1]p^5$$

Here, the numbers between brackets represents the output code sequence.

In the previous chapter, the use of state diagrams to characterise the mobile-satellite channel was presented. The same approach can be used to distinguish the output characteristics of convolutional encoders. The Figure 5.23 represents the state diagram for the encoder shown in Figure 5.22. Here, the contents of the square boxes, itemised as *a*, *b*, *c* and *d*, respectively, represent the four possible binary combinations of the memory elements S_2 and S_3. If this were to be extended to a general description, then a separate box would be needed to represent all possible combinations of the memory elements bar the first input location. The constraint length of the encoder determines the required number of boxes needed to represent each possible combination of inputs, that is 2^{K-1}. It can be seen that this mode of representation is not particularly suited to encoders of long constraint length, as diagrams can quickly become complex.

The value of S_1 can be either a "1" or a "0", the content of which will determine the value of S_2 when the next input bit enters the encoder. Similarly, the next content of S_3 is governed by the current content of S_2. At each input transition, the encoder shown in Figure 5.22 will generate a pair of code bits, corresponding to the data sequence contained within all of its memory elements. The corresponding generated code bits are illustrated on the diagram next to an associated arrow, which maps the transition in the states of S_2 and S_3 with the input. The style of the arrow is important, as it is used to differentiate between a "1" and "0" in the S_1 location. Here, a dotted line is used to represent a "1".

For example, consider the case where S_2 contains a "1" and S_3 a "0". If S_1 contains a "1", the output of the encoder will produce "01", a new input bit will enter the encoder and the contents of S_1 and S_2 will be shifted by one bit, so that S_2 and S_3 now both contain "1"s. This is shown as state *b* in Figure 5.23. On the other hand, if S_1 contains a "0", then the output of the encoder will produce "10" and the state will change from *d* to *c* on input of the next bit.

State transition diagrams can be used to characterise in a very succinct way the anticipated output sequence of an encoder, however, they cannot be used to track a code sequence with time.

Tree diagrams provide another easy to use, and readily understandable method of presenting the output sequence of an encoder for a given input stream. In this respect, tree diagrams provide the time dimension that is not available with the state diagram characterisation. Figure 5.24 illustrates the tree diagram representation for the encoder shown in Figure 5.22. Here, the convention is that shifting to the next branch on the right and stepping down represents an input bit equal to "1" and conversely, shifting to the right and stepping up represents a "0" input bit. Thus, assuming initially that the encoder is flushed with zeros, it can be seen that when the first input bit is a "1", the encoder will output "11" (to the right and down). Again, if the next input is "0", shifting to the right and climbing a branch results in the output "10".

While the tree diagram approach may be manageable for tracing input sequences of 4 or 5 bits, anything greater will result in the need for a very large tree diagram indeed. However, looking closely at the tree diagram, it can be seen that the branches of the tree repeat after the third set of branches. This is because the constraint length of the encoder is equal to 3 and therefore after the first 3 input bits, the encoder output is no longer influenced by the first input bit passing through the encoder. The repetitive nature of the tree diagram characteristics can be used to plot the output of the encoder with time in a succinct manner. This form of representation is known as the trellis diagram (Figure 5.25).

The trellis diagram provides a more convenient means of illustrating the time dependency of the encoder output. As before, using the encoder in Figure 5.22 as an example, the rows of

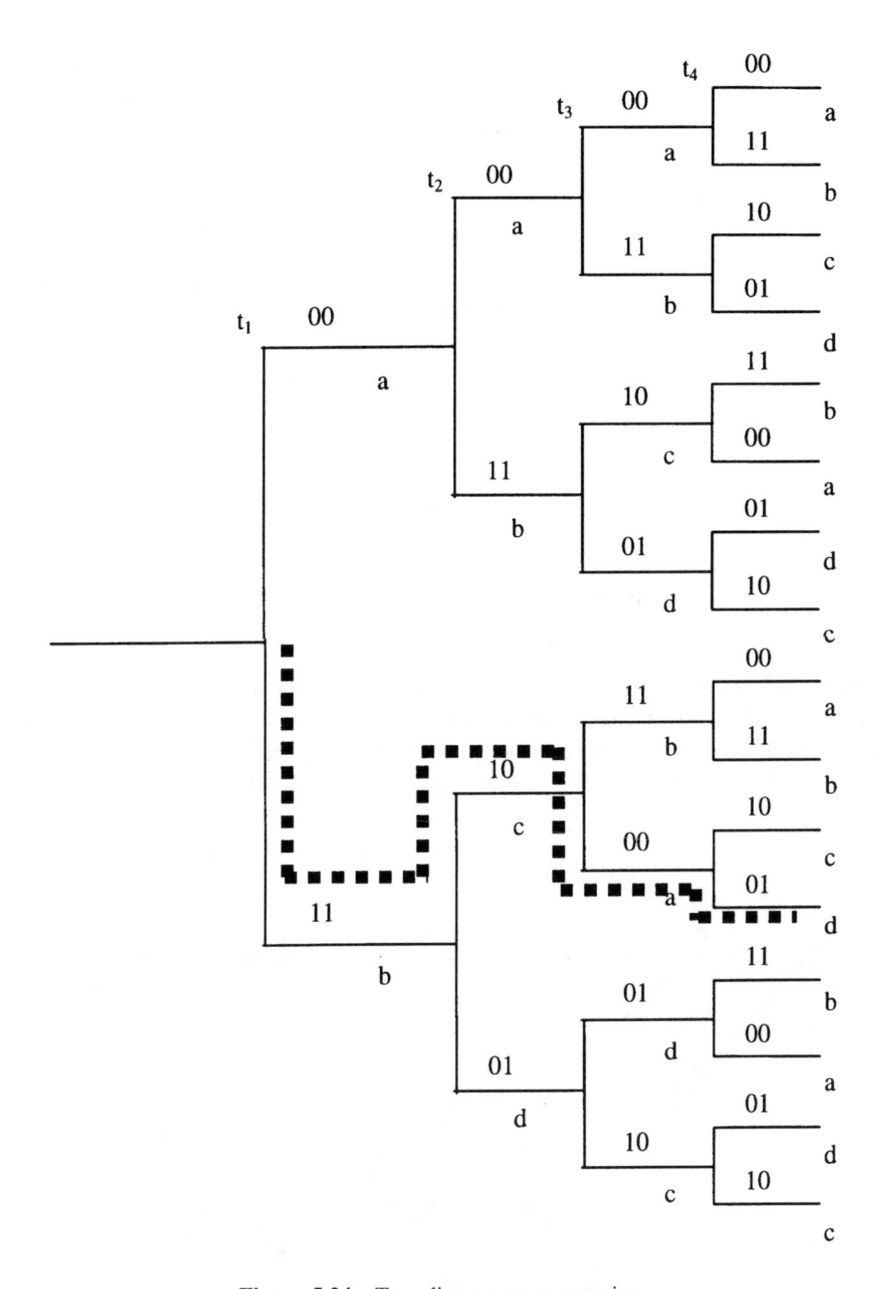

Figure 5.24 Tree diagram representation.

the trellis correspond to the four outputs states of the encoder previously identified (*a*, *b*, *c* and *d*). The columns are used to represent the progress of the output with time. Assuming the encoder is initially flushed with all "0"s, at the first input bit, as described previously, the encoder can generate one of the two pairs of digits, that is "00" or "11". As before, the concept of using a dashed line to represent an input of "1" is adopted. The output code is

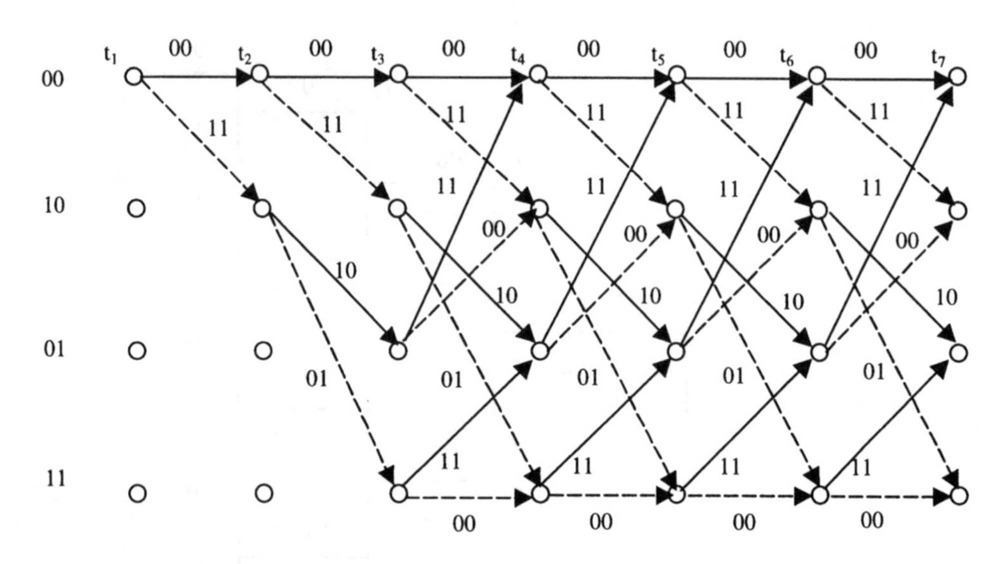

Figure 5.25 Trellis diagram representation.

shown adjacent to the arrow. At time t_2, there are four possible alternative outputs that could be generated by the encoder, as can be seen by referring back to the tree diagram. At time t_3, it can be seen that the number of possible outputs generated by the encoder increases to 8. From time t_4 onwards, the trellis repeats, with each state being entered by two possible routes and

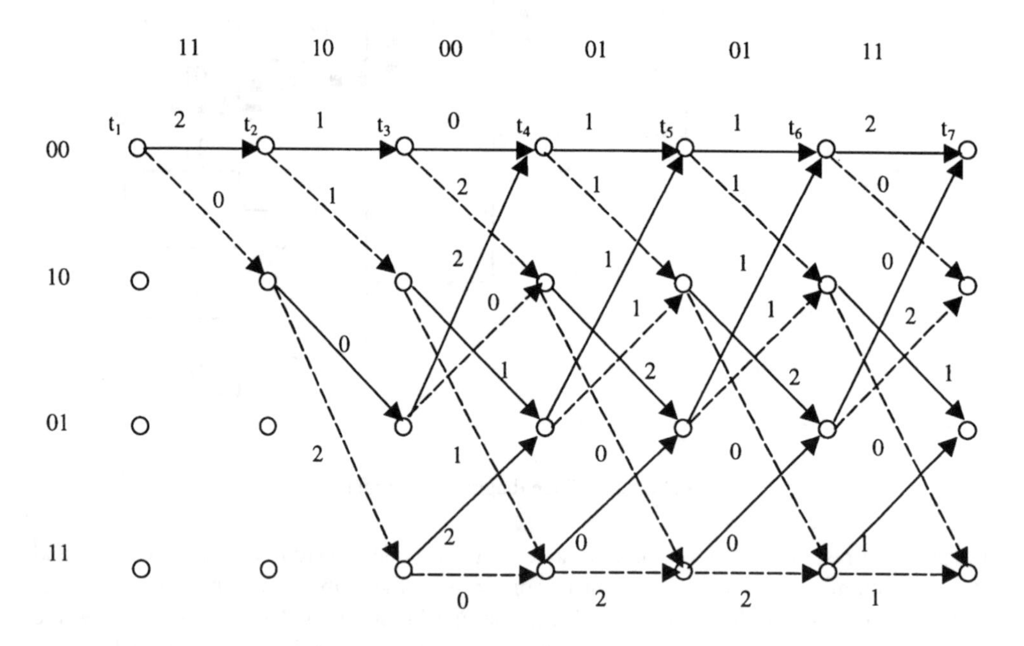

Figure 5.26 Minimum path trellis diagram.

each state generating two possible outputs. Recall that at time t_4, the first input bit no longer influences the output of the encoder, that is to say "1101" and "0101" will produce exactly the same output code. This can be verified by tracing the two paths on the trellis diagram, both of which will converge to the same node.

5.4.3.2 Decoding

Viterbi maximum likelihood decoding of convolutional codes provides the best possible results in the presence of random errors. In an attempt to match the output sequence received by the decoder, Viterbi's algorithm models the possible state transitions through a trellis identical to that used by the encoder. However, in this case, the trellis diagram rather than illustrating the possible outputs of the encoder, shows the *Hamming distance*, which in this instance is the number of bits that differ between the generated code word and the corresponding code word along a particular branch of the receiver. Referring back to the initial code sequence, the diagram shown in Figure 5.26 provides the corresponding trellis diagram. When two paths converge onto the same node, the path with the minimum Hamming distance is retained, and the other is discarded. The Hamming distance metric being the summation of all of the Hamming distances along the route. Hence at time t_4, two possible paths converge at node b, one with a Hamming distance of 0, the other 5 (2+1+2). The latter can be discarded since this is of the greatest value. Hence, it can be concluded that the first input bit is a "1", since the path emanating from the "0" first input bit has now been removed. It can be seen that the receiver introduces a processing delay associated with making a decision and requires a memory element to retain all of the code options.

Viterbi decoders make use of either hard or soft decision making. Soft decisions can be easily incorporated and this can provide an additional gain of up to 3 dB.

5.4.4 Interleaving

As was seen in the previous chapter, the mobile-satellite channel introduces errors of a bursty nature. Hence, in the short-term, the errors introduced by the channel cannot be considered to be statistically independent or memoryless, the criterion upon which most coders (block and convolutional) optimally operate. In order to mimic a statistically independent channel, a technique known as interleaving is incorporated into the transmitter chain after the output of

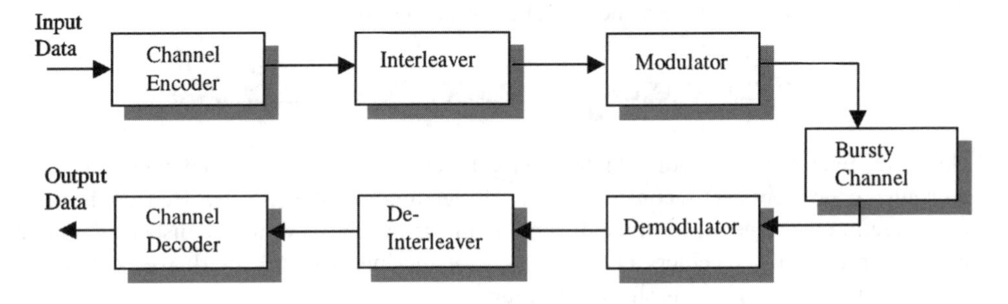

Figure 5.27 Interleaver/de-interleaver application.

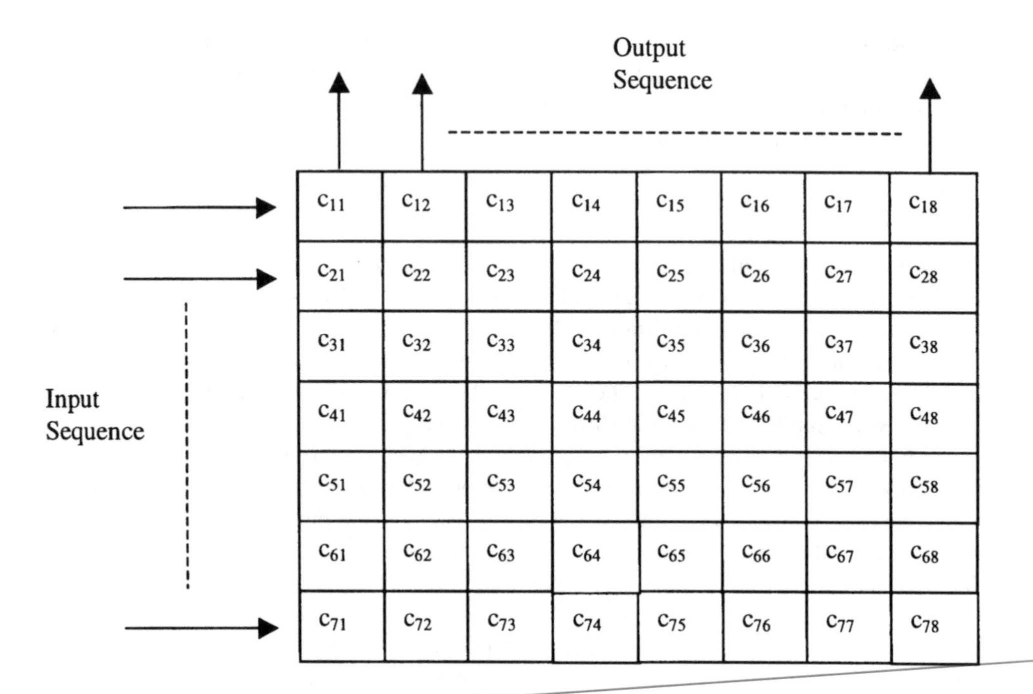

Figure 5.28 Interleaver block.

the encoder. The role of the interleaver is to re-order the transmission sequence of the bits that make up the code words in some pre-determined fashion, such that the effect of an error burst is minimised.

Interleaving can be performed for both block and convolutional coders. Block interleaving is achieved by firstly storing the output code words of the encoder into a two-dimensional array. Consider the case of an $[m \times n]$ array, where m is the number of code words to be interleaved and n is the number of code word bits. Each row of the array comprises a generated code word. Once the array is full, the contents are then output to the transmitter but in this case, data is read out on a column-by-column basis. Thus, the transmission of each symbol of a particular code word will be non-sequential, as shown below. Hence, the effect of any error bursts will have been dispersed in time throughout the transmitted code words.

In Figure 5.28, the output sequence would correspond to:

$$C_{11}C_{21}C_{31}C_{41}C_{51}C_{61}C_{71}C_{12}C_{22}C_{32}C_{42}C_{52}C_{62}C_{72}C_{13}C_{23}C_{33}C_{43}\ldots$$

At the receiver, the inverse of the interleaving function is performed by a deinterleaver and the original code words are reconstituted prior to feeding into the encoder. Hence a burst of errors affecting the transmitted bits indicated by the arrow above would be dispersed among the code words at the receiver. The placement of the interleaver/deinterleaver within the transmission/reception chain is shown in Figure 5.27.

Convolutional interleavers work along likewise concepts, achieving performance characteristics similar to block interleaving.

5.4.5 Concatenated Codes

Originally developed for deep space communications, concatenated coding occurs when two separate coding techniques are combined to form a larger code [YUE-90]. The inner decoder is used to correct most of the errors introduced by the channel, the output of which is then fed into the outer decoder which further reduces the bit error rate to the target level. A typical concatenated coding scheme would employ half-rate convolutional encoding of constraint length 7 (2, 1, 7)/Viterbi decoding as the inner scheme, and RS (255, 223) block encoding/decoding as the outer scheme.

Interleaving between the inner and outer coders can be used to further improve the performance.

5.4.6 Turbo Codes

Turbo codes were first presented to the coding community in 1993 [BER-93], and since then, they have been the subject of research throughout the world. Developed for deep space and satellite communications applications, turbo codes offer performance significantly better than concatenated codes, approaching the Shannon limit.

Turbo codes are generated using two recursive systematic convolutional (RSC) code generators, concatenated in parallel. Here, the term recursive implies that some of the output bits of the convolutional encoder are fed back and applied to the input bit sequence. The term systematic has previously been described with reference to block codes.

The first encoder takes as input the information bits. The key to the turbo code generation is the presence of a *permuter*, which performs a function similar to an interleaver described above, but in this case, the output sequence is pseudo-random. The permuter takes a block of information bits, which should be large to increase performance, for example more than 1000 bits, and produces a random, delayed sequence of output bits which is then fed into the second encoder. The output of the two encoders are parity bits, which are transmitted along with the original information bits. In order to reduce the number of transmitted bits, the parity bits are punctured prior to transmission.

The turbo decoder operates by performing an iterative decoding algorithm, resulting in the transfer of *a priori* likelihood estimates of the decoded bit sequence between the constituent decoders. Initially, the received information bits, which may be in error due to the influence of the channel, are used to perform *a priori* likelihood estimates by the respective decoders. The decoders employ a MAP (maximum a posteriori) algorithm [BAH-74] to converge on the likely sequence of data, after which the iterative interaction between decoders ceases and the output sequence is obtained from one of the decoders. An interleaver is placed between the output of Decoder 1 and the input of Decoder 2 to provide additional weighted decision input into Decoder 2, similarly, a de-interleaver is placed at the output of Decoder 2, to provide feedback to Decoder 1. The decoding time is proportional to the number of iterations between decoders.

The operation of the transmitter/receiver is shown in Figure 5.29.

5.4.7 Automatic Repeat Request Schemes

FEC schemes attempt to correct errors in transmission at the receiver, which, in some cases

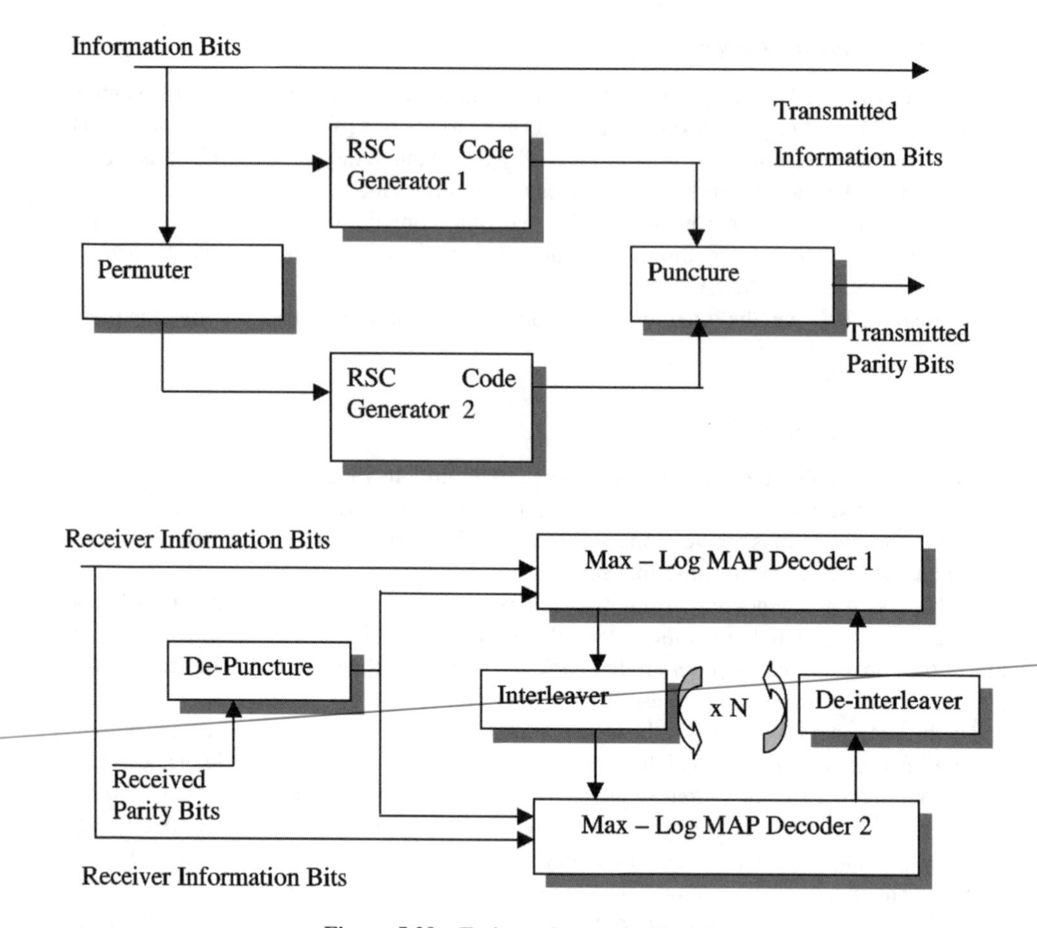

Figure 5.29 Turbo code encoder/decoder.

may not be achievable, resulting in the incorrect reception of the code word. ARQ schemes provide an alternative to FEC, where a high degree of reliability over the transmission link is required. To achieve this, ARQ schemes rely upon a re-transmission protocol that is used to inform the transmitter of whether a transmitted code sequence has been correctly received or not. Should an error be detected at the receiver, a request to re-transmit the information is made using a return channel. Here, it can be seen that the receiver technology is not as complex as that employed when using FEC. On the other hand, the transmission rate is likely to be less than that when using FEC, moreover, the transmission rate will no longer be constant but will vary in an unpredictable way. How much information is re-transmitted is determined by the ARQ protocol adopted by the network. Essentially, there are three classes of ARQ protocols: *stop-and-wait*, *continuous ARQ with repeat* (also commonly referred to as the *go-back-N ARQ*) and *continuous ARQ with selective repeat* [LIN-84].

The stop-and-wait ARQ is the most basic of the three protocols. This operates by the transmitter sending a block of coded information and then waiting for an acknowledgement (ACK) signal from the receiver, indicating that the code sequence has been received correctly. The transmitter then sends the next code block in the sequence. Should a packet of information be received incorrectly, the receiver will send a negative acknowledgement (NACK) and, subsequently, the transmitter will re-send the packet once again. Only when an ACK message is received at the transmitter will the next code word in the sequence be transmitted. In terms of throughput, this technique is not the most efficient, since the transmitter needs to be idle for relatively long periods (at least a round-trip delay of approximately 500 ms in the case of a geostationary satellite, plus processing time) before a new transmission can occur. Moreover, should transmissions occur during an error burst on the channel, it may take a considerable amount of time before all of the information can be transmitted. On the other hand, a high degree of reliability can be achieved with relatively simple receiver circuitry.

An improvement on the efficiency of the stop-and-wait ARQ is the continuous ARQ with repeat. In this case, the transmitter continuously transmits code words, which are buffered in memory, while awaiting the reception of an ACK/NACK message. In this case, both the transmitter and receiver transmit simultaneously, requiring the establishment of a full-duplex channel, unlike the half-duplex arrangement of the stop-and-wait protocol. Should a code block be received in error, the receiver will generate a NACK message, identifying the code block in question, and will discard any subsequent received packets, irrespective of whether they were received correctly or in error. Upon receiving this NACK message, the transmitter returns to the corresponding position in the message sequence and re-transmits this code block along with all subsequent blocks of code until a NACK message is received, upon which the process is repeated. Here, the inefficiency of the scheme is due to the fact that any correctly received code words may be discarded at the receiver, even if only one code word in the transmission sequence is received incorrectly. The selective-repeat ARQ solves this problem.

The selective-repeat ARQ protocol enables the transmitter to continuously send code words, as in the previous case, however, should a code word be received in error, the corresponding NACK message will only require the re-transmission of the particular code word identified by the NACK message. At the receiver, code words correctly received after a corrupted message are stored in memory until the corrupted message is received correctly, after which the message sequence is restored and output. As with continuous ARQ with repeat, the selective-repeat ARQ protocol requires the establishment of a full-duplex channel and memory at the receiver to save the correctly received blocks of code. A means of identifying each packet is also required in both cases.

The operation of the ARQ schemes is summarised in Figure 5.30.

Hybrid FEC-ARQ schemes are also applicable, whereby the receiver attempts to correct any detected errors in transmission and if not possible, requests the re-transmission of the code sequence.

5.5 Multiple Access

5.5.1 Purpose

A multiple access scheme allows many users to share a satellite's resource, that is its capacity. Essentially, there are three types of multiple access scheme employed in satellite communications, namely: frequency division multiple access (FDMA); time division multiple access

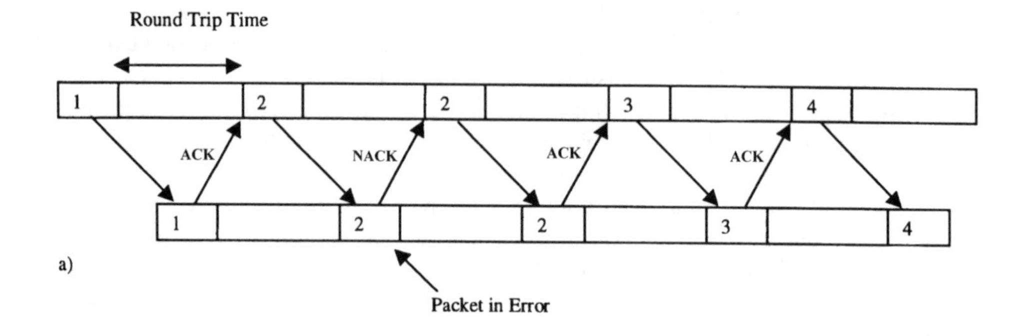

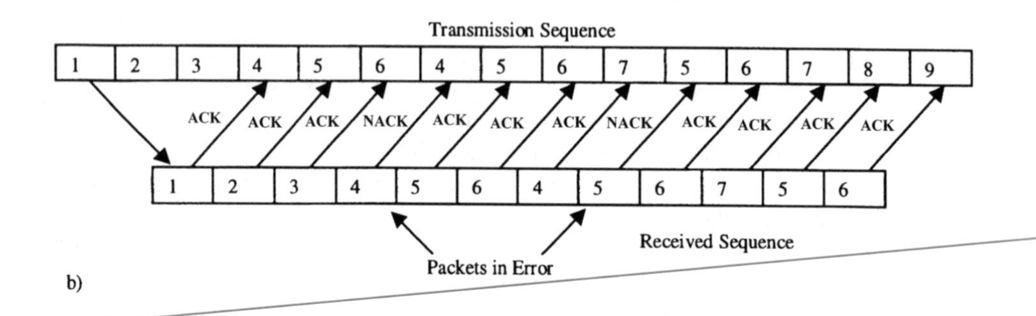

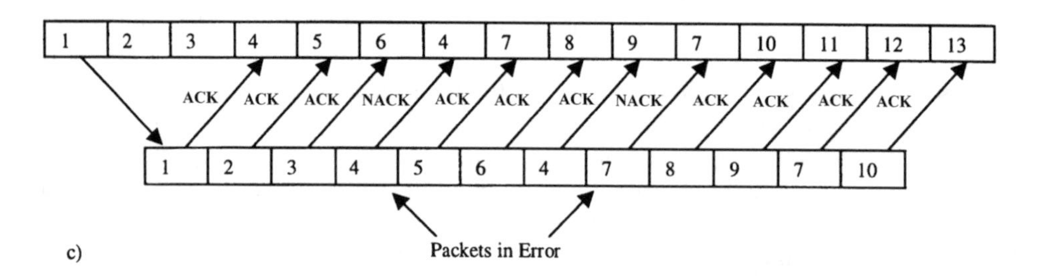

Figure 5.30 ARQ schemes: (a) stop-and-wait; (b) continuous ARQ with repeat; (c) continuous ARQ with selective repeat.

(TDMA); or code division multiple access (CDMA). Hybrid solutions are used by combining techniques such as FDMA/TDMA. In this example, the total bandwidth is divided into channels, each of which contains a TDMA frame. A user would be assigned a frequency and a time-slot upon which to transmit.

One of the main issues when considering the relative merits of an access scheme is its robustness to potential interference. Spectrum availability is limited, hence an efficient access scheme which maximises the number of available channels whilst minimising required bandwidth is desirable. As with terrestrial cellular systems, in a multi-spot-beam environment, frequency re-use is used to increase the capacity of the network. Here, spot-beams can be treated as cells, and the equations presented in Chapter 1 can be equally applied to satellite systems in order to determine the frequency re-use distance and immunity to co-channel

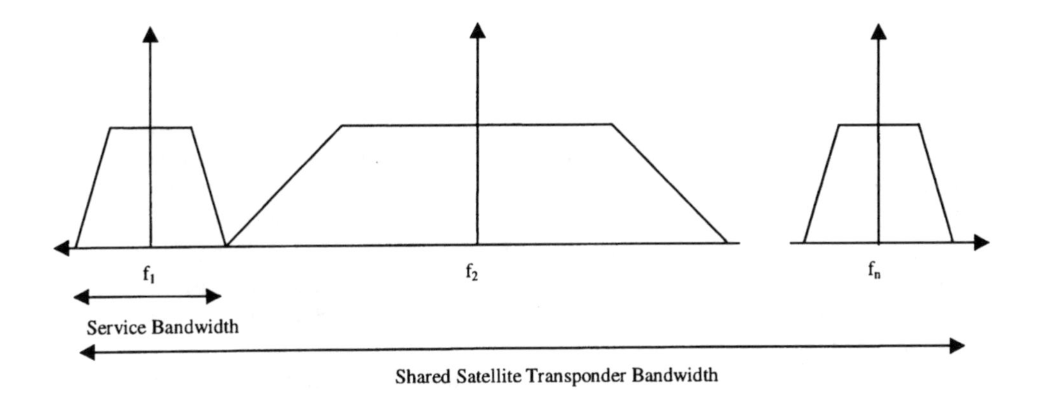

Figure 5.31 Multi-carrier usage of shared transponder bandwidth.

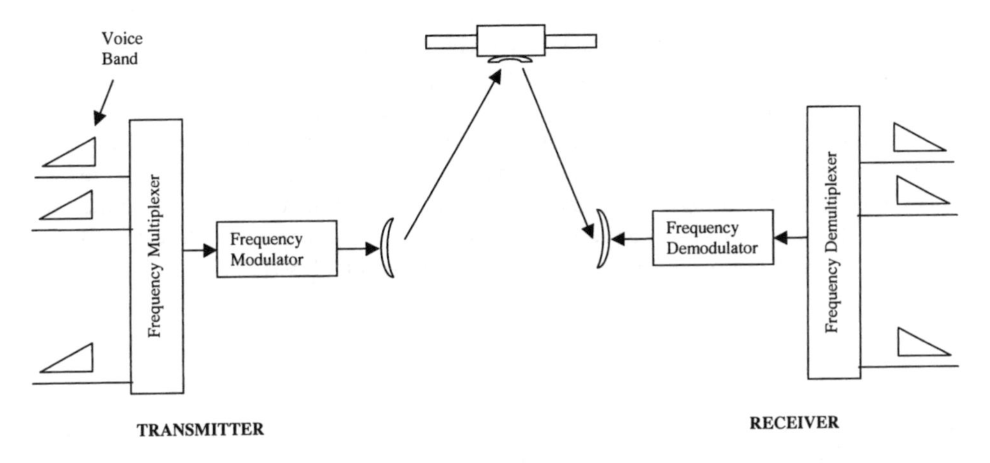

Figure 5.32 FDM/FM/FDMA application.

interference. When using FDMA or TDMA in a multi-spot-beam satellite configuration, adjacent beams cannot be configured with the same carrier frequency. This is not the case for CDMA, which can operate with a frequency re-use distance factor of 1. Consequently, the satellite antenna beam pattern will need to be considered when deriving a frequency re-use pattern. The following considers the basic characteristics of each multiple access technique prior to reviewing the current status of the standardisation of the multiple access schemes for S-UMTS/IMT-2000.

5.5.2 FDMA

FDMA is the simplest and most established technique to be employed in satellite communications. It operates by dividing the available transponder bandwidth (typically 36 or 72

MHz for a geostationary satellite) into channels, which are then assigned to users. This is shown in Figure 5.31.

One established mode of application used by fixed Earth stations for the transmission of telephony, is to multiplex several voice circuits onto an assigned channel, known as FDM/FM/FDMA (also known as Multiple Channels per Carrier (MCPC)). This tertiary expression can be explained as follows. Initially, a number of voice circuits arriving at an Earth station are combined into a single band using frequency division multiplexing (FDM). This composite signal is then frequency modulated prior to up-converting onto a network assigned carrier for transmission. At the receiver, the reverse operation is performed, that is the Earth station down converts the carrier prior to performing frequency demodulation and then demultiplexes the individual voice circuits. This is summarised in Figure 5.32.

As was noted earlier, the number of channels that can be employed per transponder bandwidth needs to take into account inter-modulation considerations. Moreover, there needs to be a guard-band between carriers to avoid mutual, adjacent channel interference. When employing non-geostationary satellites, the guard-band is governed by the Döppler shift, which increases with frequency of operation. Since the guard-band is in effect an unused resource, or network overhead, the network designer needs to carefully trade-off the need for interference protection against redundant bandwidth usage.

The other form of FDMA implementation is Single Channel per Carrier (SCPC). In this case, a carrier frequency is assigned per circuit; consequently the transmission equipment performs no multiplexing. This mode of operation allows carriers to be assigned to users either on a fixed or on a per-demand temporary basis, the latter being governed by the traffic usage. Such a scenario exists in a mobile-satellite environment, as used by Inmarsat in support of its INMARSAT-A FM telephony service (see Chapter 2). SCPC can be viewed as a more efficient use of the satellite resource than that of FDM/FM/FDMA, when traffic demand is variant.

5.5.3 TDMA

In TDMA, the available transponder bandwidth is made available to an active user for a very short period of time (known as a burst), during which its data are rapidly transmitted. The total available transponder bandwidth is shared with other users which transmit during different time-slots. A number of slots, when added together form part of a TDMA frame.

In order to ensure that each user transmits within a specific time-slot, some form of reference clock is required from which all transmitting stations can synchronise their transmissions. This reference takes the form of a reference burst, which occurs at the start of each frame. A second reference burst may also follow on from the first reference burst in order to provide a means of redundancy.

The reference burst is transmitted by a reference Earth station and is made up of three parts [FEH-83]:

- Carrier and bit timing recovery (CBR): this enables stations to lock to the carrier frequency and bit timing clock burst;
- Unique word (UW): this provides the burst reference time. This is achieved by the receiver Earth station by correlating the received UW with a stored replica. The UW can also be used to remove phase ambiguity where coherent QPSK demodulation is employed;

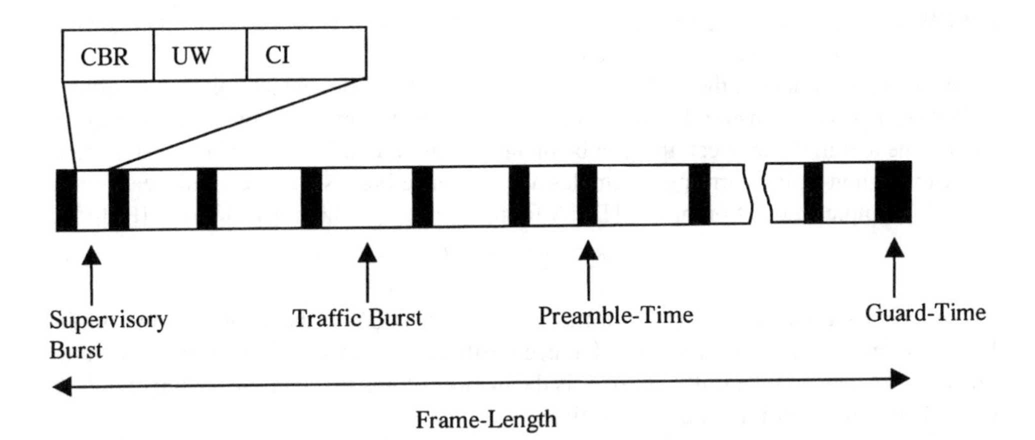

Figure 5.33 Typical TDMA frame structure.

- Control information (CI): this provides information that is used by each receiving station to control the position of its transmission bursts, as well as providing other network management information.

A traffic burst is similar in construction to a reference burst, comprising initially of CBR

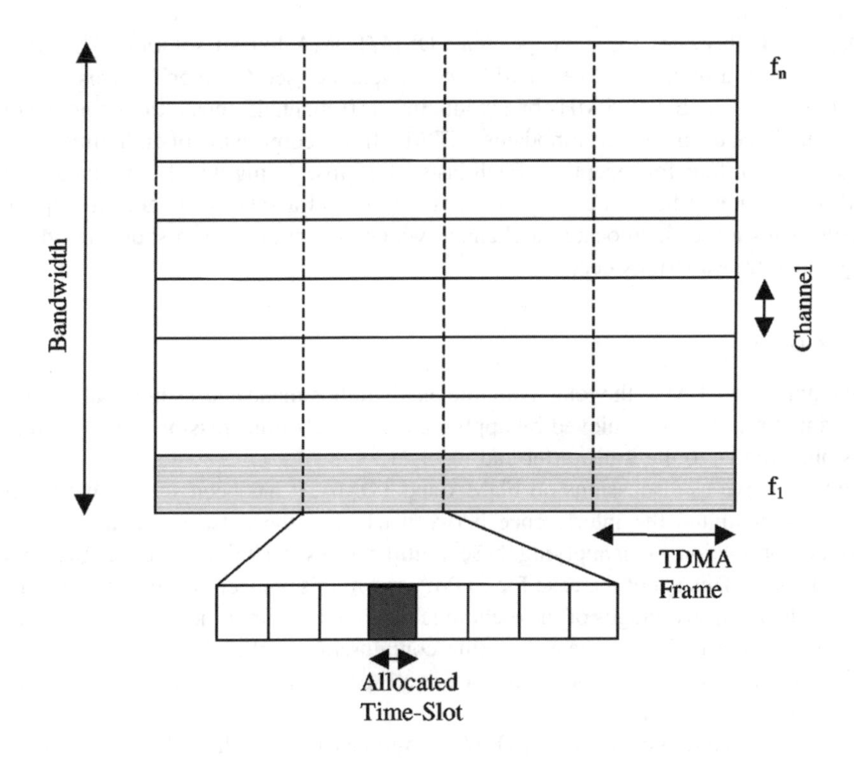

Figure 5.34 FDMA/TDMA hybrid access scheme.

and UW sequences followed by reference control information. This is collectively known as the *preamble*. A data burst follows the preamble.

The final component of the TDMA frame is the guard-time-slot, which like the guard-band in FDMA, is used to ensure that signals do not overlap. No transmissions occur during the guard-time and in this respect, this can be thought of as an overhead. Similar considerations apply to the non-traffic carrying preambles and reference burst sequences. This leads to the relatively simple calculation of the TDMA frame efficiency, which is as follows [PRI-93]:

$$\eta = \frac{R_F T_F - N_E b_p - N_R b_r - b_g(N_R + N_E)}{R_F T_F} \tag{5.48}$$

where T_F is the frame duration (s); R_F is the frame bit rate (s); N_E is the number of transmitting Earth stations; N_R is the number of reference Earth stations; b_p is the number of bits in the preamble sequence of the traffic burst; b_r is the number of bits in the reference burst; b_g is the equivalent number of bits in the guard-time.

A typical TDMA frame structure is shown in Figure 5.33.

From a satellite perspective, data will arrive in a continuous stream in the form of a time division multiplex (TDM) signal. How the satellite deals with this signal is dependent upon its payload complexity. For example, for a transparent satellite providing global beam coverage, the TDM signal is simply relayed to the ground and receiving stations are able to synchronise to the appropriate slot within the TDM stream. In more complex payload architectures, where multi-spot-beam configurations are employed, switching between beams, as discussed earlier in this chapter, allows packets of data to be directed to the appropriate beam for transmission.

Figure 5.34 illustrates the concept of a FDMA/TDMA hybrid solution. The IRIDIUM system employs a hybrid FDMA/TDMA access scheme (see Chapter 2). This is achieved by dividing the available 10.5 MHz bandwidth into 150 channels, thus introducing the FDMA component. Each channel accommodates a TDMA frame comprising of eight time-slots, four for transmission, four for reception. Each lasts 11.25 ms, during which time data are transmitted in a 50 kbit/s burst. Each frame lasts 90 ms and a satellite is able to support 840 channels. Thus a user is allocated a channel, which is occupied for a short period of time, during which transmissions occur.

5.5.4 CDMA

When employing CDMA, the total available bandwidth is made accessible to all active users at the same time. This is achieved by applying each user's transmission with a unique code that is only known to the transmitter and receiver.

From the previous discussions on FDMA and TDMA, it has been shown that in order to reduce to a minimum the interference between users of the network, there is a need for network co-ordination when applying these multiple access techniques, be it in the frequency or time domain. This is not the case for CDMA and in this respect, this can be considered to be one of the main advantages of the technique over the two other methods. Moreover, when using CDMA in a multi-spot-beam satellite configuration, unlike TDMA and FDMA, it is possible to re-use the same carrier frequency in all spot-beams. In other words, the frequency re-use factor is equal to 1.

From a commercial perspective, CDMA technology is still relatively new. Indeed, there

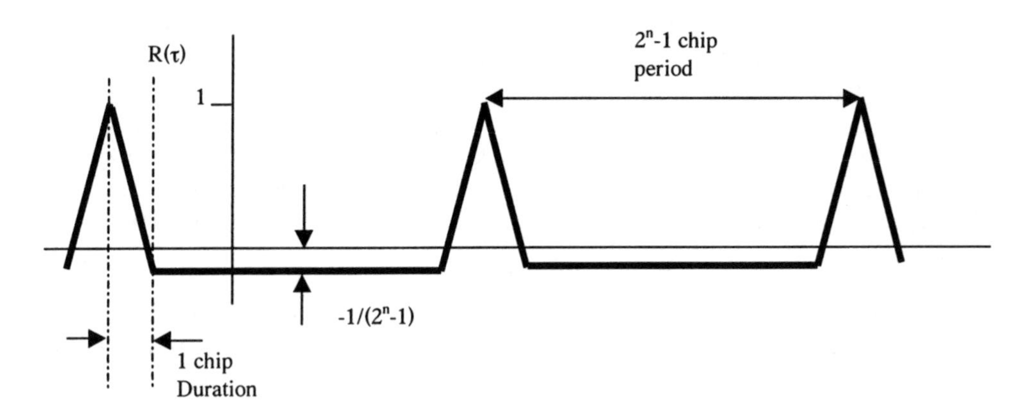

Figure 5.35 PN sequence auto-correlation function $R(\tau)$.

has been considerable debate within the satellite community regarding the relative merits of CDMA and TDMA. However, Chapter 2 has highlighted several systems that currently operate (for example GLOBALSTAR) or plan to implement (ELLIPSO, CONSTELLATION) a CDMA solution. Here, achieving the commonality with the terrestrial mobile network is of primary concern. For example, it has already been shown in Chapter 2, how the GLOBALSTAR radio interface is a derivative of the cdmaOne standard, discussed in Chapter 1. This need for commonality will be returned to when considering the candidate solutions for S-UMTS/IMT-2000 in the next section.

In theory, the maximum number of users that can be simultaneously supported by the

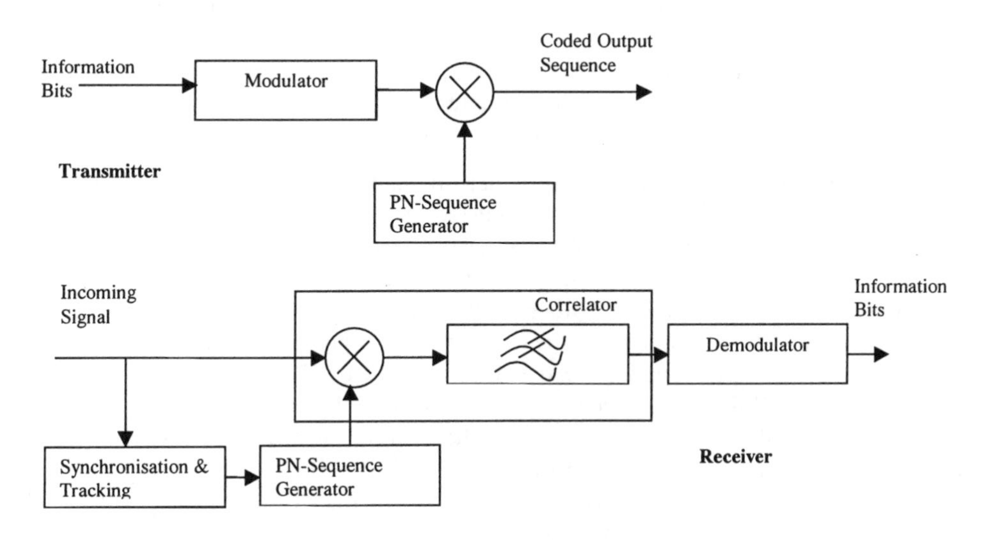

Figure 5.36 Direct sequence CDMA.

CDMA network is determined by the selection of the code sequence generator that is used to spread the user information. Code generation can be achieved by using a linear feedback multi-stage shift register and a modulo-2 adder. As with convolutional code generation, the code sequence is determined by the arrangement of the connections between the respective elements. The subsequent output produces a code sequence with noise-like properties, and since the code sequence is deterministic, it is termed *pseudo-noise*. For a cyclic shift register of n-stages, the maximum period of repetition, P, of the code sequence is given by:

$$P = 2^n - 1 \tag{5.49}$$

This is termed the *maximal length sequence*. Performing an autocorrelation function on such a sequence results in the characteristics shown in Figure 5.35. Here, it can seen that the function peaks when the codes are synchronised, otherwise the output reduces to a minimum at any other off-set.

The application of CDMA can be divided into two main techniques: *direct sequence (DS)* and *frequency hopping (FH)*.

In DS-CDMA, the spreading sequence is multiplied with the modulated signal prior to transmission. The applied code rate, referred to as the *chip rate*, is of a pseudo-random nature with a noise-like spectrum. The chip rate is very much greater than the information rate, which occupies a bandwidth approximately equal to its data rate. The subsequent convolution of the information and code sequences results in the spreading of the information bandwidth, hence this form of multiple access is also sometimes referred to as *spread spectrum* (Figure 5.36).

In order to recover the user data at the receiver, a locally generated replica of the transmitted code sequence needs to be multiplied by the incoming signal. This requires the receiver code sequence to be in synchronisation in the frequency and time domains with that of the transmitted sequence. When this condition occurs, as noted earlier, the output of the correlator is at a peak. Synchronisation is achieved in two stages:

- Initially, during the *acquisition phase*, the receiver attains a course alignment with the transmitted code;
- Having attained a course alignment, the second stage is to fine-tune this alignment and maintain synchronisation. This is generally referred to as the *tracking phase* and usually involves the application of feedback-loop circuitry.

In order to perform the acquisition phase efficiently, initially the possible boundaries over which to search in the frequency and time values of the incoming signal are defined. Such uncertainties can be the result of Döppler, multipath transmission, transmission delay and so on. This is referred to as the *time-frequency uncertainty region*. The average time it takes to search through this region in order to attain acquisition, known as the *mean time to acquisition*, is a measure of the performance of the receiver. The simplest method of achieving initial acquisition is to use a serial search. Here, the locally generated PN sequence is correlated with the received signal. If the subsequent correlation is below a detector threshold, the local PN generator is triggered, resulting in the shifting of the local code in phase by a *cell*, a fraction of chip period (usually a half) and the process is repeated. Once the output of the correlator rises above the detector threshold, the incoming and locally generated code sequences are considered to be in coarse alignment and the tracking phase is initiated. There are a number of

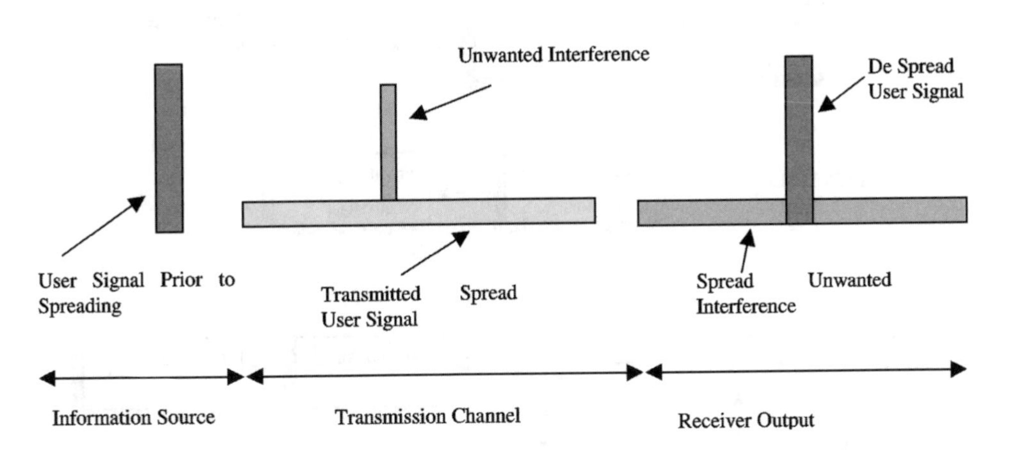

Figure 5.37 DS-CDMA interference rejection capability.

methods that have been proposed to increase the performance of the acquisition phase, some examples of which can be found in Refs. [COR-96, JOV-88, MEY-83, POL-84a, POL-84b].

Once synchronisation has been achieved, the output of the correlator can be applied to the demodulator from where the information data can subsequently be obtained. The effect of correlation at the receiver also has the benefit of reducing the level of unwanted interfering signals, by in effect, performing a similar spreading process to that at the transmitter (Figure 5.37).

The ability of the receiver to correctly receive a spread signal when in the presence of interfering signals is determined by its *processing gain*, P_G, which can be approximated by the expression:

$$P_G \approx \frac{R_C}{R_D} \tag{5.50}$$

where R_C is the chip rate (chip/s) and R_D is the data rate (bit/s).

The capacity of a CDMA network is interference limited. Hence, to maintain the maximum system capacity, it is necessary for all transmitting stations to operate with similar power levels. This is achieved by implementing a means of power control, which can take the form of open- or closed-loop. Here, open-loop power control can be used to rapidly compensate for signal fluctuations, caused by shadowing for example. The closed-loop power control can be used to compensate for the longer-term signal fluctuations. In a transparent system such as GLOBALSTAR, open-loop power control is achieved in the return link direction by the mobile terminal monitoring the fluctuation in received signal strength from the Earth station via the satellite, and adjusting its transmitted power accordingly. Here, the accuracy of the algorithm depends on how closely correlated the forward and return link propagation properties are. In the closed-loop approach, the Earth station monitors the power levels of the received signals of all mobile stations and sends control information to the mobile stations indicating the required adjustments in the level of the transmitted signal power [RAM-95].

The use of CDMA provides the opportunity to exploit the advantages that satellite diversity

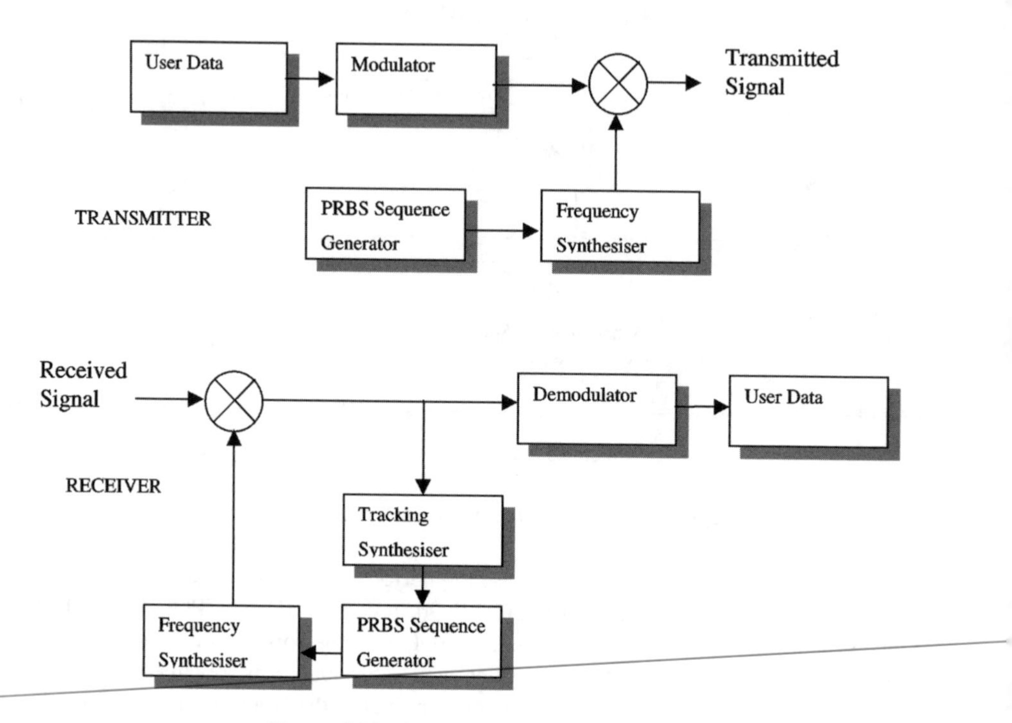

Figure 5.38 Frequency hopped CDMA.

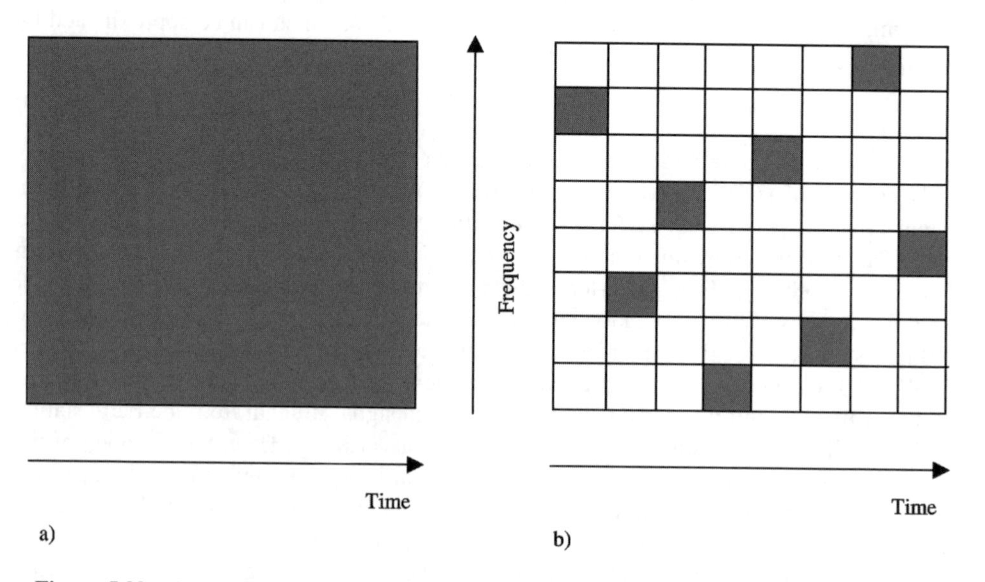

Figure 5.39 Comparison of: (a) direct sequence and (b) frequency hopping CDMA techniques.

can offer. In particular, satellite diversity can improve the quality of the received signal in terms of enhanced signal strength and link availability. Reception of the same signal from different satellites can be achieved using a RAKE receiver, whereby each "finger" of the RAKE receiver is responsible for despreading and demodulating transmissions from a particular path. The output of each finger is then combined together, resulting in increased signal-to-noise ratio.

In the frequency hopping approach, the pseudo-random sequence is used to change the transmission frequency with each change in the pseudo-code. This is achieved by driving a frequency synthesiser, the output of which is dictated by the generated code sequence. The output of the frequency synthesiser is then applied to the modulated user data sequence. Here, the modulation method could take the format of M-ary FSK or possibly PSK. Hence, the transmitted signal can be envisaged to change its transmission frequency across the available band allocated to the service. Frequency hopping can be further categorised as either fast-frequency hopping (F-FH) or slow-frequency hopping (S-FH). Whether it is fast or slow is determined by the relationship between the information rate and the frequency hopping rate. Fast frequency hopping implies that the frequency hopping rate is much faster than the information rate, conversely, slow hopping implies that the hopping rate is much less than the information rate. Hence, in slow frequency hopping, a number of information bits would be sent when using the same transmission frequency, whereas for fast hopping, the same information bit may be transmitted using different transmission frequencies. At the receiver, the same pseudo-random sequence is used to drive a similar frequency synthesiser in synchronisation with the transmitter, which can then be used to demodulate the signal. The method of transmission and reception is summarised in Figure 5.38.

Figure 5.39 compares the direct sequence with the frequency hopping spectral occupancies with time.

In addition to DS-CDMA and FH-CDMA, it also possible to apply a method of *time hopping*. Here, transmissions are sent in bursts, within random time-slots, the occupancy of which is dictated by the user's code. Further information on this technique can be found in Ref. [PRA-98].

5.5.5 Contention Access Schemes

5.5.5.1 ALOHA

The transmission of packet data is of a bursty nature. In other words, access to the transmission resource, i.e. channel, is only required at intermittent periods of time, during which packets of information are transmitted. In such an environment, the permanent allocation of channel resource to a particular transmitter can be seen to be impractical, and rather, what is known as a contention access scheme is employed. Such a scheme implies that the transmitter vies for satellite resource on a per-demand basis. In this case, provided that no other transmitter is attempting to access the same resource during the transmission burst period, an error free transmission can occur. On the other hand, there is a probability of collision of transmission packets, which in this case, will necessitate the need to re-transmit the packet after a random delay period.

The most widely used contention access scheme is ALOHA and its associated derivatives. ALOHA was developed by the University of Hawaii at the start of the 1970s. Its simplest mode of operation consists of Earth stations randomly accessing a particular resource that is used to transmit packets. Earth stations can detect whether their transmission has been correctly received at the satellite by either monitoring the re-transmission from the satellite, or by receiving an acknowledgement (ACK) message from the receiving party. Should a collision with another transmitting station occur, resulting in the incorrect reception of a packet at the satellite, the transmitting Earth station waits for a random period of time, prior to re-transmitting the packet. ALOHA is relatively inefficient with a maximum throughput of only 18.4% or $1/2e$. However, this has to be counter-weighed against the gains in simple network complexity, since no co-ordination or complex timing properties are required at the transmitting terminals.

5.5.5.2 Slotted-ALOHA

By dividing the time domain into slots, each of a duration of a single packet burst time, and allowing transmissions only at the start of a time-slot, the number of packet contentions can be reduced. This is due to the fact that collisions can now only occur during the time-slot period, and will not be a result of the partial overlapping of packets, as is the case with ALOHA. Such an approach is termed slotted-ALOHA.

The introduction of time-slots, increases the saturation capacity to 37% or $1/e$. However, the penalty for a such a gain is an increase in the network complexity. Transmitting terminals are now required to synchronise burst transmissions to particular time-slots, as in TDMA.

5.5.5.3 Slot Reservation ALOHA

This extension of the slotted-ALOHA scheme allows time-slots to be reserved for transmission by an Earth station. This can be achieved *implicitly*, by which the transmitting station initially contends for an available slot with other transmitting terminals. Available slot locations are made known to all transmitting stations within the network by the network control station using a broadcast channel. Once a transmitting station successfully gains access to a particular slot by contention, the network controller informs all other transmitting stations that the slot is no longer available, and the successful transmitting station retains the slot until transmission is complete. The network controller then informs all stations on the network that the slot is available for contention once more. The other means of slot reservation is achieved *explicitly*, whereby a transmitting station requests the network to reserve a particular slot prior to transmission.

In general terms, this mode of operation is termed a packet reserved multiple access scheme (PRMA).

5.5.6 S-UMTS/IMT-2000 Candidate Solutions

In the early 1990s, the responsibility for the standardisation of the satellite component of UMTS came under the auspices of ETSI Satellite Earth Station and Systems (SES) SMG 5, with the intention being to follow the ITU-R FPLMTS Recommendations [DON-95]. In the first few years, a few technical reports were produced by ETSI, although these addressed system issues, such as the satellite operating environments. SMG 5 was eventually closed. In 1998, the standardisation of S-UMTS was re-activated under the auspices of ETSI's SES S-UMTS Work Group.

As indicated in Chapter 1, research into the radio interface characteristics of 3G terrestrial mobile networks was the focus of significant effort around the world for most of the 1990s. As a consequence, a global harmonised set of solutions for the 3G radio interface has been agreed upon. For various reasons, activities on the standardisation of the satellite component of the 3G network have not progressed with the same degree of urgency. This may not be so surprising, however, since it is important that the satellite solution should bear a close resemblance to that of the terrestrial solution. As was discussed in Chapter 1, the harmonisation of the various terrestrial solutions, which were submitted at the same time to the ITU as their satellite counterparts, only concluded at the end of the last decade.

In fact it has been agreed that there is no need for a common 3G-radio interface for the satellite component, at least for the first phase of UMTS/IMT-2000 introduction. In the longer term, however, a standardised satellite-UMTS/IMT-2000 radio interface should be aimed for and indeed, is now the focus of the standards bodies. As was noted in Chapter 1, the ITU received five proposals for the satellite-UMTS/IMT-2000 radio interface by the proposal deadline of 20 June 1998. The five proposals included two from the European Space Agency based on the ETSI terrestrial-UMTS W-CDMA UTRA FDD and UTRA TDD solutions. ESA termed these solutions SW-CDMA, which was intended for global solutions, and SW-C/TDMA, which was for regional based solutions [TAA-99]. The Telecommunication Technology Association (TTA), South Korea, submitted a wideband-CDMA solution, based on its terrestrial proposal, termed SAT-CDMA. Inmarsat Horizons and ICO Global Communications also submitted proposals based on their proprietary solutions.

At the time of writing, work in ETSI and TTA is geared towards producing a harmonised SW-CDMA/SAT-CDMA solution with a predicted completion date of 2002.

References

[BAH-74] L. Bahl, J. Cocke, F. Jelinek, J. Raviv, "Optimal Decoding of Linear Codes for Minimizing Symbol Error Rate", *IEEE Transactions on Information Theory*, IT-20(2), March 1974; 284–287.

[BER-93] C. Berrou, A. Glavieux, P. Thitimajshima, "Near Shannon Limit Error-Correcting Coding and Decoding: Turbo-Codes", *Proceedings of International Conference on Communications, ICC'93*, May 1993; 1064–1070.

[COR-96] G.E. Corazza, "On the MAX/TC Criterion for Code Acquisition and its Application to DS-SSMA Systems", *IEEE Transactions on Communications*, 44(9), September 1996; 1173–1182.

[DON-95] P. Dondl, "Standardization of the Satellite Component of UMTS", *IEEE Personal Communications*, 2(5), October 1995; 68–74.

[EVA-99] B.G. Evans (Ed.), *Satellite Communication Systems*, 3rd Edition, IEE, London, 1999.

[FEH-83] K. Feher, *Digital Communications Satellite/Earth Station Engineering*, Prentice-Hall, Englewood Cliffs, NJ, 1983.

[ITU-93] *ITU-R Rec. M.1091*, "Reference Earth-Station Radiation Pattern for use in Coordination and Interference Assessment in the Frequency Range from 2 to about 30 GHz", 1984.

[ITU-94a] *ITU-R Rec. PI.372-6,* "Radio Noise", 1994.

[ITU-94b] *ITU-R Rec.* M.1091; "Reference Off-Axis Radiation Pattern for Mobile Earth Station Antennas Operating in the Land Mobile-Satellite Service in the Frequency Range 1–3 GHz", 1994.

[JOV-88] V.M. Jovanovic, "Analysis of Strategies for Serial-Search Spread-Spread Spectrum Code Acquisition – Direct Approach", *IEEE Transactions on Communications*, COM-36(11), November 1988; 1208–1220.

[LIN-84] S. Lin, D.J. Costello, M.J. Miller, "Automatic-Repeat-Request Error-Control Schemes", *IEEE Communications Magazine*, 22(12), December 1984; 5–17.

[MEY-83] H. Meyr, G. Polzer, "Performance Analysis of General PN-Spread-Spectrum Acquisition Techniques", *IEEE Transactions on Communications*, COM-31(12), December 1983; 1317–1319.

[POL-84a] A. Polydoros, C.L. Weber, "A Unified Approach to Serial Search Spread-Spectrum Code Acquisition – Part I: General Theory", *IEEE Transactions on Communications*, COM-32(5), May 1984; 542–549.

[POL-84b] A. Polydoros, C.L. Weber, "A Unified Approach to Serial Search Spread-Spectrum Code Acquisition – Part II: A Matched Filter Receiver", *IEEE Transactions on Communications*, COM-32(5), May 1984; 550–560.

[PRA-98] R. Prasad, T. Ojanperä, "An Overview of CDMA Evolution Toward Wideband CDMA", *IEEE Communications Surveys*, 1(1), Fourth Quarter 1998.

[PRI-93] W.L. Pritchard, H.G. Suyderhoud, R.A. Nelson, *Satellite Communication Systems Engineering*, 2nd Edition, Prentice-Hall, Englewood Cliffs, NJ, 1993.

[PRO-95] J.G. Proakis, *Digital Communications*, McGraw-Hill, New York, 1995.

[RAM-95] J. Ramasastry, R. Wiedeman, "Use of CDMA Access Technology in Mobile Satellite Systems", *Proceedings of Fourth International Mobile Satellite Conference*, Ottawa, Canada, 4–6 June 1995; 488–493.

[RAP-84] S.S. Rappaport, D.M. Grieco, "Spread-Spectrum Signal Acquisition: Methods and Technology", *IEEE Communications Magazine*, 22(6), June 1984; 6–21.

[SKL-88] B. Sklar, *Digital Communications Fundamentals and Applications*, Prentice-Hall International Editions, London, 1988.

[TAA-99] P. Taaghol, B.G. Evans, E. Buracchini, R. De Gaudenzi, G. Gallinaro, J.H. Lee, C.G. Kang, "Satellite UMTS/IMT2000 W-CDMA Air Interfaces", *IEEE Communications Magazine*, 37(9), September 1999; 116–125.

[YUE-90] J.H. Yuen, M.K. Simon, W. Miller, F. Pollara, C.R. Ryan, D. Divsalar, J. Morakis, "Modulation and Coding for Satellite and Space Communications", *Proceedings of the IEEE*, 78(7), July 1990; 1250–1266.

6

Network Procedures

6.1 Introduction

Satellite personal communication networks (S-PCN) have two main objectives [ANA-94]:

- To extend existing and possibly future services provided by public networks to mobile users.
- To complement terrestrial mobile networks by providing analogous services in areas where satellite technology is more effective and economic.

With the rapid development in Internet protocol (IP) technologies in recent years, the first objective requires satellite systems to interwork with both public circuit-switched networks, such as the PSTN and ISDN, and packet-switched networks such as the Internet, and to incorporate intelligent network capabilities to support mobility.

The second goal sets the objective of providing a truly universal personal communication system to mobile users. This can be achieved by the provision of dual-mode user equipment which communicates with both the satellite and terrestrial mobile networks so that when users roam outside of the terrestrial coverage, their requested services can still be supported via the satellite segment.

The concept of mobility support is central to the development of personal communication systems. Mobility encompasses personal mobility and terminal mobility. Personal mobility is supported by associating a logical address with a user, instead of with a physical terminal [ANA-94]. In this way, a user can access services through different networks irrespective of their geographical location and terminal type (fixed, portable or mobile), i.e. users can use their services independent of the access network technology. Terminal mobility refers to the mobile terminal's capability of accessing the network from different geographical locations or physical access points within the service coverage, i.e. a logical address is associated with a mobile terminal. To achieve a truly universal personal communication system, both personal mobility and terminal mobility, with wireless access, have to be supported.

For S-PCN, the main network functions that need to be implemented are envisaged to be the same as those in terrestrial mobile networks, although the procedures may be slightly different in order to take into account the satellite system characteristics. Such functions include call handling, mobility management, resource management, security and network management. Second-generation terrestrial mobile networks have now reached maturity, the development of current and future generation mobile systems, including both satellite and terrestrial need to be compatible with such networks. In Europe, it is envisaged that the underlying network of future generation mobile systems will be based on that of the GSM/GPRS, and therefore the network of satellite systems need to be designed on a par with the GSM/GPRS network protocols so that interworking between the two segments (terrestrial and satellite) can be more efficient and less complicated. Bearing this in mind, network protocols and functions for call handling, mobility management and resource management in this chapter are studied in parallel with the GSM network. The extension and modifications of such network functions required for a GSM compatible S-PCN are discussed.

6.2 Signalling Protocols

6.2.1 Overview of GSM Signalling Protocol Architecture

GSM communication protocols are structured on the ISO/OSI reference model (ITU-T Recommendations X.200–X.219) with other protocol functions specific to cellular radio networks being developed. The layer 1 to layer 3 GSM signalling protocol architecture and its distribution among the network nodes is shown in Figure 6.1. Interfaces Um and Abis are GSM specific while interfaces A, B, C, and E are based on common channel signalling systems No. 7 (SSN7) reported in ITU-T Series Q.700–795. The Um-interface is defined between an MS and a BTS, whereas the Abis-interface is located between a BTS and a BSC. Signalling exchange between a BSC and an MSC is through the A-interface. Communication between an MSC and the VLR, HLR and other MSCs are via the B-, C- and E-interfaces, respectively.

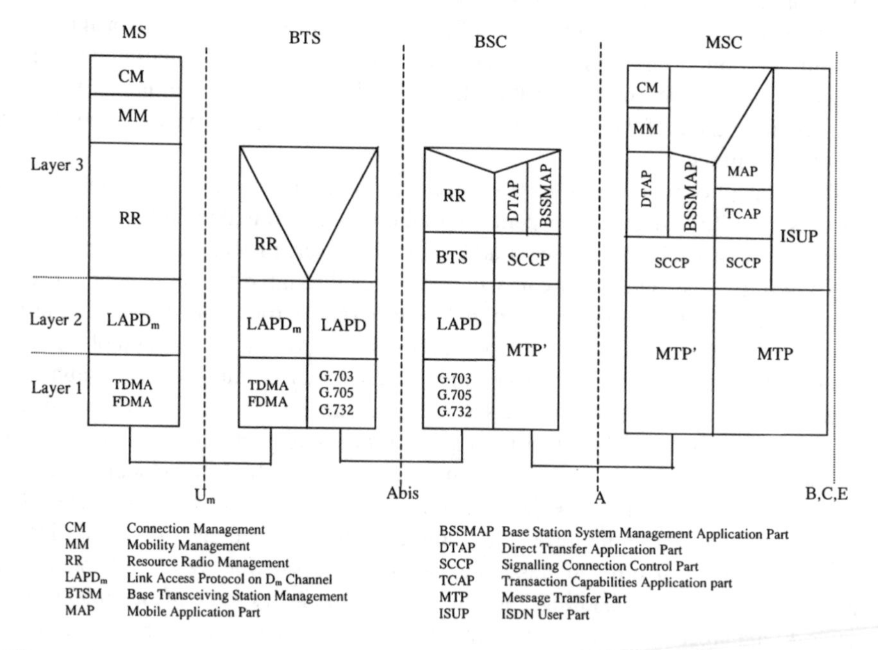

Figure 6.1 GSM signalling protocols and distribution among network elements.

In layer 2 of the MS and BTS protocol stacks, a modified link access protocol on the D channel ($LAPD_m$) is used. $LAPD_m$ is a modified version of the ISDN LAPD protocol specifically for use in mobile applications. It protects the data transfer between the MS and BTS over the radio interface.

A mobile application part (MAP) is specifically developed in order to accommodate radio signalling in GSM networks. It is implemented in all the switching centres directly linked to the mobile network. MAP groups a number of protocols which are able to support mobility control functions and is specified in GSM Recommendation 09.02 [ETS-94]. It consists of several application service elements (ASE) necessary for registration transaction and database inquiry and for the determination of a mobile station's current location.

Of particular interest to this chapter is the functions defined in the *radio resource* (*RR*) *management*, *mobility management* (*MM*) and the *connection management* (*CM*) layers. These three layers are sub-layers to the network layers or layer 3 of the ISO/OSI reference model.

The RR layer handles the administration of frequencies and channels. It is responsible for the set-up, maintenance and termination of dedicated RR connections, which are used for point-to-point communication between the MS and the network. It also includes cell selection when the mobile station is in idle mode (the term idle mode refers to the state when the mobile is switched-on but is not in the process of a call) and in handover procedures. It also performs monitoring on the broadcast control channel (BCCH) and common control channel (CCCH) on the downlink when there is no active RR connection.

The MM layer is responsible for all functions that support mobility of the mobile terminal. It includes registration, location update, authentication and allocation of new temporary mobile subscriber identity (TMSI).

The CM layer is responsible for the set-up, maintenance and termination of circuit-switched calls. It provides the transport layer with a point-to-point connection between two physical subsystems.

6.2.2 S-PCN Interfaces and Signalling Protocol Architecture

In analogy, the GMR specifications (ETSI GEO-Mobile Radio Interface Specifications), which describe the requirements and network architecture for a geostationary satellite system to interwork with GSM, use the GSM protocols as the baseline protocols for carrying out the satellite network control functions. ETSI has produced a series of recommendations for two GMR configurations, known as GMR-1 and GMR-2, respectively. Figure 6.2 shows the network interfaces defined in the GMR network architecture [ETS-99a, ETS-99b]. A brief description on the interfaces follows:

- *S-Um-interface* – similar to the Um-interface as defined in GSM protocols, this is used for signalling between a Gateway Transceiver System (GTS) and an MS.
- *A-interface* – this is the interface between the Gateway Subsystem (GWS) and the Gateway Mobile Switching Centre (GMSC), although strictly speaking, it lies between the Gateway Station Control (GSC) and the GMSC. This interface is used to carry information on GWS management, call handling and mobility management.
- *Abis-interface* – this is an internal GWS interface linking the GTS part to the GSC part.

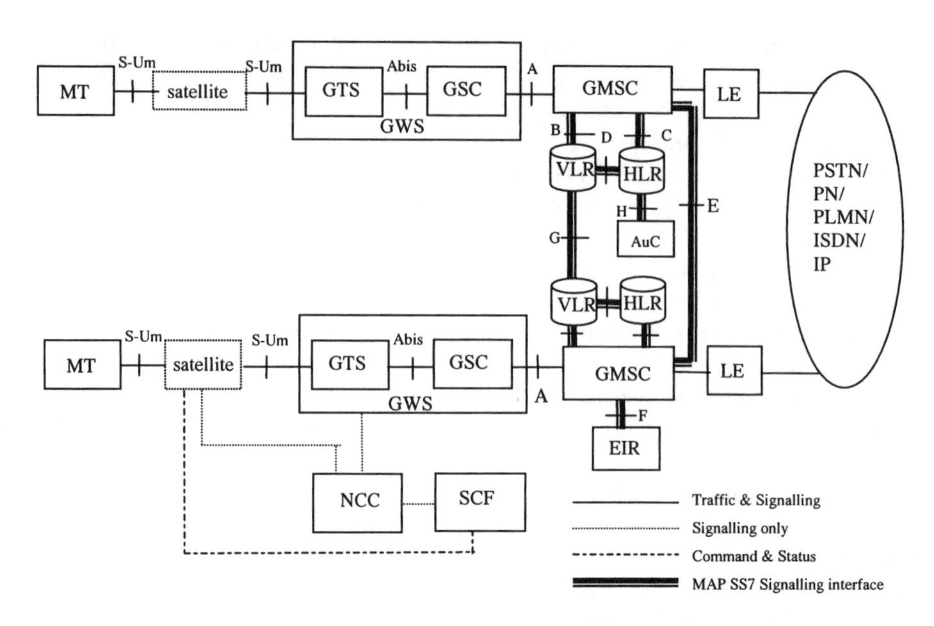

Figure 6.2 Functional interfaces of a GMR system.

This interface is used to support the services offered to the GMR users and subscribers. It also allows the control of radio equipment and radio frequency allocation in the GTS.

- *B-interface* – this interface uses the MAP/B protocol allowing the GMSC to retrieve or update local data stored in the VLR. When an MS initiates a location updating procedure with an GMSC, the GMSC informs its VLR which stores the relevant information. This procedure occurs whenever a location update is required.
- *C-interface* – this interface uses the MAP/C protocol allowing the GMSC to interrogate the appropriate HLR in order to obtain MS location information. Additionally, the GMSC may optionally forward billing information to the HLR after call clearing.
- *D-interface* – this interface uses the MAP/D protocol to support the exchange of data between an HLR and VLR of the same GMSC. It also supports the MAP/I protocol for the management of supplementary services.
- *E-interface* – this interface uses the MAP/E protocol to support the exchange of messages between the relay GMSC and the anchor GMSC during an inter-GMSC handover.
- *F-interface* – this is the interface between the GMSC and the AuC/EIR. It uses the MAP/F protocol for user authentication.
- *G-interface* – this interface uses the MAP/G protocol between VLRs of different GMSCs in order to transfer subscriber data.
- *H-interface* – this is the interface between the HLR and the AuC. When an HLR receives a request for authentication and ciphering data for a mobile subscriber and if the data requested is not held at the HLR, it will send a request to the AuC to obtain the data.

The GMR signalling protocol architecture is also similar to that of the GSM. Although most of the functionalities in each layer of the signalling protocol stack remain the same, some modifications have been made in order to accommodate satellite specific functional-

Table 6.1 Additional logical channels in the physical layer

Group	Channel	Descriptions
DCCH	Terminal-to-terminal associated control channel (TACCH)	Dedicated for terminal-to-terminal call set-up. It may be shared among several such calls
BCCH	GPS broadcast control channel (GBCCH)	Broadcast of GPS time and satellite ephemersis information in the forward link
	High-margin synchronisation channel (S-HMSCH)	Provides time and frequency synchronisation and spot-beam identification in the forward link
	High-margin broadcast control channel (S-HMBCCH)	Contains information for an MS to register in the system. This channel provides an alternative to the MS when the S-BCCH cannot be detected
CCCH	High-power alerting channel (S-HPACH)	A special paging channel to provide high penetration alerting when an MS cannot be reached through normal paging
	BACH	Reserved for alerting messages

ities. In particular, additional functions need to be included in the physical layer (layer 1), the data link layer (layer 2) and the network layer (layer 3) of the MS and the GTS to take into account the satellite channel characteristics and the satellite network architecture. Specifically, the location information of both satellites and MSs need to be measured and reported. Additional logical channels have been specified to take into account such characteristics, as described in Table 6.1.

6.3 Mobility Management

6.3.1 Satellite Cells and Satellite Location Areas

Mobility management consists of two components: location management and handover management. Mobility management strategies have been extensively researched in the past in both land mobile and mobile-satellite networks, some of which are studied in conjunction with resource management and call control strategies [BAD-92, DEL-97, EFT-98a, EFT-98b, HON-86, HU-95a, HU-95b, JAI-95, MAR-97, SHI-95, WAN-93, WER-95, WER-97]. A fundamental part in deriving mobility management strategies is the definition of a satellite cell and a satellite location area. In cellular land mobile networks, such as the GSM, a cell is defined as the locus where a broadcast channel transmitted by a BTS is received with signal quality at or above a pre-defined threshold level [DEL-97]. With this definition, a cell is suitable for a given MT if the relevant broadcast signal received by the MT is at or above this pre-defined threshold quality. In essence, in land mobile networks, a cell is characterised by the presence of a broadcast channel, the basic functions of which are as follows [DEL-97].

- Broadcast of data concerning the cell organisation, such as the cell identifier, the location area identifier for a given cell, the service type in a given cell, the MT identifier and resource organisation.

- Broadcast of paging messages, for example in mobile terminated call set-ups.
- Provision of a reference signal for the MT to carry out measurements in cell re-selection and handover procedures.

More specifically, a one-to-one correspondence among cells, broadcast channels and BTSs exists. In other words, in each cell there exists a single BTS which transmits on a single broadcast channel.

However, in a multi-FES, multi-spot-beam satellite system, the definition of a satellite cell is more complicated. There is no one-to-one correspondence among satellite cells, satellite spot-beams and FESs [DEL-96]. More than one FES can access a given satellite spot-beam. In [DEL-97], a satellite cell is identified in association with both FES i and spot-beam j at a given time t, whereby FES i transmits on a broadcast channel toward spot-beam j at time t. In a full FES-to-spot connection, each FES can be connected to any spot-beam. In this case, a satellite spot-beam can be considered as the overlap of N satellite cells, N being the total number of FES. Each of these N satellite cells has the same coverage, the spot-beam coverage, served by a different FES [DEL-94a].

With the above satellite cell definition and denoting $A_{\mathrm{FESpot}}(i, t)$, $A_{\mathrm{FESact}}(i, t)$, $A_{\mathrm{FEStar}}(i, t)$ as the potential coverage area, the actual coverage area and the target coverage area of FES i at time t, respectively, it follows that [DEL-97]:

$$A_{\mathrm{FESpot}}(i,t) = \bigcup_{j=1}^{j=N} A_{\mathrm{sat}}(j,t)C(i,j,t) \tag{6.1}$$

$$A_{\mathrm{FESact}}(i,t) = \bigcup_{j=1}^{j=N} A_{\mathrm{sat}}(j,t)B(i,j,t) \tag{6.2}$$

$$A_{\mathrm{FEStar}}(i,t) \subseteq A_{\mathrm{sat}}(j,t)B(i,j,t) \tag{6.3}$$

$$A_{\mathrm{FESpot}}(i,t) \supseteq \bigcup_{j=1}^{j=N} A_{\mathrm{FEStar}}(j,t) \tag{6.4}$$

where $A_{\mathrm{sat}}(j, t)$ is the area covered by spot-beam j at time t; $C(i, j, t)$ is a binary function equal to 0 when FES i cannot be connected to spot-beam j at time t and equal to 1 otherwise; and $B(i, j, t)$ a binary function equal to 1 when FES i is actually connected to spot-beam j at time t and equal to 0 otherwise.

6.3.2 Location Management

6.3.2.1 Operations

Location management is concerned with network functions that allow mobile stations to roam freely within the network coverage area. It is a two-stage process that allows the network to locate the current point of attachment (in public land mobile network (PLMN) terms, the point of attachment refers to the base station) of the mobile for call delivery [AKY-98]. The first stage is location registration or location update, while the second stage is the call delivery as shown in Figure 6.3 [AKY-98].

In the location registration stage, the mobile station periodically notifies the network of its new point of attachment, allowing the network to authenticate and to update the user's

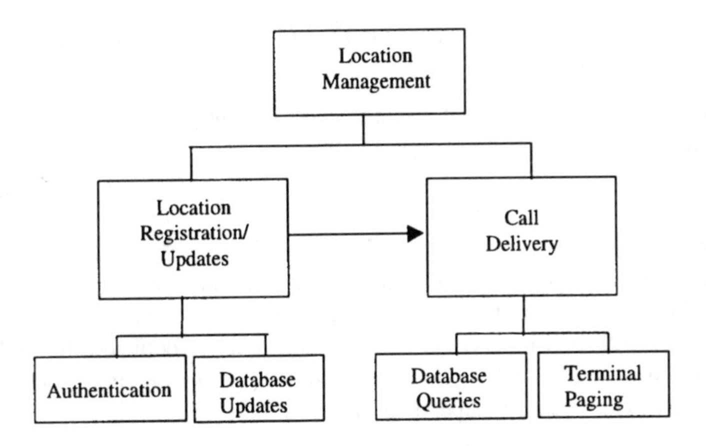

Figure 6.3 Location management operations.

location profile. In the call delivery stage, the network queries the user's location profile and locates the current position of the mobile terminal by sending polling signals to all candidate access ports through which an MS can be reached.

An important issue for designing a location management strategy is the amount of signalling load and delay involved in both location registration/update and call delivery processes. The definition of satellite location areas, in conjunction with the definition of a satellite cell, has a direct impact on the signalling load.

6.3.2.2 Location Update and Terminal Paging Strategies

Location registration is a location monitoring procedure which is invoked automatically by the mobile terminal. This procedure ensures that the network can identify the location of a mobile terminal as accurately as possible. Traditionally, location areas are used for identifying the MT's position. A location area (LA) is the smallest area unit that can be used to locate the MT when it is in idle mode.

In cellular land mobile networks, an LA is usually defined as the area wherein an MT can roam without having to perform a location update. Such an area is normally bounded by a cluster of cells. A possible parameter used to trigger a location update could be the signal strength quality received by the MT from the base station. The current location of the MT is constantly compared with the location information broadcast by the surrounding base stations. The mobile terminal continuously monitors the signal quality of the broadcast control channel in order to decide whether it is out of reach of the current LA. When the MT decides to access a new location area, a location update is required.

Similar location update procedures need to be implemented in mobile-satellite networks. For non-geostationary satellite systems, the satellite network has to cope with both the satellites' motion and that of the terminal. This has an impact on the location registration/update and handover (which will be discussed in a later section) procedures. If inter-satellite links (ISL) are deployed, it adds further dimensions to the definition of location areas and mobility management in satellite networks. In the sections that follow, it is assumed that there is no ISL. There are two basic approaches [CEC-96] for defining the location of the MT in relation to a location area:

- The guaranteed coverage area (GCA) approach [CEC-96].
- The terminal position (TP) approach [CEC-96].

GCA Approach

In the GCA based approach, an LA is the static area surrounding an FES where for a guaranteed 100% of the time, the FES can reach an MT with an elevation angle at both sides of the link (FES-satellite and MT-satellite) above a pre-defined threshold value. This approach uses the satellite cell definition discussed in the previous section. With this approach, $A_{\mathrm{FEStar}}(i,\ t)$ becomes independent of t such that:

$$A_{\mathrm{FEStar}}(i) = A_{\mathrm{FEStar}}(i,t)\ \forall t \tag{6.5}$$

and that $A_{\mathrm{FEStar}}(i)$ contains all the points on the Earth's surface such that

$$A_{\mathrm{FEStar}}(i) = \{P|\forall t\ P \in A_{\mathrm{FESpot}}(i,t)\} \tag{6.6}$$

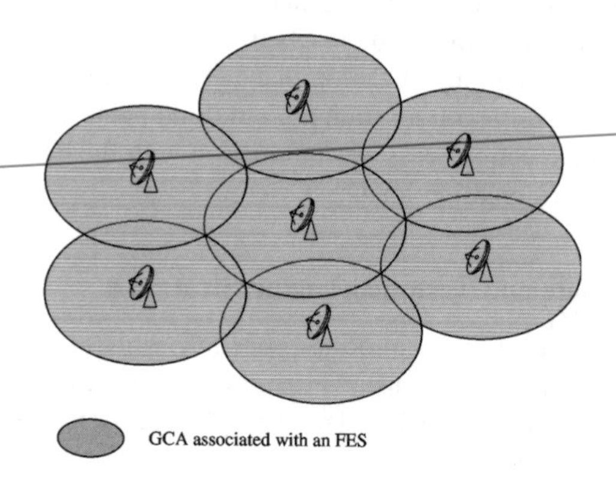

Figure 6.4 Location area under guaranteed coverage area based approach.

With this approach, the location information to be stored in the network database is that of the FES, with which the MT is registered. Since this LA definition guarantees that the FES will be able to contact the MT at all times, the paging of the MT through the registered FES will be guaranteed to be successful. In this way, an MT terminated call can be directly routed to the registered FES. This approach is quite similar to the approach adopted in terrestrial mobile networks. The GCA based approach is shown in Figure 6.4.

The GCA approach requires a continuous service area coverage, resulting in a large number of FESs. In order to reduce the number of FESs, and consequently the number of GCAs, a possible scenario is to provide GCA in regions where the traffic density is high and to provide intermediate coverage area (ICA) in between GCAs [CEC-96]. This scenario is termed the partial GCA, as shown in Figure 6.5. With partial-GCA, the system has to implement extra functionalities in order to page the MT. The traffic signalling associated with tracking MTs outside a GCA is greater than that inside a GCA. However, the traffic densities on the ICAs are much lower than those in the GCAs, thus the extra signalling should

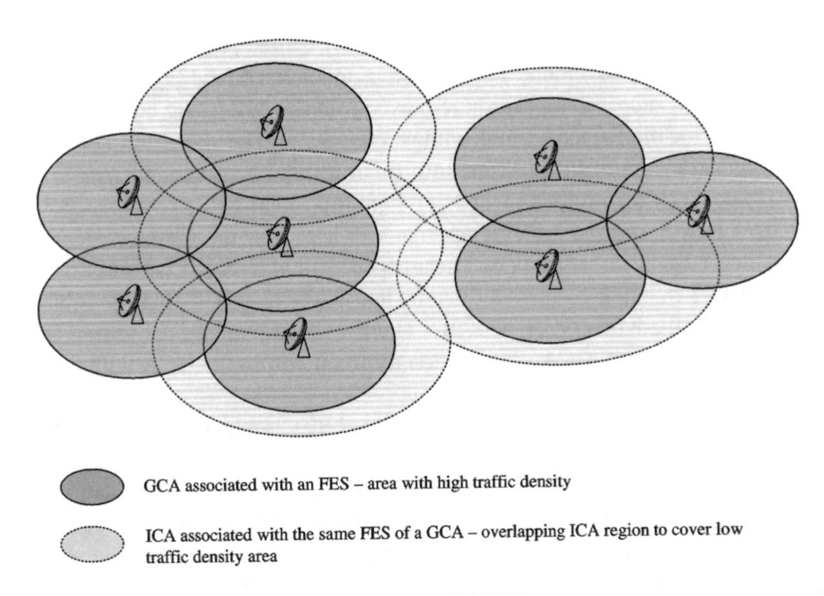

Figure 6.5 Partial GCA.

not have a great impact on the system performance. The GCA approach can be terminal position based or BCCH based.

GCA-Terminal Position (GCA-TP) Based Approach In this approach, a location update is triggered by the position of the MT, i.e. when the MT roams outside of a specific GCA, a location update is performed. The location information may be computed by the MT or by the FES, depending on the intelligence of the MT and the distribution of functionality among the network elements. If the FES is in charge of the location computation of the MT, any location update decision will be made by the FES.

The MT requests its position from the network at regular intervals. A suitable FES will be selected among a group of FESs on the basis of the signal strength of their broadcast signal received by the MT. This group of FESs includes the serving FES and all the other adjacent FESs. When the serving FES detects that an MT has reached the border of its controlling area, it will inform the MT of its new FES identification, which will take over responsibility for communicating with the MT. The MT then contacts the new FES, which updates its databases accordingly. Once this location update procedure is completed, the MT will communicate with the new FES to enable subsequent position identification and paging. The connection between the MT and the old FES will then be released. However, if the MT is unable to contact the new FES due to network congestion, the databases of the new FES cannot be updated. In this case, the connection between the MT and old FES will be maintained until the MT can contact the new FES.

Since the FES measures the MT position on a regular basis, this information is preferably stored in the FES and can be re-used to optimise the paging procedure. The FES identifier must be stored in the network database. Since the FES has the knowledge of the satellite ephemerides, it will be able to map the position of the MT onto the <satellite, spot-beam> co-ordinates pair. This mapping can be performed by the FES during the paging process.

The main disadvantage of this approach is that it requires periodic position calculations independent of the MT's movement. This will have an impact on the signalling load. The time interval between two successive position requests from the MT has to be traded-off between the frequency of location updates and paging efficiency. The more frequent the position request, the more accurate the position information, which in turn increases the paging efficiency. However, it also implies a heavier signalling load.

GCA-BCCH Based Approach This approach is based on the monitoring of the BCCH channel broadcast by each spot-beam. The FES identifier is broadcast only over its GCA through the BCCH channels of the spot-beams that cover the GCA. It is possible that more than one satellite will cover a GCA. The BCCH channels containing the same FES identifier are bounded within the FES GCA. As result, a terminal within the GCA will always receive the same FES identifier.

When an MT is switched on, it selects the best BCCH channel and decodes the FES identifier associated with that channel. Registration with the FES then occurs. Location update occurs when the MT receives a BCCH with a new identifier with better quality. In this case, the MT triggers the location update procedure. This should only happen when the MT is approaching the border of the GCA. Normally, there are overlapping areas between two adjacent GCAs. When the MT is within these overlapping areas, the MT detects more than one FES identifier. When the BCCH with a new FES identifier is received with better quality than the one with the old FES identifier, the MT triggers a location update request with the new FES. The new FES then updates the network databases accordingly. This involves the updating of the HLR associated with the MT and the deletion of the registration information in the old FES. Once this is completed, the new FES acknowledges the MT and the MT will start to listen to the BCCH of the new FES.

Spot-beams covering more than one GCA are required to broadcast all the corresponding FES identifiers. In this situation, more than one FES controls the spot-beams.

In order to optimise the paging procedure, two techniques can be used to restrict the area where the MT is located.

1. Terminal position: in this technique, the MT's position will be measured on a regular basis either by the FES or by the MT which then reports to the FES of its most up-to-date position. The area surrounding the latest measured MT position will be paged first. If the MT does not respond, a wider area is paged.
2. Spot-beam area: the FES stores the spot-beam footprint area through which the MT made its last contact. Spot-beams that overlap that particular area are paged first.

The main disadvantage of the GCA-BCCH approach is that the borders of a GCA are difficult to be precisely defined by an FES. This may lead to frequent location updates between two adjacent GCAs.

Partial GCA-TP Based Approach This should be used in conjunction with the GCA terminal position based approach. When the MT roams inside a GCA, the GCA-TP approach is used. When the MT roams outside of the GCA, the last FES associated with the MT will continue to measure the MT's position. After each measurement, the FES decides and informs the MT of the best FES for it to be associated with. Location updates then occur with the new FES. The FES decision should also take into account the satellite ephemerides.

Partial GCA-BCCH Based Approach In this approach, the FES identifier is broadcast beyond its GCA, but only to those areas that are not covered by any GCAs. When an MT roams outside of a GCA coverage area, it continuously listens to the BCCH of the last FES with which it was associated. If the MT detects a BCCH with better quality, it informs the old FES. Subsequently, location update occurs with the new FES and the databases are updated accordingly. For regions with more than one ICA overlap, numerous location updates may occur due to the ICA dynamic coverage characteristics.

Terminal Position Based Approach

In the TP based approach [CEC-96], an LA is defined by the triplet co-ordinates (latitude, longitude, variable radius) of an MT. Both the network and the MT must store this location information, which corresponds to the current location of the MT. However, this information does not indicate the FES through which the MT can be reached. Hence, the location information of the MT has to be associated with a set of FESs so that the MT can be paged. This implies that the MT needs to be paged with more than one FES. With this approach, the routing procedures will be different from that adopted in the terrestrial mobile network. Figure 6.6 shows the TP based approach.

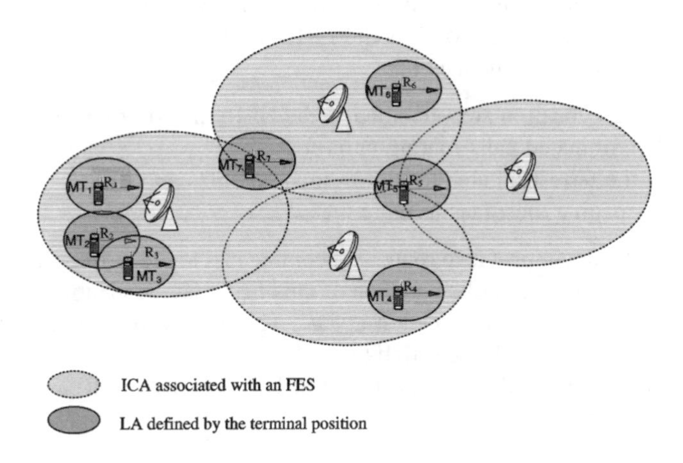

Figure 6.6 Location area for terminal position approach.

In this approach, the MT must be able to determine its position so that it can decide whether a location update is necessary. In order to determine the third co-ordinate, the variable radius, the following approaches have been proposed in Ref. [CEC-96]. The variable radius should associate with a virtual terminal instead of the MT and the MT should register to a virtual terminal. The virtual terminal defines the location area. This can avoid defining a location area for each MT. This would reduce the operation of location updates for each MT and hence reduce signalling traffic. The following strategies have been identified in defining the variable radius in association with the virtual terminal.

(a) All the virtual terminals have a specified fixed, variable radius regardless of the mobility characteristics of the MTs.
(b) A virtual terminal has a specified variable radius according to its mobility class.
(c) The variable radius of a virtual terminal can be updated by the network according to the mobility behaviour of the MT.

Strategy (a) is inflexible since different terminals with different mobility characteristics co-exist. Strategy (b) requires a set of MT mobility classes be made available. Strategy (c) is an extension of strategy (b) and can be efficient. However, it requires extra functionalities in the network.

Assuming that strategy (b) is adopted, during the session set-up procedure, the MT registers with a virtual terminal. The virtual terminal then informs the network of its mobility class which defines the variable radius. The network then associates the variable radius with a temporary mobile terminal identity (TMTI). When the MT roams beyond the location area associated with the variable radius, a location update must be performed regardless of the number of users registered on that virtual terminal in that location area. Location update of the MT is performed by changing the terminal data in the visited area. If a new TMTI is generated, i.e. the mobility class of the virtual terminal in the visited area is different from the previous mobility class, then the user registration data needs also be updated.

In order to deliver a call, the area defined under the variable radius must be paged. All spot-beams that overlap with that area must page the MT. It is possible that more than one satellite and more than one FES are involved during the paging process. An example is shown in Figure 6.6, where in order to page MT2, both FES1 and FES2 have to be used. This approach has a considerable impact on the network.

- Instead of storing a location area identifier in the HLR, the co-ordinate triplets have to be stored for MT terminated calls routing purposes.
- Translation of the terrestrial area to spot-beams, satellites and FES.
- The sending of paging messages to more than one FES.

A way of reducing the impact is to route an incoming call to the most probable FES, where the MT has a high probability of being located. This FES may be identified by the FES that was used to perform the last MT location update. This FES will then page the MT in the uncertainty area where the MT is located. If this is unsuccessful, this FES will have to re-route the paging message to a second most probable FES. This procedure will carry on until the MT responds to the paging message.

Paging Strategies Associated with the TP Based LA Approach In the TP based LA approach, a location update is required when a mobile terminal travels a distance exceeding the variable radius since the last location update with the network. The location area may be covered by a number of spot-beams. In order to reduce paging signalling, the coverage area of a spot-beam with the best visibility from the MT is selected as a paging area. The paging area is evaluated by a satellite paging algorithm using the satellite constellation ephemerides and paging related information stored in a VLR. However, paging signals may be subject to loss due to the channel fading effect as well as the inaccuracy in the predicted paging area algorithm. If an LA is covered by a large number of spot-beams, without an accurate prediction of a paging area, paging signalling will increase. As a result, the paging delay will also increase. [HE-96] identified two paging techniques used in conjunction with the two-step paging strategy in order to reduce the paging delay in a TP based location update approach. The following summarises the standard two-step paging strategy, together with the two supplementary paging techniques.

1. Standard two-step paging strategy: in the first step, the spot-beam whose coverage area is identified as the paging area issues a paging command to the MT. If the paging response is not received within a pre-defined paging interval, the whole LA is paged. Note that this pre-defined interval may involve more than paging commands being sent to the MT.
2. Diversity satellite paging: the propagation of satellite signals relies on line-of-sight radio transmission. If the line-of-sight transmission path is blocked, the paging signal may be lost. The diversity satellite paging is proposed to counteract the channel fading effect. This technique makes use of two spot-beams from two different satellites. The coverage of one of the spot-beams is identified as the paging area and the satellite of the other spot-beam is the next best visible satellite. In the first step paging, both satellites issue paging commands to an MT. If the paging signal from the best visible satellite is lost such that no paging response from the MT is received by this satellite within a pre-defined paging interval, the MT will find the next best visible satellite and synchronise itself with this second satellite. In this way, the MT may be successfully paged in the first paging step.
3. Redundant paging: in the first paging step, the paging signal is duplicated within the paging cycle. The objective is to increase the first step paging success rate in the presence of fading and shadowing.

6.3.2.3 Design of Database Architecture for Location Management

In PLMN networks, database architecture design plays an important role in minimising unnecessary signalling in mobility management. Location registration involves the updating of databases and call delivery requires the querying of the location databases in order to identify the current location of a called MT. Currently, there are two basic approaches to the design of database architecture.

Centralised Database Architecture This approach is well adopted in the SS7-MAP used by GSM900/1800 and IS-41 used by the IS-136/cdmaOne networks. These two location management standards are very similar to each other. They make use of a two-tier database structure to define the functions of the two network location databases, the home location register (HLR) and the visitor location register (VLR), for locating an MT. Each network has its associated HLR, which contains user profile information such as the types of subscribed services, location information, etc. The user is permanently associated with the HLR of their subscribed network. The distribution of VLRs varies among networks. Each VLR stores information on the MTs which visit its associated location area. Incoming calls, (calls that an MT receives), are routed to the MT by the VLR via its HLR. Outgoing calls will only go through the VLR so that access to the HLR is not required. Signalling exchanges for location registration and call delivery will always have to go through the HLR regardless of how far away the HLR position is from the current location of the MT. This may result in an undesirably high connection set-up delay. However, an advantage of this approach is that the number of database updates and queries for location registration and call delivery is relatively small.

Research into optimisation of the centralised database architecture to reduce the signalling delay has been proposed in the past. The following summarises a few techniques, which have been reported in Ref. [AKY-98], for minimising the signalling delay and traffic in PLMN networks. More detailed explanation can be obtained from the referenced papers.

1. *Dynamic hierarchical database architecture* [HO-97]: in this architecture, an additional level of databases called the directory registers (DR) is added onto the IS-41 standard. Each DR covers the service area of a number of MSCs. In addition, each MT will be assigned with a unique location pointer configuration. Three location pointers are available at the DR:

   - A local pointer stored at the MT's serving DR indicating the current serving MSC of the MT.
   - A direct remote pointer stored at a remote DR indicating the current serving MSC of the MT.
   - An indirect remote pointer stored at a remote DR indicating the current serving DR of the MT.

   These pointers are set-up so that database queries by a calling MSC can be done at the DR level and that incoming calls can be forwarded immediately to the current serving MSC of the MT without querying the HLR of the MT. The HLR may also be set-up with a pointer to the serving DR so that only the local pointer of the serving DR needs be updated when the MT moves to another area. The main purpose of this approach is to avoid querying at the HLR, which may have been located too far away from the current position of the MT, and hence reducing the signalling delay and traffic. However, there may be a possibility that it is more costly to set-up these pointers. Under this circumstance, the original centralised database architecture should be used.
2. *Per-user location caching* [JAI-94]: in this approach, a cache of location information of MTs is maintained at a nearby service transfer point (STP). Whenever the MT is accessed through the STP, an entry is added to the cache which contains the mapping of the MT's ID with its serving VLR. When there is an incoming call to the MT, the STP checks the entry of the MT in the cache. If no entry exists for the called MT, the usual signalling procedure will be performed. If the entry of the MT exists, the STP will query the VLR directly. However, there may be a possibility that the MT has already moved to another location area of a different VLR. In this case, the usual signalling procedure will be used. This scheme is efficient only when the STP finds an entry of the MT in the cache. Otherwise, the signalling cost of implementing this scheme will be even higher than that of the conventional schemes. Strategies have been derived in order to increase the entry success, such as imposing a threshold in the time interval of an entry that should be attained/updated in the cache.
3. *User profile replication* [SHI–95]: this approach replicates the user profile at selected local databases. Whenever an incoming call for an MT arrives, the network checks whether the user profile is available locally. If it is available, the location information of the MT can be determined from the local databases, hence, no HLR query is necessary. If not, the network will perform the conventional procedure to identify the MT location. It has to be noted that when an MT moves to another location, the network will have to update the MT's user profile in all of the selected databases. This may result in higher signalling for location registration. If the mobility rate of the MT is low and the call arrival rate is high, then this method may significantly reduce the signalling overhead. This approach may not be feasible if the mobility rate of the MT is high, which leads to frequent location updates of the user profiles among different network providers and to significantly high signalling traffic.

4. *Pointer forwarding* [JAI-95]: pointer forwarding allows the old VLR to communicate with the new VLR by using a forwarding pointer at the old VLR when an MT moves to a new location area of a different VLR from the current VLR. The serving MSC will then follow the pointer chain in order to identify the serving VLR. This eliminates the querying of and reporting to the HLR about the location change of the MT. However, a predefined limit must be set for the length of the pointer chain. Once this predefined limit is reached, further pointer forwarding will be prohibited and a location update must be reported to the HLR when the next move occurs. As with the user profile replication scheme, this approach may not be efficient if the mobility rate of the MT is high.
5. *Local anchoring* [HO-96]: under this scheme, a VLR close to the MT will be selected as its local anchor for reporting any location changes. This will eliminate the reporting to the HLR. Since the local anchor is close to the MT, the signalling cost incurred in location registration can be reduced. When an incoming call arrives, the HLR then queries the local anchor of the MT. The local anchor will then query the serving VLR for the routing address of the MT. There are two strategies in selecting a local anchor: dynamic and static. Under both strategies, the serving VLR of the MT during its last call arrival becomes the local anchor. The local anchor changes when the next call arrival occurs. However, dynamic local anchoring allows the network to make decisions on whether the local anchor should be changed to the serving VLR according to the mobility and call arrival pattern. Static local anchoring completely eliminates the involvement of the HLR for location registration. However, it involves the updating of the local anchor whenever there is a call arrival. This may lead to poorer performance than the conventional centralised database architecture. On the other hand, it has been demonstrated that the signalling cost for dynamic local anchoring is always better or equal to that of the conventional scheme.

Distributed Database Architecture Under this database architecture, the database is distributed physically over different locations but remains logically centralised. An example of this type of database architecture is that based on the standardised register service ITU-T X.500. The database is arranged in a tree-like structure which represents the directory information base. All the objects are stored in the hierarchy directory information tree. The branch of the tree will be filed in a physically separated directory system agent. Individual objects will be accessed by a directory user agent, which searches through the tree in order to retrieve the data. The time taken to search for the data is a critical factor in this method. Different schemes in employing this type of database architecture have been studied in the past and are summarised as follows.

1. *A fully distributed scheme* [WAN-93]: in this scheme, the two level HLR/VLR will no longer be used. Instead, the location databases are organised as a tree with the root of the tree at the top, the leaves at the lowest level and the branch will identify the subtree that the leaves belong to. Each MT is associated with a leaf database, which contains the location information of the MT that is residing in its subtree, as shown in Figure 6.7. For each database along this subtree, an entry exists for the MT. As shown in Figure 6.7, when a call arrival occurs for MT2 initiated by MT1, the network follows the entry of the database. If MT2 belongs to a different location area, the databases in the subtree that MT1 resides will

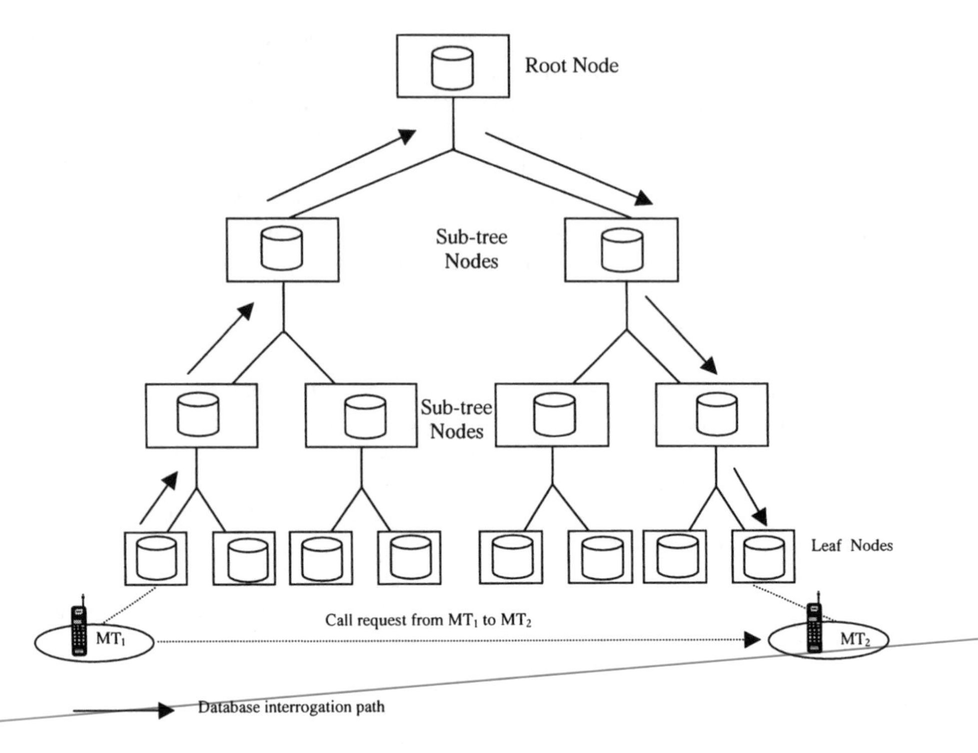

Figure 6.7 A fully distributed database architecture.

not contain any entry of location information of MT2. The network will then have to trace the entry of MT1 from the root database and then follow the path of the subtree that MT2 resides. Location registration occurs when MT1 moves to an LA that belongs to a different leaf database. Although this scheme reduces the distances that the signalling messages have to travel, the number of database updates and queries increases resulting in the delay for location registration and call delivery.

2. *Database partitioning* [BAD-92]: in this scheme, the database is partitioned according to the mobility pattern of the MT. Partition is achieved by grouping location servers within areas that the MT moves most frequently. Location registration is performed only when the MT enters a partition that belongs to different location servers (LS), as shown in Figure 6.8. Depending on the mobility rate and the call arrival pattern, this scheme may minimise location registration in areas where the mobility rates of the MT are high, hence the signalling cost can also be reduced.
3. *Database hierarchy* [ANA-95]: this is similar to the fully distributed scheme apart from the fact that not every node in the subtree contains a location database. Furthermore, the MT can be associated with any node in the tree (in comparison with only the leaf in the fully distributed scheme). For a node in the subtree where the MT resides, a pointer will be set-up pointing to the next database along the path of this subtree. Thus, the location of an MT can then be located by following the pointers, as shown in Figure 6.9. The placement of databases has to be determined in order to minimise database access and updates [AKY-98].

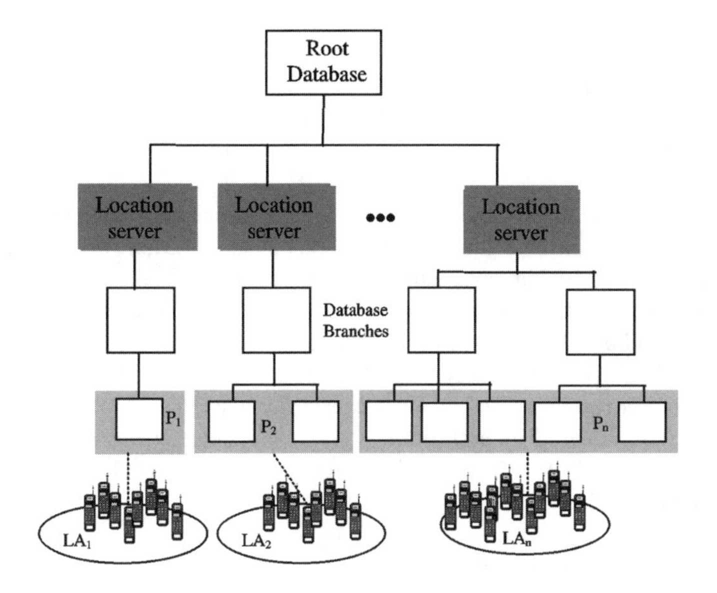

Figure 6.8 Database partitioning.

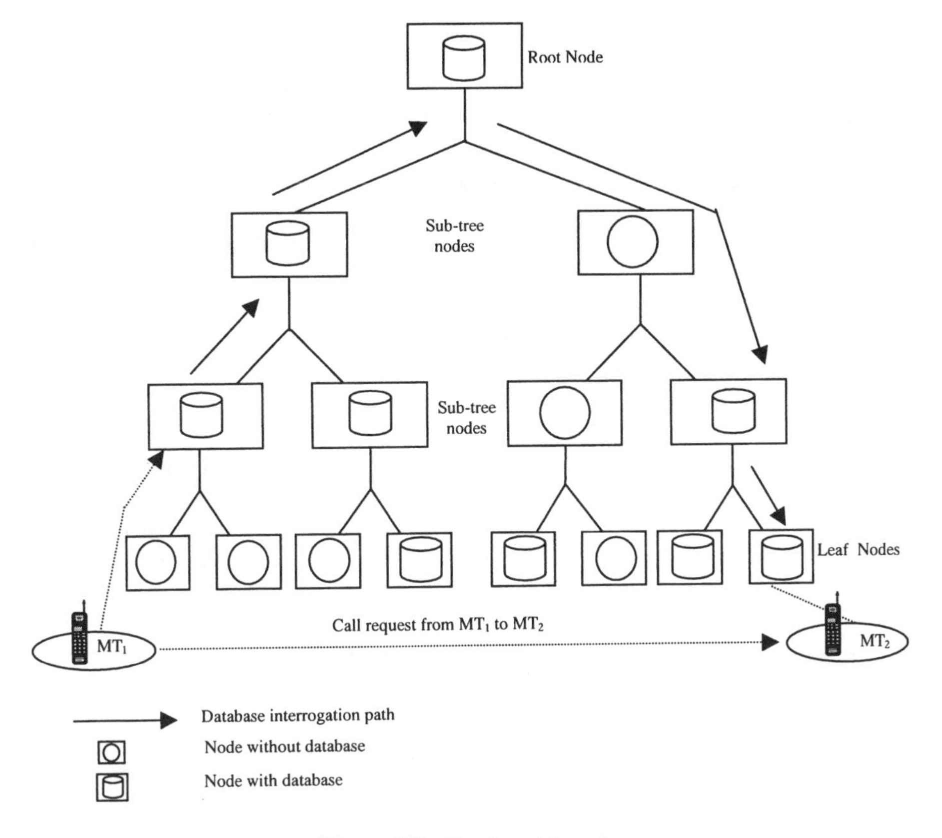

Figure 6.9 Database hierarchy.

6.3.3 Handover Management

6.3.3.1 Phases of Handover

Handover management involves network functions that allow mobile stations to change their current access point or base station during a connection. It ensures the continuity of an on-going connection.

Handover management ensures that an active call connection is maintained when the mobile terminal moves and changes its point of attachment to the network. Three main phases are involved in handover: handover initiation, handover decision and handover execution [EFT-98b]. The main task involved in the handover initiation phase is gathering of information such as the radio link measurements. If the radio link quality falls below a predefined threshold, a handover will be initiated. Based on measurements, the handover decision phase will select the target resources. In handover execution, new connections are established and old connections are released by performing signalling exchanges between the mobile terminal and the network. Table 6.2 shows the different phases in handover and the strategies associated with each phase.

Table 6.2 Handover phases and strategies

Handover (HO) management			
Functional phases	Processes	Main tasks	Handover strategies
Handover initiation	Monitoring, data collection and processing	QoS report generation, e.g. radio link measurements	HO controlling schemes • Network controlled HO (NCHO) • Network assisted HO (NAHO) • Mobile controlled HO (MCHO) • Mobile assisted HO (MAHO)
Handover decision	Decision making	Checking of new connection availability and selection of target cell/spot-beam	
Handover execution	Path creation and switching, handover completion and route optimisation	Establishment of new signalling and traffic connections	Connection establishment Schemes (signalling channel connection) • Backward handover • Forward handover Connection transference Schemes (traffic channel connection) • Soft handover • Hard handover

6.3.3.2 Handover Initiation

A handover can be initiated due to poor radio link performance or other QoS degradation. In addition, a network can also initiate a handover for operations and maintenance purposes. A user can also initiate handover which arises from the user's performance requirements. In general, the following three types of handover initiation can be distinguished:

- *QoS parameters initiated handover*: the most common form of QoS initiated parameters is the radio link quality – the signal strength and the carrier-to-interference ratio. Other forms of QoS parameters are also possible for use in handover initiation such as the delay and the BER. These parameters are monitored continuously, either by the terminal or the network or both, and are compared with predefined threshold values in order to determine whether a handover should be initiated.
- *Network parameters initiated handover*: this type of handover is initiated due to network management criteria such as system resource utilisation or maintenance issues. It is not directly related to QoS parameters.
- *User profile initiated handover*: this type of handover is initiated due mainly to user service profile and tariff structure. It is not directly related to QoS parameters. However, this type of handover initiation is more applicable in an integrated network environment, i.e. integrated satellite and terrestrial network, such that a user may choose to switch to another network for cheaper call charges, for example.

Stand-alone Satellite Network Scenario For a stand-alone satellite network scenario, handover occurs in non-geostationary satellite constellations, due mainly to the motion of the satellite. Neither network parameters nor the user profile are the cause for this type of handover. Due to the dynamic features of the non-geostationary satellite constellation, there are two main handover categories in this scenario [CEC-97].

Intra-FES Handover

This type of handover occurs due to the change of spot-beams caused by the motion of the satellite. This is particularly common in a satellite-fixed cell system. The satellite motion results in a change or degradation in the radio link quality, which is then used to determine whether a handover should be initiated. This type of handover is deterministic and periodic in nature according to the visibility period, and prediction can be made to assist in the measurement in the link quality. This type of handover is further divided into inter-beam handover and inter-satellite handover.

Inter-beam handover refers to the transfer of a call from one spot-beam to another of the same satellite. Figure 6.10 shows the inter-beam handover scenario. Such handover is due mainly to the satellite motion. During the handover process, the serving satellite and FES remain unchanged, hence re-synchronisation is not required. Since both the old and new link follow more or less the same path, it implies that their shadowing characteristics will be highly correlated. A pure signal quality based handover criterion may not be sufficient in this case. Since this type of handover is predictable, the positional information of the MT in relation to the satellite location can be used as the handover criterion. Due to the relatively short overlapping region between adjacent spot-beams, the whole handover process has to be fast. The predictable nature of this type of handover offers radio resource management flexibility and allows a fast resource re-establishment during the handover execution process.

Inter-satellite handover refers to the transfer of a call from one satellite to another, as shown in Figure 6.11. This type of handover is due to the low elevation angle as a result of the satellite motion. As the elevation angle becomes lower, the propagation path loss and the depth of shadowing increase, resulting in a decrease in the received power. In contrast with the inter-beam handover, the old and new links are with different satellites and follow different paths. A delay difference occurs between the two paths. In this case, re-synchronisation is

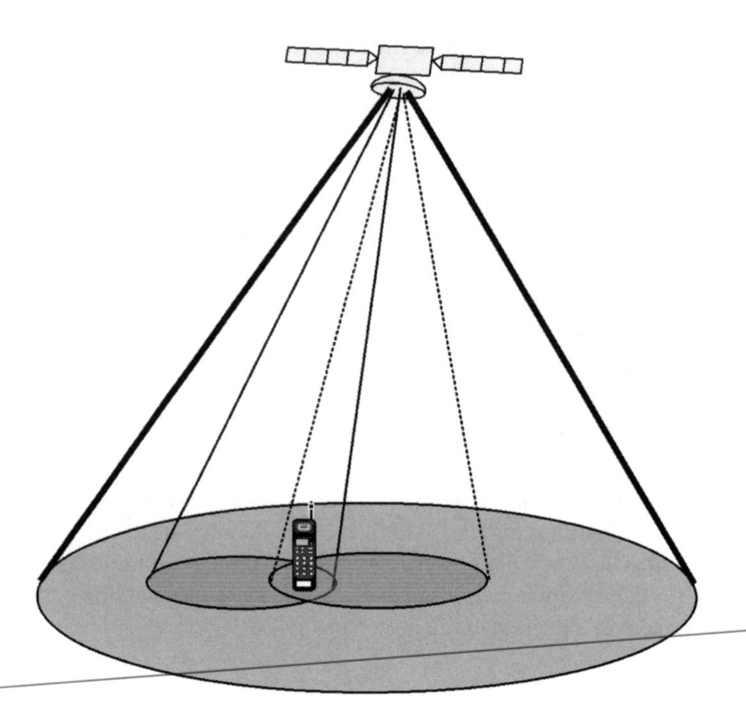

Figure 6.10 Intra-satellite inter-spot-beam handover.

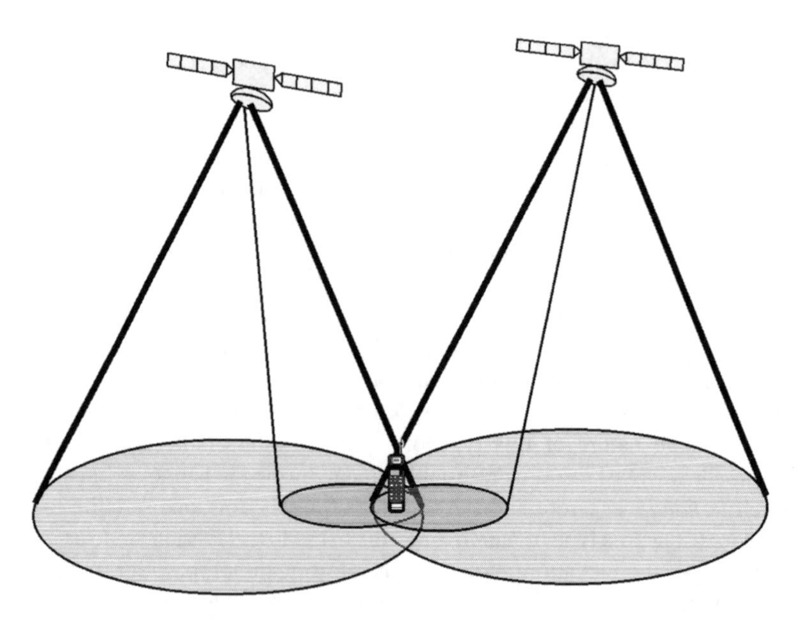

Figure 6.11 Inter-satellite handover.

required. The path difference also implies that shadowing from these two paths is uncorrelated. Handover can be initiated using radio link related parameters. Since the overlapping region between two satellite footprints is relatively large in comparison with that between two spot-beams, the allowable handover time can be longer.

Inter-FES Handover

Inter-FES handover refers to the change from one FES to another during a call. Only the feeder link is involved. Thus, the handover is transparent to the mobile terminal. Furthermore, in order to avoid frequent changes in the signalling and traffic link, it is usual that the call would still carry on with the original FES. This FES is called the *anchor FES* during the handover process. Figure 6.12 shows the mechanism of the inter-FES handover. Although the original FES may be at a distance far away from the new FES, which implies an increased cost for calls suffering from this type of handover, inter-FES handover is rare – only mobile terminals associated with a high degree of mobility, such as in high speed trains or aeroplanes, may experience this type of handover. Unlike intra-FES handover, which affects mainly on-going calls, inter-FES has a great impact on the network as it implies a transfer of routing control from one FES to another. This handover is equivalent to an inter-BTS or inter-MSC handover in terrestrial mobile networks. In designing a satellite network, the need for inter-FES handover should be kept to a minimum.

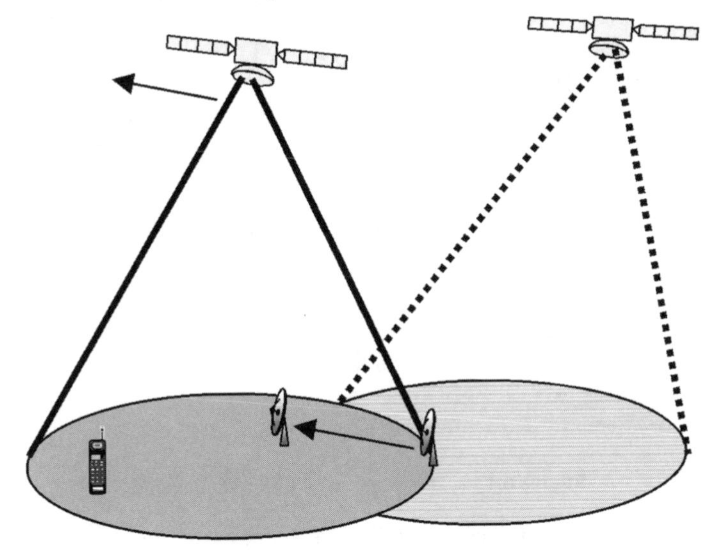

Figure 6.12 Inter-FES handover.

Integrated Satellite-Terrestrial Network Scenario In an integrated network, an on-going call may be handed over from a satellite network (serving network) to a terrestrial network or *vice versa* due to one of the reasons identified during the handover initiation phase. Handovers involved in two different but integrated networks are normally referred to as inter-segment handover. Since the propagation delays of the two networks are different, the synchronisation procedure during handover requires careful consideration. Furthermore, different measures may be used for link qualities and resource control procedures. Two main types of inter-segment handover are involved in an integrated satellite-terrestrial network: satellite to terrestrial handover and terrestrial to satellite handover.

Terrestrial to Satellite Handover This type of handover is triggered due to one of the following reasons [CEC-97].

1. A mobile terminal, which has an on-going call served by a terrestrial network, is at the boundary of a terrestrial cell and is moving outside of the terrestrial coverage but within a satellite spot-beam coverage. A handover from the terrestrial network to the satellite network under this situation can be regarded as an extension of an inter-cell handover to allow the on-going call to be continued while the mobile continues to roam between the two networks. This type of handover is normally triggered by radio link parameters. Inter-working between the terminal and the network is required.
2. A mobile terminal, which has an on-going call served by a terrestrial network, is moving towards a neighbouring cell in which no channels are available due to traffic overflow but is within a satellite-spot-beam coverage. This type of handover is triggered by the network.

Figure 6.13 shows the procedure of terrestrial to satellite handover.

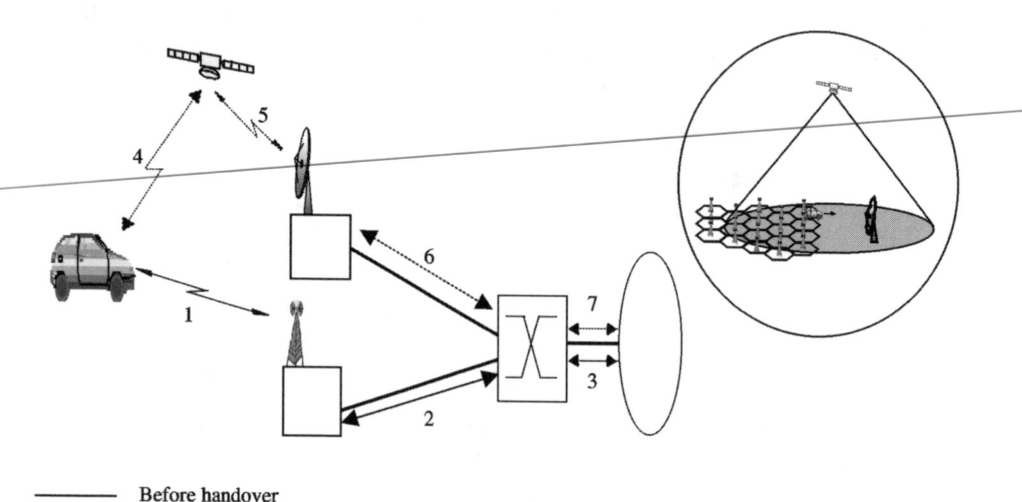

Figure 6.13 Terrestrial to satellite handover.

Satellite to Terrestrial Handover Satellite resources are more expensive than the terrestrial network and are scarce. Hence this type of handover can sometimes be viewed as a network management process in order to release the satellite resources and in some cases to improve the service quality. The following briefly describes the situation in which this type of handover will occur.

- In an integrated satellite-terrestrial network environment, the user may prefer to use the terrestrial segment as the default segment on economic grounds. Then handover may occur when a terrestrial channel becomes available in an area where a mobile terminal is communicating through a satellite link. This can be seen as a user profile initiated handover.

- When a mobile is communicating through the satellite link and enters into an area where the service quality is poor, for example in urban areas. This is a radio link initiated handover.
- When a mobile is originally communicating through a terrestrial link, but the call is handed over to the satellite link due to being out of terrestrial coverage or due to traffic overflow in the neighbouring terrestrial cell to which the mobile is approaching. In the former situation, once the mobile terminal re-enters a terrestrial coverage, the call should be handed over to the terrestrial network. In the latter situation, once the mobile leaves the overloaded terrestrial cell, the link should be switched back to the terrestrial network. Figure 6.14 shows the routing of calls before and after handover.

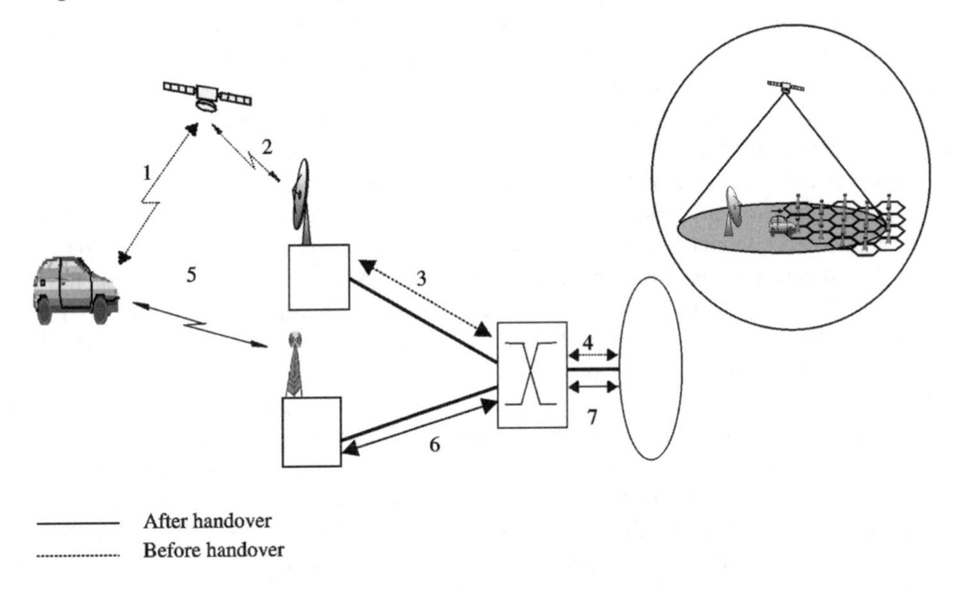

Figure 6.14 Satellite to terrestrial handover.

In order to make a proper decision on whether a handover should be initiated, the following information will be useful.

- MT position: this information will enable the satellite network to know whether the MT is in the proximity of the terrestrial coverage and then to decide on whether measurements on the terrestrial link quality should be made. The information can also be used to initiate a handover. This will require mobile terminals to be incorporated with a GPS or an equivalent type receiver (see Chapter 9).
- Terrestrial link signal quality: the signal quality of the terrestrial link has to be monitored when the MT is in the proximity of the terrestrial coverage and when channels are available in the terrestrial network.
- Availability of resources in the terrestrial link: the availability of resources in the terrestrial network in the area of interest (i.e. the handover zone) should be made known to the satellite network before the handover. This information can either be supplied by the terrestrial network or reported by the MT through information transmitted from the terrestrial base station to the MT.

6.3.3.3 Handover Strategies

Handover strategies refer to the schemes adopted in each of the three phases of handover. As shown in Table 6.1, there are four different *handover controlling schemes* [EFT-98a, EFT-98b] governing the handover initiation and decision phase. For handover execution, different connection establishment schemes and connection transference schemes can be used, respectively, for handing over the signalling channel and the user traffic channel.

Handover Controlling Schemes There are four handover controlling schemes: mobile controlled handover (MCHO), network controlled handover (NCHO), mobile assisted handover (MAHO) and network assisted handover (NAHO). Each scheme differs in the location of the functional entities which perform the radio link measurements and make the handover decision. Either the MT or the network can perform the radio link measurement functions. However, only one of them will make the handover decision.

Mobile Controlled Handover Under this scheme, the MT performs measurements on both the current radio link as well as the surrounding links. Based on these measurements, a handover decision is made by the MT. With this scheme, the network is not involved in the initiation and decision process. This type of handover has the advantages of fast handover initiation and decision since only the MT is involved. However, since the handover decision is made based purely on the downlink measurement at the MT, the handover decision may not be reliable if the uplink and downlink are uncorrelated. Furthermore, the complexity of the MT will be high since intelligence has to be incorporated into the MT to handle link measurements and handover decision making.

Network Controlled Handover In contrast to MCHO, the radio link measurements and handover decision are both performed by the network. The network will make the handover decision and handover command will then be passed on to the MT. In comparison with MCHO, the handover initiation and decision process is slightly slower than MCHO because of the additional signalling for the network to inform the MT of handover. However, the MT can remain simple since no intelligence for handover has to be incorporated. Again, the handover decision may not be reliable since only the uplink is measured.

Mobile Assisted Handover With MAHO, both the MT and the network are involved in the radio link measurements, the MT for the downlink and the network for the uplink. The MT will periodically report to the network on the downlink measurements. The network will make a decision on handover based on both the uplink and downlink information. This type of controlling scheme ensures that the handover decision is reliable. The interval between when the MT reports to the network has to be carefully selected. If the interval is too small, i.e. the rate that the MT reports to the network is too frequent, the signalling load will be too high. However, too large an interval may inhibit a fast response to the need for handover. In comparison with MCHO, the MT is simpler. However, the signalling load is higher since the MT needs to report to the network periodically. In comparison with NCHO, the MT is more complex as it needs to be equipped to take the downlink measurements and report back to the network. The signalling load is also higher for the same reason as explained before.

Network Assisted Handover This is the same as the MAHO case, both the MT and the network will monitor the radio link quality in NAHO. However, in this case, the network will report the uplink measurement to the MT and the MT will make the handover decision based on both the downlink and uplink measurements. Since the MT will make the decision, the handover process will be faster than that of the MAHO at the expense of a higher MT complexity. The signalling load will also be higher than the MCHO and NCHO since reporting of uplink measurements to the MT from the network is required.

Table 6.3 summarises the advantages and disadvantages of each handover controlling scheme.

Table 6.3 Advantages and disadvantages of different handover controlling schemes

Handover controlling scheme	Advantages	Disadvantages
MCHO	Fast handover initiation Low handover signalling	Handover decision unreliable High level of complexity in MT design
NCHO	Relatively fast handover initiation in comparison with MAHO and NAHO Lower handover signalling in comparison with MAHO and NAHO but higher than MCHO Low level of complexity in MT design	Handover decision unreliable
MAHO	Handover decision more reliable	High handover signalling resulting in slower handover initiation
NAHO	Moderate complexity in MT design Handover decision more reliable Handover process functionality distributed between MT and network	High handover signalling resulting in slower handover initiation in comparison with MCHO and NCHO but faster than MAHO High level of complexity in MT design

Handover Connection Establishment Schemes Handover connection establishment schemes are responsible for the setting up of new signalling channels in the new target cell (or FES) and the disconnection of the old signalling links in the old cell (FES) in the handover execution process. There are two main schemes for connection establishment: backward handover and forward handover.

Backward Handover In backward handover, the old signalling link is used for the exchange of handover signalling messages. This approach is usually adopted in conventional cellular radio systems since in most cases the variation in the signal level at the cell boundary is smooth, especially so when the cell size is large. Hence, there will not be a sudden drop in the signal level. However, if the signal level received from the old link suffers from serious shadowing effect before a new link can be established, the handover process will have to be terminated as signalling exchange will not be possible through the old connection, resulting in dropping of the call. Thus, backward handover is not reliable.

Forward Handover In order to increase the reliability of handover, a different approach can be adopted. In the forward handover scheme, a new signalling link will be established and will be used for the exchange of handover signalling messages. Information on the old connection will be used to transfer the connection from the old BTS (or FES) to the new BTS (or FES). Once everything is firmly established, the new BTS (or FES) will instruct the old BTS (or FES) to release the network resources.

Connection Transference Schemes There are three different connection transference schemes for use in establishing the user traffic channels, namely, hard handover, soft handover and signalling diversity.

Hard Handover Hard handover is a "break before make" handover in which the old traffic connection is released before the new one is established. With this scheme, the MT maintains connection with the best transceiver station until the signal deteriorates below the threshold level. When this happens, the original connection will be released and the MT searches for another suitable transceiver station to re-establish the connection. An interruption to the service will be unavoidable and hence seamless handover cannot be performed (Figure 6.15).

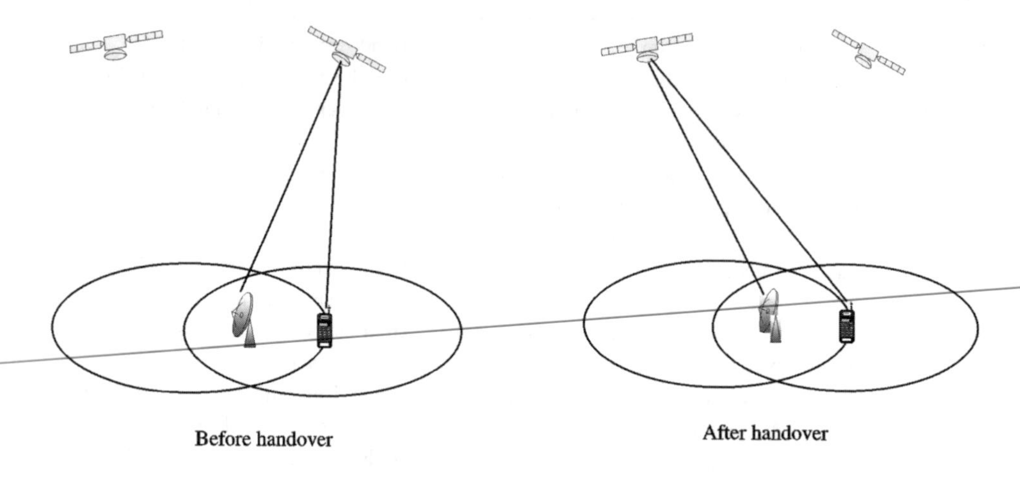

Figure 6.15 Hard handover.

Soft Handover In contrast, soft handover maintains the call connection through the old link until a new link is firmly established, i.e. a "make before break" handover. Soft handover is always associated with diversity. Hence, it is sometimes known as *diversity handover*. With soft handover, the service will not be interrupted since the old connection is still used for communication during the handover procedures. As a result, seamless handover can be achieved. However, since both the old and new links are established during the handover process, extra network resources are required. Soft handover can be further divided into switched diversity and combined diversity. The system complexity required to implement the two different types of soft handover differs.

- Switched diversity implies that communication is carried out through only one link, not both simultaneously. Switched diversity consists of three main stages: (1) the call connection is maintained through the old links; (2) the establishment of new links is under preparation, e.g. channel allocation synchronisation, etc. while the call connection is still maintained through the old link; (3) after the new links are firmly established, the call connection will be switched over to the new links. Synchronisation between the old and new links is required in order that the data packets can be switched to a new link. Figure 6.16 shows the three stages of the switched diversity process, assuming a single MT, two satellites and single FES configuration.

- Combined diversity implies that communication is carried out through both old and new links during a handover. There are also three stages similar to the switched diversity. The difference is that in stage (2), both links will be used for communication even when the new links are not fully established. Both the MT and the FES will require combining functions to be implemented in order to perform combined diversity. Furthermore, synchronisation between the old and new links is required so that the data packets can be combined. Figure 6.17 shows the combined diversity process with the same configuration as in switched diversity.

Signalling Diversity With signalling diversity, the signalling procedures are performed through both new and old signalling links, while the user traffic continues to be transmitted through the old link. When the new traffic channel is firmly established, the user traffic is then switched to the new link and the old link is then released. This scheme is especially useful for inter-segment handover between the satellite and terrestrial segments, in which the propagation delay difference between the two links is large. For inter-segment handover, it is difficult to use soft handover because it requires the use of old and new links for user traffic

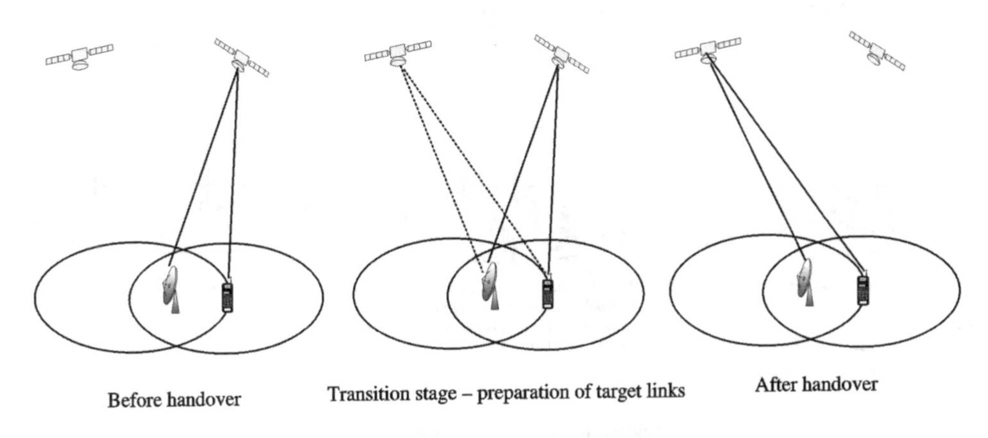

Figure 6.16 Switched diversity handover.

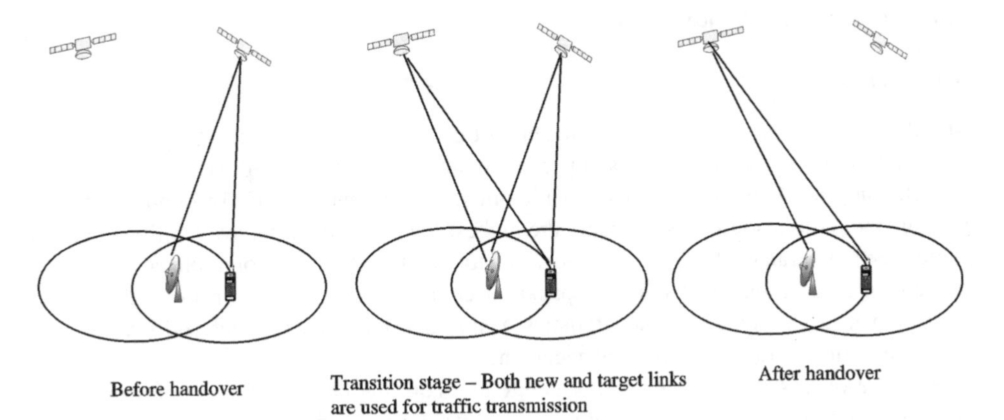

Figure 6.17 Combined diversity handover.

simultaneously during the handover stage. However, synchronisation between the two connections will be difficult due to the large propagation delay difference between the two links. Thus, signalling diversity provides one possible solution for inter-segment handover to establish the target user traffic connection while the original one is still in use, although simultaneous signalling transmission and reception of both links from the MT is required. Figure 6.18 shows the signalling diversity scheme for terrestrial to satellite handover. In (a), a call is in progress through the terrestrial link. In (b), a handover from the terrestrial link to satellite link is identified and the handover signalling procedure proceeds in both satellite and terrestrial links. In (c), a traffic channel is established and the call is switched from the terrestrial to satellite connection. For real-time traffic connection, the user will experience a slight interruption, which is equal to the delay difference between the two connections. This scheme can be regarded as an asynchronous diversity handover. It is a "make before break" scheme, therefore it is different from hard handover. It differs from the soft handover because no synchronisation is required on the traffic channel since only one link is used for traffic transmission at any one time.

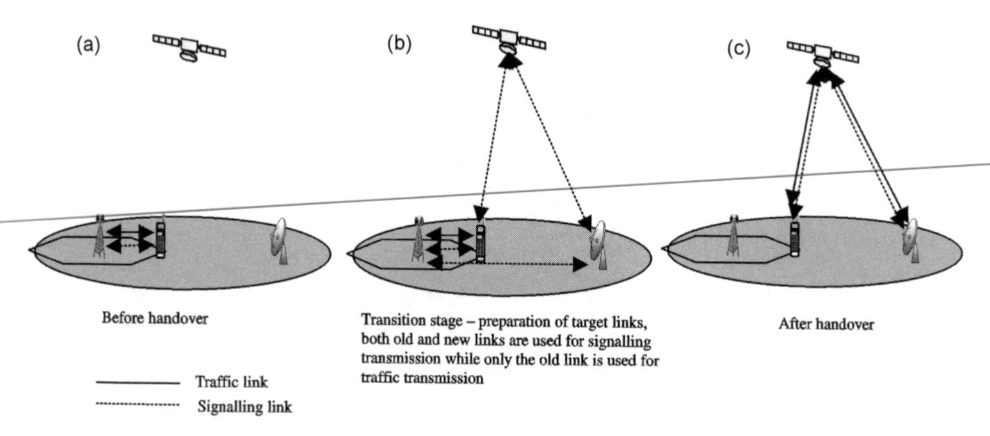

Figure 6.18 Signalling diversity.

6.4 Resource Management

6.4.1 Objectives

Mobile communication systems need to support divergent services with a large range of bit rates and traffic characteristics. It is important to optimise the use of radio resources in such a way that no one channel is over used while others remain under used. An important task to manage the radio resources is to ensure the highest level of reliability, availability and at the same time to guarantee the QoS being provided to the users. Satellite communication systems operate in a large range of radio propagation environments, ranging from rural to urban environments. The dynamic characteristics of non-geostationary systems add several challenges in terms of radio resource management.

Radio resources will have to be allocated to users during the full call duration without any interruption, regardless of the type of satellite constellation and the mobile's movements. In order to ensure end-to-end communications, radio channels must be established, maintained and released. To perform such tasks, dedicated signalling channels are required, which in turn require extra radio resources for non-user traffic transmission. Satellite resources are limited, hence, optimisation of their use is of prime importance. Even when a mobile terminal is in idle mode, i.e. no call is in progress, radio resources have to be used for continuous position monitoring of the MT such that the MT can be paged to receive incoming calls and transmit outgoing calls at any time and anywhere.

The movement of the mobile and the satellites in the case of non-geostationary satellite systems, although the latter will have a more significant effect for satellite communications, require that radio resource management strategies should take into account handover. Two issues need to be considered in handling handover: (a) The signalling exchanges for handover should be kept to a minimum in order to optimise radio resource usage; (b) handover calls should have priority over ordinary incoming calls. These two issues require careful radio resource management strategies with an aim to minimise the radio resource usage but at the same time not to impose too complex a radio resource management scheme.

6.4.2 Effects of Satellite System Characteristics

Alongside quality, two important aspects that a satellite radio resource management strategy has to consider are: availability and mobility. In the past, adaptive power control techniques have been used to combat the propagation channel fluctuation so that link availability can be enhanced. However, the limitation in the transmit powers of both the satellite and mobile terminals makes these techniques not fully compatible when high system capacity is required in a telecommunication system. Instead, geometric considerations have been taken into account by using the minimum elevation angle and satellite diversity techniques so as to increase the range of operational environments.

Minimum Elevation Angle In most cases, the location of the FESs in a satellite system is carefully chosen, normally in areas where there are no obstacles surrounding the FESs, e.g. hill tops, so as to permit maximum satellite visibility and communication even with very low elevation angle. However, the situation with mobile terminals is more complicated due to the continuous and somewhat random movement of the mobile, which can be in wooded areas, urban areas with lots of high rise buildings and so on. These will cause interruption to the service as a result of shadowing. Therefore, a minimum elevation angle for both the FES and mobile terminal has to be defined in order to ensure acceptable system availability. In the worst case, the minimum elevation angles are typically 5° for an FES and 7–10° for a mobile terminal [ETS-93]. The system performance can be assured for minimum elevation angles greater than these values, assuming that there are no blockages in the radio path, which is random in nature. The high minimum elevation angle of the mobile terminal will decrease the effect of shadowing on the link. In particular, with respect to a geostationary satellite, a higher minimum elevation angle corresponds to a reduced propagation length and hence a shorter transmission delay will be introduced. However, the coverage area surrounding the mobile terminal will also decrease resulting in a higher number of FESs for a given system coverage.

Satellite Diversity Satellite diversity is used in order to reduce the probability of satellite invisibility. Consider that if only one satellite is within the visibility range from the mobile, the probability that the satellite visibility is interrupted due to shadowing or other effects may not be negligible in certain operational environments. However, if two satellites are used, the probability of a satellite being invisible to a mobile terminal will be the product of the probabilities of satellite invisibility due to each individual satellite, resulting in a lower probability. This is based on the assumption that the two events are statistically independent, which is achievable if the azimuth distribution of both satellites is large enough [CEC-96]. However, this requires the system to track both satellites simultaneously which leads to high system complexity. Furthermore, the management of radio resources will be more complex and the associated signalling will increase with the number of diversity paths.

6.4.3 *Effects of Mobility*

The dynamic nature of non-geostationary systems and to a lesser extent, the user's movement with respect to geostationary satellites, requires the consideration of handover procedures during transmission. In comparison with the non-geostationary satellite's velocity, a mobile terminal's velocity can be regarded as negligible and sometimes is regarded as fixed for resource management purposes. Concentrating on a stand-alone satellite network scenario and in intra-FESs, handover can be grouped into two categories: intra-satellite handovers and inter-satellite handovers.

Intra-satellite handovers This is also referred to as inter-beam handover in a multi-spot-beam system. This type of handover involves only one FES and hence is a sub-branch of intra-FES handover. The occurrence of intra-satellite handover is due to: (a) the highly dynamic non-geostationary satellite constellation such that spot-beam footprints are moving fast on the Earth during a call resulting in frequent spot-beam handovers; (b) the movement of the terminal from one spot-beam coverage to another in a geostationary satellite constellation. If a GLOBALSTAR-like 16-spot-beam LEO constellation is considered, the visibility duration for each spot-beam is around 2 minutes for inner beams and 5 minutes for outer beams. For a MEO constellation with a 120-spot-beam coverage, the visibility duration is between 7 and 10 minutes.

Hence, handover is necessary for a 3-minute call duration through a LEO satellite system. For a call duration of greater than 10 minutes, handover is required for all types of non-geostationary satellite constellations.

Inter-satellite handover Inter-satellite handover is more common in LEO satellites, with an average interval of 10 minutes between two handovers and about 2 hours for MEO satellite constellations. Hence, for a 3-minute call duration, the probability of requiring an inter-satellite handover is only about 15% for LEO satellites and less than 2% for MEO satellite constellations. Hence, most of the calls in MEO satellite systems will not suffer from inter-satellite handover [CEC-97].

Handover occurrence in non-geostationary satellite systems is deterministic since the satellite motion follows Newton's three Laws of Motion. For MEO satellites, proposals have been made on the use of steerable spot-beams, which require complex satellite antenna arrays but will greatly simplify the radio resource management process for handover, which rarely happens in MEO satellite systems. For LEO satellites, satellite diversity can be used. In switched diversity, the resource is time-shared among satellites in visibility. The transmission is switched from one satellite to another. In combined diversity, the signal power is combined in order to optimise power consumption and to improve link quality. However, it requires more complex resource management and demodulation techniques. Dual satellite diversity involves the use of two different paths for signal transmission. Techniques to control interference between the two paths will have to be imposed in order to optimise the system capacity and performance. Satellite diversity is inherent in the downlink, radio resources have to be reserved for every satellite within the visibility of the mobile terminal. This implies that the same amount of radio resources has to be reserved also in the uplink. Diversity techniques are normally studied in association with multiple access techniques. The complexity of radio resource management should be traded-off against the diversity gain, taking into account the available bandwidth.

6.4.4 Resource Allocation Strategies

6.4.4.1 Introduction

As a cellular mobile radio system seeks to re-use its time, frequency and power resources to enhance its resource utilisation efficiency, so does a satellite communication system. While FDMA re-uses the frequency bandwidth, TDMA re-uses the time-slots and CDMA the power resources. The use of multi-spot-beam antennas in satellite systems makes use of the frequency re-use concept and each spot footprint can be regarded as equivalent to a terrestrial cell. The concept of re-use distance in terrestrial cellular systems to arrange the frequency re-use pattern can be applied in a multi-spot-beam architecture. Figure 6.19 shows the application of the cellular principle to a multi-spot-beam satellite system.

Multiple access techniques have been discussed in Chapter 5, this chapter aims at studying the relevance of multiple access techniques in relation to satellite resource allocation. There are several approaches to assigning radio resources: fixed channel assignment, dynamic channel allocation (DCA) and hybrid channel allocation. The tasks for resource allocation are to assign a satellite/spot-beam and a traffic channel to the users in as an efficient way as possible.

6.4.4.2 Multiple Access Considerations

There are three main multiple access schemes: FDMA, TDMA and CDMA. Under the universal transmission radio access (UTRA) concept, the resource management strategy has to be as independent as possible of the multiple access scheme. However, there are technical aspects such as satellite diversity and interference issues which will have a big impact on the design.

TDMA This scheme is highly sensitive to differential propagation delay between users sharing the same carrier frequency. A timing advance will have to be carried out to correct the transmission time-slot after each frame transmission. In addition, TDMA requires efficient frequency re-use planning in order to avoid interference between two adjacent spot-beams of all the satellites which are visible to the user. Therefore, frequency re-use cannot be achievable within a short distance. For most TDMA satellite systems, a seven-cell cluster size is normally used. TDMA is more suitable to GEO satellite constellations in which

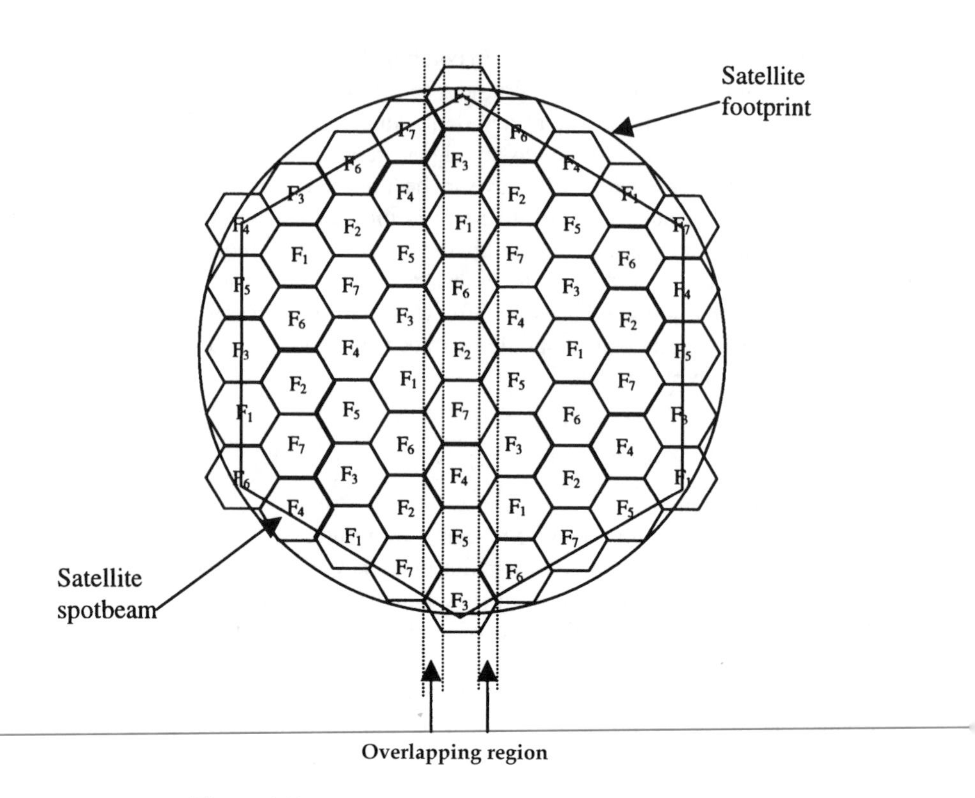

Figure 6.19 Satellite spot-beam cellular concept.

handover rarely occurs. Most of the timing advance algorithms will work efficiently under TDMA but synchronisation when satellite diversity is employed is still a complex matter. A pseudo-cell concept will have to be imposed, otherwise acceptable link performance can only be dependent upon a large link margin for a TDMA system.

CDMA Bandwidth efficiency is the main driving force in the use of CDMA since frequency re-use planning is not required. All available frequencies can be re-used in every single spot-beam. CDMA makes use of the pseudo-code concept in order to distinguish between different channels. It transmits modulated data onto wideband carriers that are distinguishable from each other by different pseudo-noise (PN) sequences. Receivers retrieve their intended data by searching for their PN sequence. In order to avoid interference, the traffic carriers must be spread with synchronised and orthogonal PN sequences.

Although synchronous-CDMA (S-CDMA) proves to be the most efficient to eliminate interference arising from other users sharing the same carrier and the same spot-beam, interference from other spot-beams which overlap the coverage of the intended spot is still considerable. The synchronisation process to ensure orthogonality between all links requires signalling to adjust the transmission in both the time and frequency domains for every user independently. If dual satellite diversity is deployed, the timing advance will be addressed to only one satellite. Half of the users sharing the same frequency band will statistically be synchronised to this one particular satellite while generating intrinsic noise to the other. The system capacity is subsequently reduced. If orthogonality between PN sequences is not required, i.e. asynchronous CDMA, synchronisation is not necessary. Under this situation, the number of available PN sequences will increase tremendously. However, this implies that interference levels generated by co-channel users cannot be suppressed as efficiently which may reduce the system capacity. However, since the number of PN sequences is increased, such a reduction in system capacity may still be well within the resource utilisation efficiency of that offered by S-CDMA. In the case of multi-satellite diversity, longer codes may be required in order to discriminate between different links and consequently, synchronisation among different satellites will be more complex.

6.4.4.3 Fixed Channel Allocation (FCA)

Channel assignment in first- and second-generation terrestrial mobile networks is performed according to a fixed frequency re-use plan where a fixed set of channels is permanently assigned to each cell site. This is to ensure a large enough frequency re-use distance in order to eliminate co-channel interference. As noted earlier, in satellite systems, each spot-beam coverage is equivalent to a terrestrial cell.

There are two approaches where radio resources can be assigned in a satellite system. The first approach is to assign each spot-beam a set of channels from a pool of all possible radio channels. A channel search is performed only over the subset of channels allotted to each spot-beam assigned to serve the call. The first available channel encountered during the channel search is assigned to serve the call. If no channel is available in that spot-beam, the call will be blocked or lost, regardless of whether channels are available in other spot-beam coverage areas. For an evenly distributed FCA scheme, the number of channels, N, permanently allocated to each cell is given by [DEL-94b]

$$N = \frac{M}{K} \tag{6.7}$$

where M is the number of system resources and $K = D^2/(3R^2)$ is the frequency re-use factor with D and R being the re-use distance and the cell radius, respectively (see Chapter 1).

Whenever an inter-beam handover occurs, a change of radio resource from one spot-beam to another is required and re-routing of the call is also necessary. As a consequence, signalling to inform the mobile terminal of the available radio resource information is also required. For non-geostationary satellite constellations where the satellite movement is high, it is difficult to assign a fixed frequency re-use plan to a continuously moving satellite coverage. Furthermore in non-geostationary satellite systems, where inter-beam handover is frequent, a large amount of signalling is required.

Another approach is to associate the radio resources with fixed ground areas. In this method, a steerable spot-beam antenna can be used such that its footprint coverage can be fixed all of the time. This approach will always ensure that the mobile is under the same spot-beam and hence beam-to-beam handover is not required.

Either approach requires a fixed frequency re-use plan that permanently assigns channels to cells based on worst-case interference scenarios. In order to take into account the variation of traffic demand in different geographical locations, some cells may be assigned with more channels than others. However, this approach lacks the flexibility to adapt uneven traffic demand by allowing unused traffic channels in one cell to be used in another at any given time. In general, fixed channel assignment is inefficient for satellite systems where the resource is limited and the variation of traffic demand is high.

6.4.4.4 Dynamic Channel Allocation (DCA)

DCA was first studied by Cox and Reudink in order to improve network efficiency in personal communication systems [COX-72a, COX-72b, COX-73]. In a DCA scheme, each channel can be used in any cell of the network provided that the constraint on the re-use distance D is respected. Instead of permanently assigning channels to each cell, channels are only assigned for the duration of a call. A new call at any cell x may use any channels which are not used in cells x nor in any interfering cell of x, $I(x)$, which is at a distance less than D. Figure 6.20 shows the DCA concept.

In order to allocate the best available channel for cell x at an instant when a call arrival occurs at cell x, a suitable cost function $C_x(i)$ has to be defined such that the best channel j is selected according to the following criteria [DEL-95]:

$$C_x(j) = \min\{C_x(i)\} \;\; \forall i \in \Lambda(x) \tag{6.8}$$

where $\Lambda(x)$ is the set of available channels at cell x at the instant of the call arrival.

Two approaches can be used to evaluate the cost function $C_x(i)$ of any channel $i \in \Lambda(x)$. The first approach assumes a pure DCA scheme such that a single pool of channels from which all mobile terminals and FESs can select any traffic channels satisfying the re-use distance constraint. Under this approach, the allocation cost function is defined as [DEL-94b, DEL-95, NAN-91]:

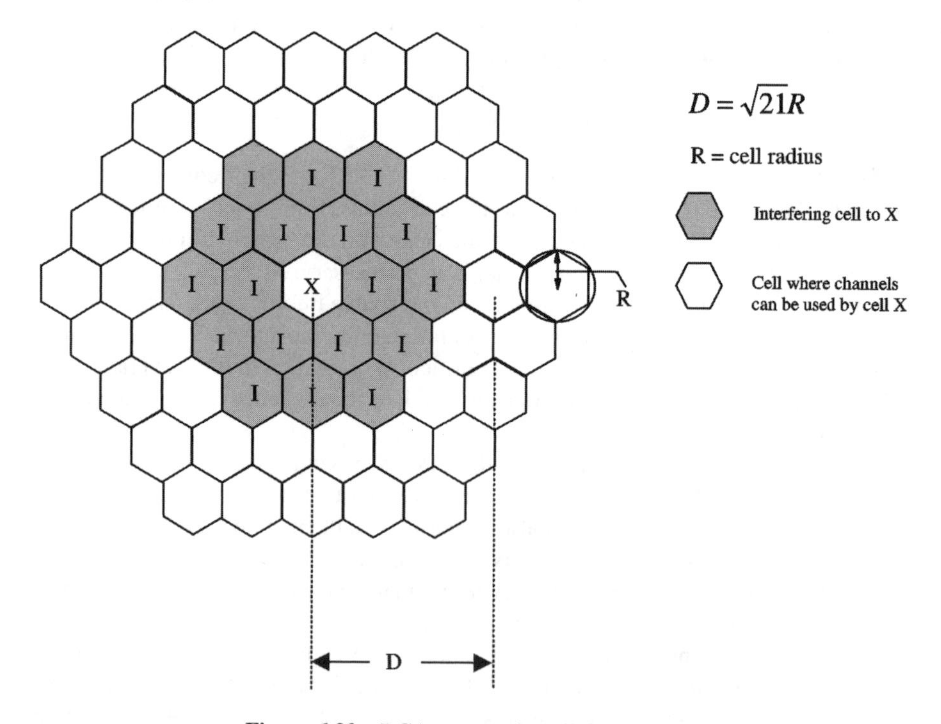

Figure 6.20 DCA concept in cellular systems.

$$C_x(i) \triangleq \sum_{k \in I(x)} \{u_k(i)\}, \ \forall i \in \Lambda(x) \tag{6.9}$$

where

$$u_k(i) = \begin{cases} 1 & \text{if } i \in \Lambda(k) \\ 0 & \text{otherwise} \end{cases} \tag{6.10}$$

This approach sometimes leads to inefficient channel distribution in the network, especially under heavy traffic conditions since the minimum re-use distance can be much greater than D. Call blocking probability can substantially increase, sometimes even greater than that of the FCA scheme.

Another approach [DEL-95] is to allocate a set of channels $F_D(x)$ to cell x according to a FCA scheme satisfying the minimum re-use distance constraint. DCA is then used to select channels for use in cell x with priority set on those channels belonging to $F_D(x)$. If no channel is available on $F_D(x)$, channels in other cells will be selected provided that the constraint on D is respected. Thus, the allocation cost function for this approach contains two parts: one part is the availability of channels in $F_D(x)$ and the other is the availability of channels in other non-interfering cells if no channel is available in $F_D(x)$. Mathematically, the allocation cost function is expressed as:

$$C_x(i) \triangleq q_x(i) + \sum_{k \in I(x)} \{C_x(k, i)\} \ \forall i \in \Lambda(x) \tag{6.11}$$

where

$$q_x(i) = \begin{cases} 0 & \text{if } i \in F_D(x) \\ 1 & \text{otherwise} \end{cases} \tag{6.12}$$

and

$$C_x(k, i) = u_k(i) + 2[1 - q_k(i)] \ \forall k \in I(\text{x}) \tag{6.13}$$

$C_x(k, i)$ denotes the allocation cost contribution for channel $i \in \Lambda(x)$ due to the interfering cell $k \in \mathrm{I}(x)$. The term $q_x(i)$ in equation (6. 12) implies that priority is set to allocate channels available in $F_D(x)$.

Both DCA approaches require channel re-arrangement to reallocate the channel assigned to x once the call is terminated. Without this re-arrangement, the performances of both DCA approaches degrade significantly. DCA is particularly effective in areas where traffic demand is not uniformly distributed so that resources can be concentrated on the most heavily used spot-beam.

6.4.5 Network Operations and Procedures

6.4.5.1 Overview

In order to play a complementary role to terrestrial networks, the design of the network operations and procedures in satellite networks will, in as far as possible, re-use those of the terrestrial networks, to facilitate integration. Several network procedures need to be considered in order to establish and to release a call connection including.

- Terminal switch on/off
- Session set-up/release
- Call set-up/release
- User registration/de-registration
- Location update
- Handover

While the last two procedures have been discussed in detail in previous sections, this section will concentrate on the other procedures in conjunction with location update and handover procedures whenever necessary.

6.4.5.2 Terminal Switch On/Off

Users of mobile-satellite communications systems are normally international business travellers, who would often travel from one part of the world to another with their mobile terminal to receive incoming calls and to make out-going calls wherever they are. Since this may often involve long-distance travel, keeping the terminal in idle mode will incur significant signalling, delay and information exchange between different databases. Hence, efficient switch on and off procedures have to be derived in order to reduce such network overheads.

Terminal Switch On The main purpose of the switch on procedure is to prepare the terminal for later possible use. Although the state of a terminal after the switch on procedure is often compared with a terminal in idle mode, there are significant differences. When the terminal is in idle mode, the terminal is connected to the network but there is no user session and no active calls. The terminal will periodically "listen" to the broadcast channels in order to detect any incoming calls. Furthermore, procedures such as location update, spot-beam and satellite handovers to allow terminal roaming have to be performed so that registration to the network is not lost. When the terminal is switched on, the terminal is not actually registered to the network, i.e. procedures such as location update will not be performed. The terminal will just receive broadcast information from the satellite. Hence, the switch on procedure is a purely internal procedure within the terminal. No interaction between the network and the terminal will occur.

The switch on procedure consists of two actions: power on and personal identification number verification. Broadcast information received from the fixed Earth station can possibly include the reachable location in the satellite network and information on the reachable fixed Earth station within that location area. In order to provide this type of broadcast information, information on the terminal position is required. The reachability information is used essentially during a user session set-up and location update procedures such that the location information of the network is known.

Terminal Switch Off The main function of the switch off procedure is to completely terminate the connections between the terminal and the network and to stop all the processes in progress in the terminal. The actions required for the switch off procedure depend on the state of the mobile terminal.

- When the terminal is just switched on, the switch off procedure merely switches the power off.

- When the terminal is in idle mode, the terminal must terminate any contacts with the network. There are mainly two active mobility management processes when the terminal is in idle mode: satellite/spot-beam handover and location update. The former is due to the satellite movement and will only be performed at the transmission level; consequently it is transparent to the switch off procedure. The latter involves interaction between the terminal and the network. Hence, all processes which activate these procedures will be terminated and should not be initiated.
- Another on-going process when the terminal is in idle mode is the detection of incoming calls through the broadcast channels. Hence, during the switch off process, incoming calls must be prevented. In order to do this, either the terminal does not respond to the network when it detects an incoming call on the broadcast channels or the network does not inform the terminal of the incoming call. However, the latter approach has to be performed in conjunction with the detach procedure, which is optional. If the detach procedure is not implemented, a careful design of the latter approach is required.
- Furthermore, all the user registrations have to be deleted. The terminal has to send a message to inform the network that the terminal will be switched off to enable the network to delete all the registrations in order to remove all the access agreements between the network and the terminal. If the message is not received by the network, there will be inconsistency between the data stored in the terminal and in the network. One possible solution is to add a timer in the network and in the mobile terminal. The timer can be reset when the virtual terminal (see next section) and the network starts contact. However, the timer in the terminal must expire before the timer in the network. When the timer in the terminal expires, a message is sent from the terminal to the network informing it of its switch off state. If acknowledgement from the network is received, the terminal timer is reset. If no acknowledgement from the network is received, all the registration data in the terminal will be deleted. Likewise, when the timer in the network expires, all the registration data concerning the terminal are deleted and the access agreement between the terminal and the network is then removed. In the case where the terminal cannot inform the network of the switch off process, the terminal deletes all the data as soon as the timer expires. The network will subsequently delete all the registration data.
- The data generated as a result of such processes in the terminal have to be deleted, unless such data are kept for specific purposes. For example, information on the last location area can be kept and used in the terminal for another session set-up.
- When the terminal is in an active mode and a session has just been opened but with no call in progress yet, the session will then be released automatically. When the session is released, it leaves the terminal in the idle mode state and the procedure of switch off will then follow the previously mentioned terminal idle mode procedure.

6.4.5.3 Session Set-Up/Release

Session set-up can be initiated by the user or the network. A network initiated session can sometimes be considered as part of a call set-up process. Session set-up is a procedure to prepare the users to use their desired networks. The procedure may include obtaining access to the network, user authentication, terminal authentication and information collection on user profile. One may compare such procedures with logging into a computer. In the following, the session set-up/release procedures are derived from Refs. [CEC-95, CEC-96, CEC-97]. Four logical entities may be identified during a session set-up: user role selection, access control, user session and terminal session.

User Role Selection This logical entity provides the possibility that one subscriber identity device (SID) may contain several user roles. Users can specify their user roles according to, for example, their professional status, their personal status, etc. Each user role is distinguished by an international mobile user identity (IMUI). The IMUI has a one-to-one correspondence with the service provider identity (SPI). Only one user role can be chosen per SID during a session set-up according to the user's preference.

Access Control Access control consists of two parts: (1) creation of an access agreement between a terminal and a network; (2) the retrieval of the TMTI when the access agreement exists. When these two procedures are completed, communication between the network and the terminal can start. Depending on the complexity of the terminal, it can have simultaneous access to multiple networks. For example, in an integrated satellite and terrestrial network environment, a dual-mode terminal can access both the satellite and terrestrial network at the same time. This creates the *virtual terminal* concept in which a single physical terminal can provide users with multiple virtual terminals, each virtual terminal accesses one particular network environment.

Before the creation of an access agreement between a terminal and a network, several information and verification procedures are required. The user has to select a network to which they would like to have access and can be reachable in the case of a multi-mode terminal. Furthermore, the user should have access rights to the selected network. Thus, the access control function will first select a network according to the access mode selected by the user. Given the selected access mode, it will then select the user's preferred network, checking whether the selected network is reachable by the terminal and whether the user has access rights to the selected network.

Once the access to the network is verified, the best location area for the user will be selected according to some predefined criteria. In terrestrial networks, the choice of the best location area can be based on the signal strength. However, in a satellite environment and in particular for non-geostationary satellite networks, the use of signal strength will not be appropriate. Discussion on the definition and the selection of location areas can be referred to Section 6.4.1.

Following the identification of the network and location area, resources should be selected for communication between the terminal and the network. For a location area using a terminal position approach, resource selection in the satellite environment identifies the {FES, satellite and spot-beam} or {satellite, spot-beam} set if it uses the GCA approach. Furthermore, resources available in one FES will be shared among different operators.

If a terminal position based approach is used, the selection of an appropriate FES constitutes part of the resource selection process. The following outlines some of the FES selection strategies.

- Nearest FES landing: the FES which is nearest to the mobile terminal's geographical location will be selected.

- Originating FES landing: the FES which belongs to the administrative area where the MT is registered is selected.
- Home FES landing: the FES closest to the MT's home network is selected.
- Idle FES landing: the least busy FES is selected to minimise call blocking probability.
- Terminating FES landing: the FES nearest to the fixed user involved in the call is selected. It has to be noted that in both the idle FES and terminating FES approaches, different FESs can be used to serve the MT on a call-by-call basis. However, this requires that the MT tunes to the ever changing FES paging channel even before it is called.
- Centralised FES landing: the FES which also acts as the network control centre will be selected. This strategy has the advantage of providing optimum route selection and it reduces the complexity of paging since the MT only needs to listen to the paging channel from the centralised control FES.

Access control is established once the resource is selected. A TMTI is then allocated to the terminal by the network. The terminal can have several TMTI associated with the networks to which it has access.

User Session A *user session* is a continuous period of time during which a user has a secured association with the desired service provider in order to prepare them for the use of services. The service provider identifies the user through the SID. A user session can be divided into three phases [CEC-95]:

1. User session set-up
2. User session active
3. User session release

In the user session set-up phase, the user notifies the service provider with whom they would like to set-up a session. The service provider can accept or refuse this attempt. Mutual authentication and encryption keys will be exchanged in order to have a secured association. The user profile will be retrieved. This association with the service provider will be maintained during the whole session. In summary, the user session set-up procedure includes the following functions:

- Identification of the user to the network.
- Authentication of the network to the user and *vice versa*.
- Establishment of an authentication channel over the air interface for terminal encryption keys and encryption initiation to allow terminal sessions.
- Access control such as user profile verification.

The user session active phase involves user registration, receiving and making incoming and outgoing calls. During the user session, several consecutive services can be handled by the service provider. The procedure during this phase depends on the criteria for user session release. The user session is released after a specified condition is met, for example, a specified duration of time, or after each call or after the user registration. When the user finishes using the service and does not make a request to maintain the session, the session will be released by default a few seconds after the completion of a service. For a group usage terminal, access to the network may still be maintained so that other users registered on this terminal can still access the network.

The user session is related to call handling in the following manner: a user session should be set-up before a call is invoked. During the whole active call phase and call release phase, the user session should have already been set-up. However, user session and registration are somewhat different. A user session should be in place before and after making user registration. During the actual user registration, the user session is released. When user de-registration occurs, a new user session is set-up again. For other procedures which do not require the involvement of the user, such as handover and location update, a user session is not required.

Terminal Session Terminal session is introduced in order to support the universal personal telecommunication (UPT) concept so that simultaneous registration of the same user on different terminals for different services can be made possible. A terminal session is a continuous period of time during which it has a secured association with the service provider for the simultaneous use of multiple services for multiple users. However, the service provider will not distinguish among different users associated with that terminal.

A terminal session will start only after a user registration, which associates the terminal with a specific user. The user registration process also ensures that the TMTI assigned to the terminal is updated and terminal authentication keys are exchanged. This also implies that a user session should have already been established before the terminal session, since a user session is required prior to user registration. A terminal session will last until the user performs a de-registration procedure. Similar to a user session, a terminal session also consists of three phases.

1. Terminal session set-up: authentication of the terminal takes place in this phase in which security parameters and service profiles are retrieved so that a secure connection between the terminal and the network can be established.
2. Terminal session active: during this phase, one or more consecutive calls can be performed and completed.
3. Terminal session release: in this phase, the terminal encryption key will be destroyed.

Before information exchange can take place between the user and the service provider, the corresponding terminal should first have access to the network. In an integrated satellite and terrestrial network environment, the user may have preferences in selecting a particular segment. Once the network is accessed, the terminal will be allocated with a TMTI so that the terminal and the network can associate with each other. The TMTI will be valid for as long as an active session or a user registration takes place on the terminal. The TMTI will only be withdrawn when the last user of the terminal ends the session.

6.4.5.4 User Registration/De-registration

The registration procedure enables a user to inform the network of their location area where they can be reached through a mobile terminal in order to receive incoming calls. This is normally referred to as the in-call registration. Registration can only occur when the user session is active. Procedures such as mutual authentication, user profile verification, terminal capability to support the selected services, identification of location areas, resource selection will be carried out first to enable the service provider to store the information in the database for user registration. For a multi-mode terminal which can support different network environments (e.g. both satellite and terrestrial), an environment capability check needs also be carried out. Some systems may implement a time limitation for which a user can register on a terminal.

Registration can be performed on a per-service basis, i.e. a user can register for different services on different terminals. Similarly, more than one user can register on the same terminal. This brings about a distinction between the first user registration and subsequent user registrations. When the first user registers on a terminal, no TMTI would have been previously assigned to the terminal. When subsequent users register on the same terminal, a TMTI check has to be carried out in order to ensure the correct identity of the terminal. Depending on the call handling procedure, different users registered on the same terminal may not be able to use their services at the same time. When the registration procedure is completed, the user session will be released and the user will have access to the network at all times until a de-registration procedure is in place.

The de-registration procedure informs the network of a change in the user registration information. It requires the establishment of a new user session before the de-registration process can be carried out. All the information such as the terminal, the location area, the registered services will be deleted.

Figures 6.21 and 6.22 show examples of the information flow for the first and subsequent user registrations, respectively.

6.4.5.5 Call Handling

Call set-up/release are part of the call handling procedures. The most fundamental call handing service is the provision of point-to-point voice communication services. As technology advances, other call handling services for multimedia calls, multi-party calls and so on will require different call set-up/release procedures, which may be derived from that of the point-to-point speech communication. This section will only concentrate on the call set-up/release procedures for point-to-point voice communication services.

Similar to session set-up/release, call handling consists of three consecutive phases [CEC-97].

1. Call set-up: this involves the establishment of signalling and bearer connections.
2. Call in progress: this is the phase where the actual communication takes place. Signalling exchanges may take place for in-call negotiation or for handover.
3. Call release: this involves the release of signalling and bearer connections should one of the parties decide to end the call.

Before a call can be set-up, a session should have already been established. The session will be maintained until after the call is released. While session set-up is responsible for all the service independent procedures, a call set-up procedure will concentrate on service dependent actions.

In order to account for possible integration with terrestrial mobile networks, a logical approach in deriving the call handling procedures will be to re-use, as much as possible, those defined in terrestrial mobile networks. However, there are satellite specific constraints when considering the design or the re-use of such protocols. Two important factors need to be considered: the high propagation delays and the satellite motion. According to Ref. [CEC-97], at least ten one-way signalling exchanges are required between the BSS and the mobile terminal for a call set-up in a terrestrial network such as GSM. For a satellite network, and if a

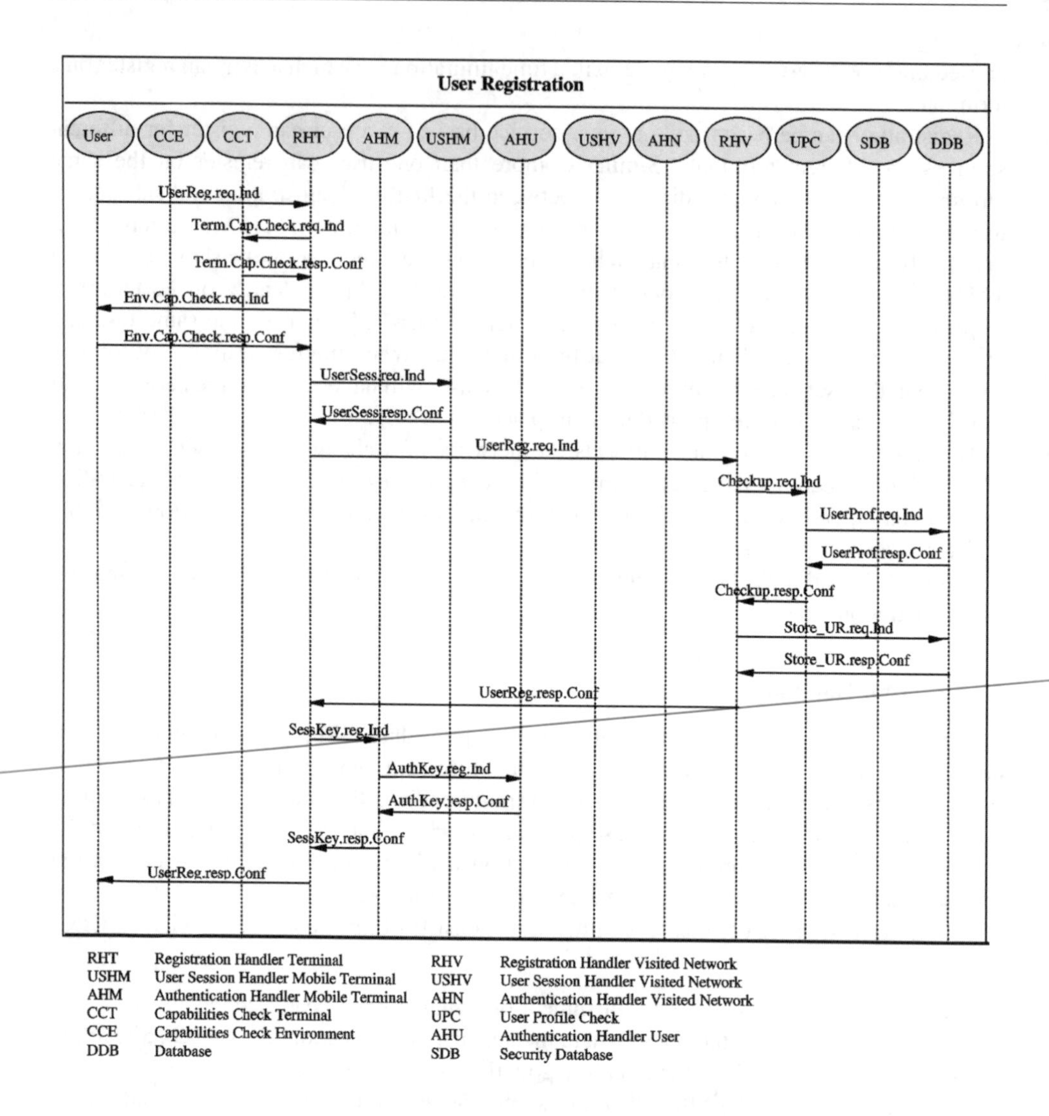

Figure 6.21 First user registration information flow [CEC-95].

geostationary satellite is considered, each one-way signalling exchange takes one round-trip delay of about 270 ms. This will have a significant impact on the QoS. In addition, the requirement of burst synchronisation and tracking in satellite systems also impose constraints on the choice of the access techniques, in particular, those that are time division based. The much larger spot-beam coverage of a satellite system compared to that of a terrestrial cell, coupled with the satellite movement introduces a much larger range of propagation delay for synchronisation and tracking purposes. Hence, adaptation of terrestrial procedures is required.

Call Set-up For out-going calls, i.e. when a user requests to set-up a call for a particular service, the following procedures are carried out.

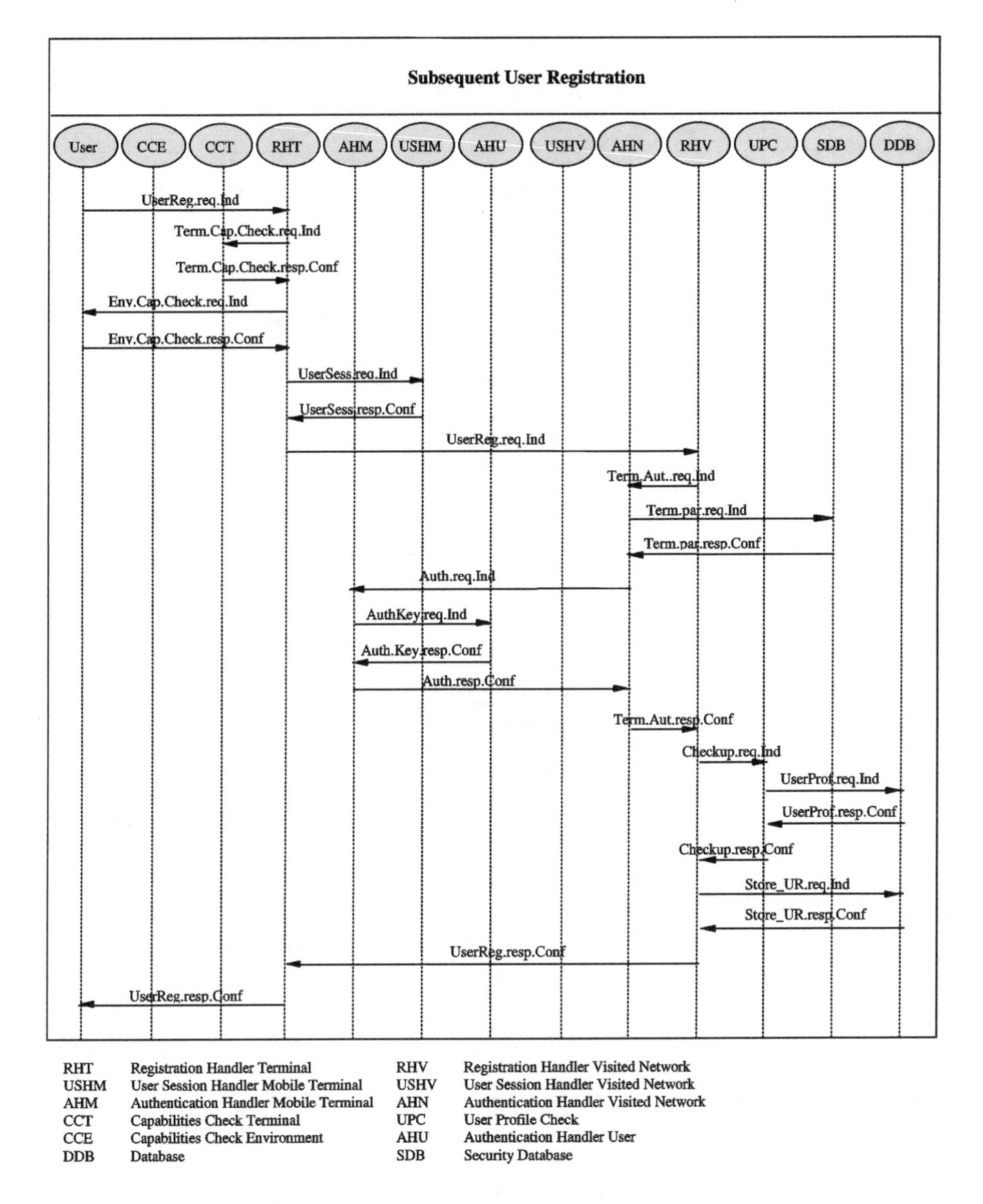

Figure 6.22 Subsequent user-registration information flow [CEC-95].

- Verification of active terminal session: before a call can be set-up, the terminal session should be in active mode. If this is not the case, a terminal session set-up is initiated.
- Terminal capability verification: the terminal capability to support the requested service will be checked. Identification of terminal capability can be carried out in two ways: (1) the terminal capability is stored in the network database and identification can be

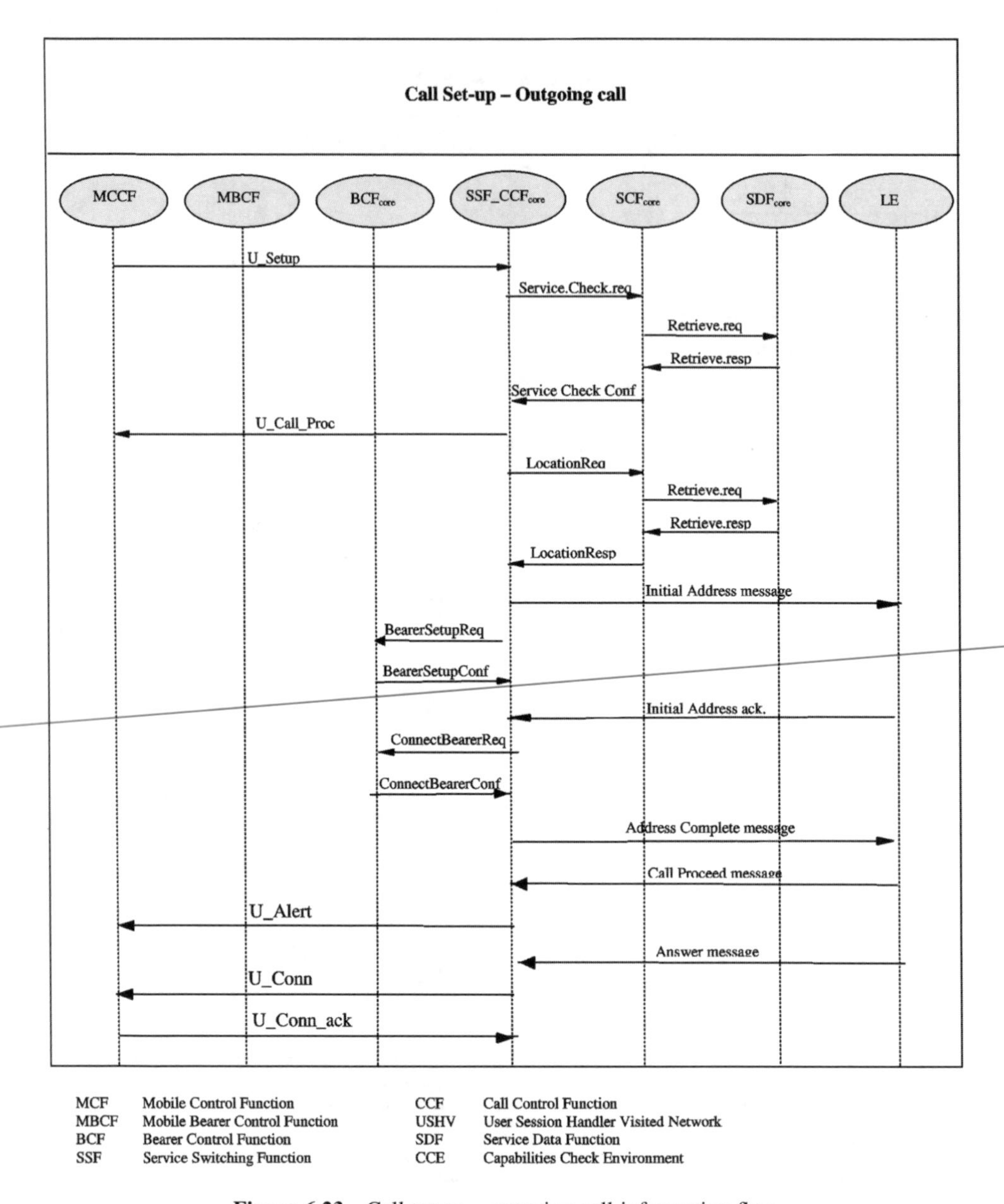

Figure 6.23 Call set-up – outgoing call information flow.

performed via a database query; (2) the terminal itself stores its capability information, which can be retrieved in the call set-up procedure.

- User profile verification: the service description will be checked against the user profile in order to determine whether the calling and called parties have the access rights to the requested service. This is carried out through a human machine interface (HMI) procedure, in which the calling and the called partys' identities, in terms of the IMUI and the international mobile user number (IMUN), and the requested service will be submitted.

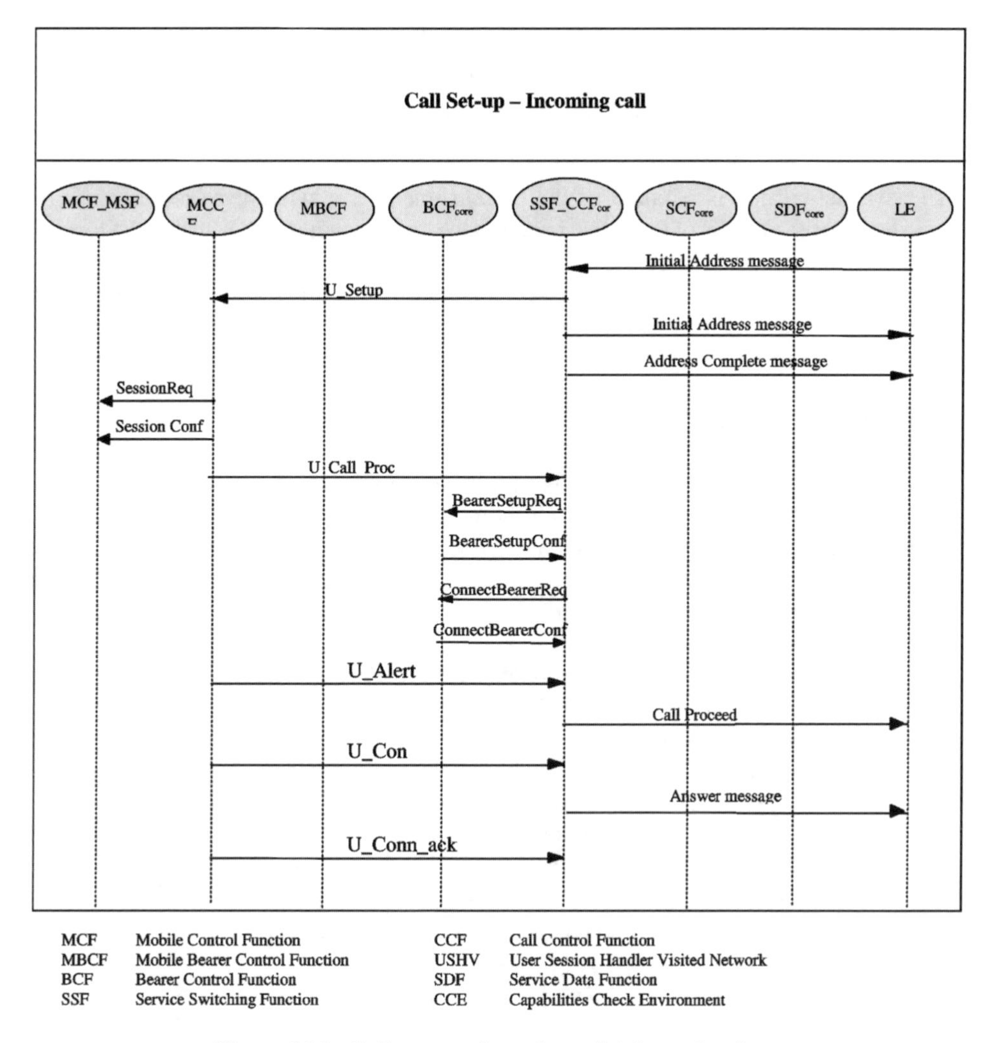

Figure 6.24 Call set-up – incoming call information flow.

The user profile information database will then be retrieved, which should at least contain data sufficient to verify whether the users have subscribed to a particular service. Other information such as access rights to the requested network may also be included.

- Network capability check: once affirmative answers are obtained from both the terminal capability and the user profile checks, the capability of the network to support the requested service will be verified. If the network cannot support the requested service and if the user terminal is only a single mode terminal such that only one network is supported, the call request has to be rejected. However, if the user terminal is a multi-mode terminal, i.e. the terminal supports more than one network, the possibility of selecting another network may be included.
- Once the network capability check is accomplished, the call set-up phase for service

negotiation, bearer allocation connection, etc. is initiated. A signalling association will be established between the calling and called parties.

When the called user, assuming also a mobile user, receives a call (incoming call), the following procedures are carried out.

- The first action that is required of the establishment of a signalling association is the translation of the IMUN of the called user into the corresponding IMUI so that the location of the called user can be identified and a terminating session (a trusted session between the called user and their service provider) can be established. In order to locate the called user, the IMUI and the service description will be used as input parameters in return for the location triplet of the called user consisting of a location area identifier, an {FES, satellite, spot-beam} or a {satellite, spot-beam} set, and a terminal identity. This defines the location area in which the called user can be reached.
- When the location area of the called user is successfully identified, a paging control process is carried out in order to select the resource to use in communication between the network and the terminal on which the user is registered. Paging control is closely related to location management, as set out in Section 6.4.1, in order to identify the location area and the network to which the user is attached. It is also part of the call control process such that suitable resources required for communication between the network and the terminal can be identified.
- A terminating session is then set-up between the called user and its service provider. Terminating session set-up is network initiated and is a lot less complicated than a normal session set-up since most of the relevant data, such as the access data and user registration information, are already available in the network during the call set-up process. The called user's service provider is identified through service provider mapping as in the session set-up process. Mutual authentication of the called user and their service provider takes place so that the service provider can inform the called user of the incoming call.

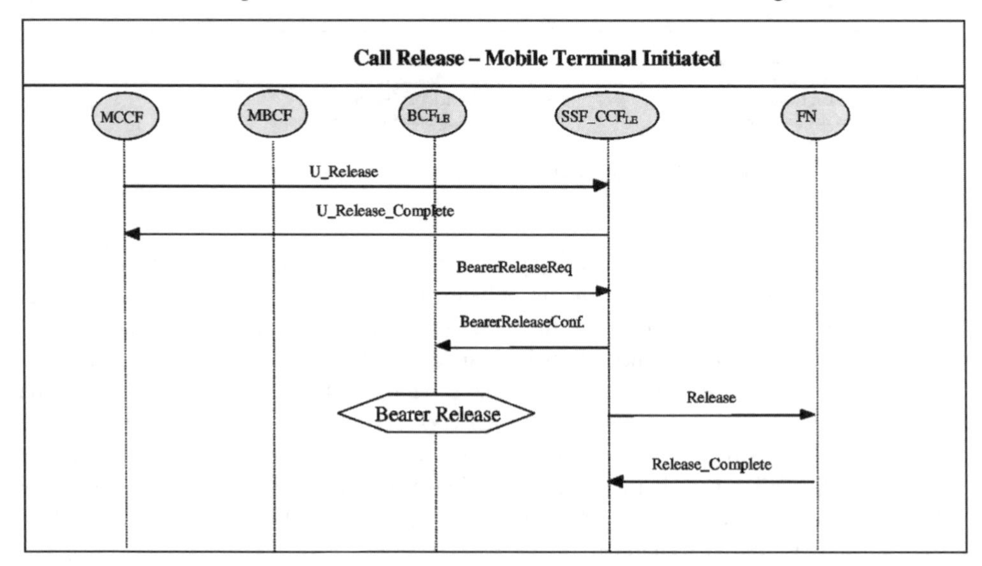

Figure 6.25 Mobile terminal initiated call release information flow.

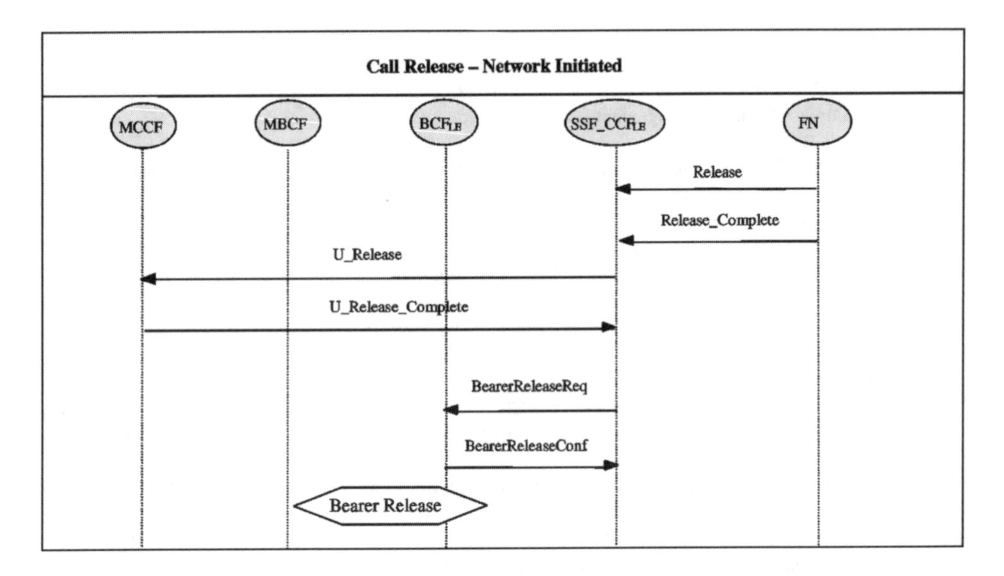

Figure 6.26 Network initiated call release information flow.

- When the terminating session is successfully established, the call set-up phase is initiated for service negotiation, bearer allocation connection, etc.

Figures 6.23 and 6.24 [CEC-96] show examples of the information exchanges for incoming and outgoing call set-up procedures, respectively.

Call Release There are two types of call release: one is initiated by the mobile terminal and the other is initiated by the network. Figures 6.25 and 6.26 [CEC-97] show examples of the message exchanges for the mobile terminal initiated and network initiated call release procedures.

References

[AKY-98] I.F. Akyildiz, J. McNair, J.S.M. Ho, H. Uzunalio Lu, W. Wang, "Mobility Management in Next Generation Wireless Systems", *Proceedings of IEEE*, 87(8), August 1999; 1347–1384.

[ANA-94] F. Ananasso, M. Carosi, "Architecture and Networking Issues in Satellite Systems for Personal Communications", *International Journal of Satellite Communications*, 12(1), January/February 1994; 33–44.

[ANA-95] F. Ananasso, F.D. Priscoli, "Issues on the Evolution Toward Satellite Personal Communication Networks" *Proceedings of Globecom '95*; 541–545.

[BAD-92] B.R. Badrinath, T. Imielinski, A. Virnami, "Locating Strategies for Personal Communication Network", *Performance Evaluation*, 20(1–3), May 1994; 287–300.

[CEC-95] EC RACE Project SAINT, "Interworking with other Networks", *CEC deliverable R2117 SRU/CSR/DR/P216-B1*, November 1995.

[CEC-96] EC RACE Project SAINT, "Operation and Procedures in Integrated UMTS Satellite UMTS Network", *CEC deliverable, R2117/AES/DAS/DR/P215-A1*, 1996.

[CEC-97] EC ACTS Project SINUS, "System Architecture Specifications", *CEC deliverable AC212 AES/DNS/DS/P222-B1*, October 1997.

[COX-72a] D.C. Cox, D.O. Reudink, "A Comparison of Some Channel Assignment Strategies in Large Scale Mobile Communications Systems", *IEEE Transactions on Communications*, COM-20(2), April 1972; 190–195.

[COX-72b] D.C. Cox, D.O. Reudink, "Dynamic Channel Assignment in Two Dimensional Large Scale Mobile Radio Systems", *Bell System Technical Journal*, 51(7), September 1972; 1611–1629.

[COX-73] D.C. Cox, D.O. Reudink, "Increasing Channel Occupancy in Large Scale Mobile Radio Systems: Dynamic Channel Reassignment", *IEEE Transactions on Vehicular Technology*, VT-22(4), 1973; 218–222.

[DEL-94a] F. Delli Priscoli, F. Muratore, "Assessment of a Public Mobile Satellite System Compatible with the GSM Cellular Network", *International Journal of Satellite Communications*, 12(1), January/February 1994; 13–24.

[DEL-94b] E. Del Re, F. Fantacci, G. Giambene, "Performance Analysis of a Dynamic Channel Allocation Technique for Satellite Mobile Cellular Networks", *International Journal of Satellite Communications*, 12(1), January/February 1994; 25–32.

[DEL-95] E. Del Re, R. Fantacci, G. Giambene, "Efficient Dynamic Channel Allocation Techniques with Handover Queuing for Mobile Satellite Networks", *IEEE Journal on Selected Areas in Communications*, 13(2), February 1995; 397–405.

[DEL-96] F. Delli Priscoli, "Mobility Issues for a GEO Multi-Spot Satellite System with OBP Capabilities", *Proceedings of 2nd European Workshop in Mobile and Personal Satcoms(EMPS)*, Springer-Verlag, Rome, October 1996: 108–114.

[DEL-97] F. Delli Priscoli, "Functional Areas for Advanced Mobile Satellite Systems", *IEEE Personal Communications*, 4(6), December 1997; 34–40.

[EFT-98a] N. Efthymiou, Y.F. Hu, R.E. Sheriff, "Performance of Intersegment Handover Protocols in an Integrated Space/Terrestrial UMTS Environment", *IEEE Transactions on Vehicular Technology*, 47(4), November 1998; 1179–1199.

[EFT-98b] N. Efthymiou, Y.F. Hu, A. Properzi, R.E. Sheriff, "Inter-Segment Handover Algorithm for an Integrated Terrestrial/Satellite-UMTS Environment", *9th IEEE International Symposium on Personal, Indoor and Mobile Radio Communications,* PIMRC '98, Boston, MA, 8–11 September 1998; 993–998.

[ETS-93] ETSI Technical Report "Satellite Earth Stations & Systems (SES), Possible European Standardisation of certain aspects of S-PCN, Phase 2: Objectives and Options for Standardisation", *ETSI TR DTR/SES-05007*, September 1993.

[ETS-94] ETSI Technical Report "Mobile Application Part (MAP) Specifications V.4.8.0", GSM 09.02, 1994.

[ETS-96] ETSI Technical Report "Satellite Earth Stations and Systems (SES), Possible European Standardisation of Certain Aspects of S-PCN, Phase 2: Objectives and Options for Standardisation", *ETSI TR DTR/SES-00002*, June 1996.

[ETS-99a] "GEO-Mobile Radio Interface Specifications: GMR-2 General System Description", *ETSI TS101377-01-03*, GMR-2 01.002, 1999.

[ETS-99b] "GEO-Mobile Radio Interface Specifications: Network Architecture", *ETSI TS101377-03-02*, GMR-2 03.002, 1999.

[HE-96] X. He, R. Tafazolli, B.G. Evans, "Paging Signalling in Non-Geosatellite Mobile Systems", *Proceedings of the Fifth International Conference on Satellite Systems for Mobile Communications and Navigation*, London, 13–15 May, 1996; 139–142.

[HO-96] J.S.M. Ho, I.F. Akyildiz, "Local Anchor Scheme for Reducing Signalling Cost in Personal Communication Networks", *IEEE/ACM Transactions on Networking*, 4(5), October 1996; 709–726.

[HO-97] J.S.M. Ho, I.F. Akyildiz, "Dynamic Hierarchical Database Architecture for Location Management in PCS Networks", *IEEE/ACM Transactions on Networking*, 5(5), October 1997; 646–661.

[HON-86] D. Hong, S.S. Rappaport, "Traffic Model and Performance Analysis for Cellular Mobile Radio Telephone Systems with Prioritised and Non-Prioritised Handoff Procedures", *IEEE Transactions on Vehicular Technology*, VT-35(3), August 1986; 77–92.

[HU-95a] L.-R. Hu, S.S. Rappaport, "Personal Communication Systems Using Multiple Hierarchical Cellular Overlays", *IEEE Journal on Selected Areas in Communications*, 13(2), February 1995; 407–415.

[HU-95b] Y.F. Hu, R.E. Sheriff, E. Del Re, R. Fantacci, G. Giambene, "Satellite-UMTS Traffic Dimensioning and Resource Management Technique Analysis", *IEEE Transactions on Vehicular Technology, Special issue"Universal Mobile Telecommunications System*", 47(4), November 1998; 1329–1341.

[JAI-94] R. Jain, Y.B. Lin, "A Caching Strategy Employing Forwarding Pointers to Reduce Network Impact of PCS", *IEEE Journal on Selected Areas in Communications*, 12(8), October 1994; 1434–1444.

[JAI-95] R. Jain, Y.B. Lin, "An Auxiliary User Location Strategy Employing Forwarding Pointers to Reduce Network Impact of PCS", *ACM-Baltzer Journal Wireless Networks*, 1(2), July 1995; 197–210.

[KRU-98] N.E. Kruijt, D. Sparreboom, F.C. Schoute, R. Prasad, "Location Management Strategies for Cellular Mobile Networks", *Electronic & Communication Engineering Journal*, 10(2), April 1998; 64–71.

[LIN-95] Y.B. Lin, S.K. DeVries, "PCS Networking Signalling using SS7", *IEEE Communications Magazine*, 33(6), June 1995; 44–55.

[MAR-97] J.G. Markoulidakis, G.L. Lyberopoulous, D.F. Tsirkas, E.D. Sykas, "Mobility Modelling in Third Generation Mobile Telecommunications Systems", *IEEE Personal Communications*, 4(4), August 1997; 41–56.

[MOD-90] A.R. Modarressi, R.A. Skoog, "Signalling System 7: A Tutorial", *IEEE Communications Magazine*, 28(7), July 1990; 19–20, 22–35.

[NAN-91] S. Nanda, D.J. Goodman, "Dynamic Resource Acquisition: Distributed Carrier Allocation for TDMA Cellular Systems", *Proceedings of Globecom '91*, December 1991; 883–888.

[SHI-95] N. Shivakumar, J. Widom, "User Profile Replication for Faster Location Lookup in Mobile Environments", *Proceedings of ACM/IEEE Mobicom '95*; 161–169.

[TEK-91] S. Tekinay, B. Jabbari, "Handover and Channel Assignment in Mobile Cellular Networks" *IEEE Communications Magazine,* 29(11), November 1991; 42–46.

[WAN-93] J.Z. Wang, "A Fully Distributed Location Registration Strategy for Universal Personal Communication Systems", *IEEE Journal Selected Areas in Communications*, 11(6), August 1993; 850–860.

[WER-95] M. Werner, A. Jahn, E. Lutz, A. Böttcher, "Analysis of System Parameters for LEO/ICO Satellite Communication Networks", *IEEE Journal Selected Areas in Communications*, 13(2), February 1995; 371–381.

[WER-97] M. Werner, C. Delucchi, H.-J. Vogel, G. Maral, J.-J. De Ridder, "ATM-based Routing in LEO/MEO Satellite Networks with Intersatellite Links", *IEEE Journal Selected Areas in Communications,*15(1), January 1997; 69–82.

[WIL-92] D.R. Wilson, "Signalling System No. 7, IS-41 and Cellular Telephony Networking" *Proceedings of IEEE*, 80(4), April 1992; 664–652.

7

Integrated Terrestrial-Satellite Mobile Networks

7.1 Introduction

Satellite integration into terrestrial mobile networks may take two paths: the evolutionary or the revolutionary approach. In the evolutionary scenario, existing standards are gradually evolved towards a new technical standard. In contrast, the revolutionary policy necessitates a completely new approach to the problem, disregarding existing standards and consequently new standards will emerge. At the start of the last decade, it was not clear which path the future development of mobile communications would take. However, by the end of the decade, it became clear that any new system would have to take into account the investments made by industry and the technological developments that had taken place. Moreover, the level of market take-up requires the need to ensure that there is some form of backward compatibility with existing systems.

We have already seen how GSM is evolving towards GPRS and EDGE, while cdmaOne is following a similar path towards cdma2000. With this in mind, it is clear that there would be very little support for introducing a revolutionary new system at this stage of the mobile communications development. It could be argued that W-CDMA is indeed a revolutionary new system, however, it is important to realise that the underpinning core network technology is still that of GSM. However, with the introduction of an all-IP core network, certainly the move from circuit to packet-oriented delivery can be considered to be a significant change in the mode of delivery.

As far as satellites are concerned, the future requirement to inter-work seamlessly with terrestrial mobile networks is paramount. Hence, there is a need to be able to adapt the terrestrial mobile standards to those of the space segment. With this in mind, this chapter will concentrate on the evolutionary approach and in particular, issues in relation to satellite-personal communication network (S-PCN) integration with the fixed public switched telephone network (PSTN), GSM and GPRS will be discussed.

In order to determine the level of integration between the S-PCN and different terrestrial networks, the requirements imposed by the mobile users, the service providers and the network operators have to be identified. Such requirements will enable the identification of the required modifications or adaptation of terrestrial mobile network functions for the support of interworking between space and terrestrial networks.

This chapter is partly based on the work carried out in the EU RACE II SAINT project, which paved the way for satellite-UMTS studies in Europe and ETSI's GMR specifications, which define the requirements for integration between a geostationary satellite and the GSM network.

7.2 Integration with PSTN

7.2.1 Introduction

Mobile satellite networks are required to interwork with both fixed and mobile networks, including PSTN and GSM. Due to different national PSTN implementation standards, interoperability between PSTN and S-PCN relies on the use of ITU-T Signalling System No. 7 (SSN7) both within national networks and at the international interconnection point between national networks. This is achieved using SSN7 signalling at the international switching centre (ISC) of the PSTN, and an interworking function at the fixed Earth station (FES), or the gateway, of the satellite network to adapt SSN7 to an S-PCN compatible format, as shown in Figure 7.1 [CEC-95]. A two level coupling between the PSTN and S-PCN has been proposed.

- Gateway functions: the gateway functions ensure end-to-end interoperability between the PSTN and S-PCN subscribers and *vice versa*.
- Access functions: the access functions allow S-PCN subscribers to access the S-PCN via an FES using a transit link between the FES and the S-PCN.

The signalling configuration as shown in Figure 7.1 has the advantage that little or no modifications to existing SSN7 procedures implemented in the PSTNs are required.

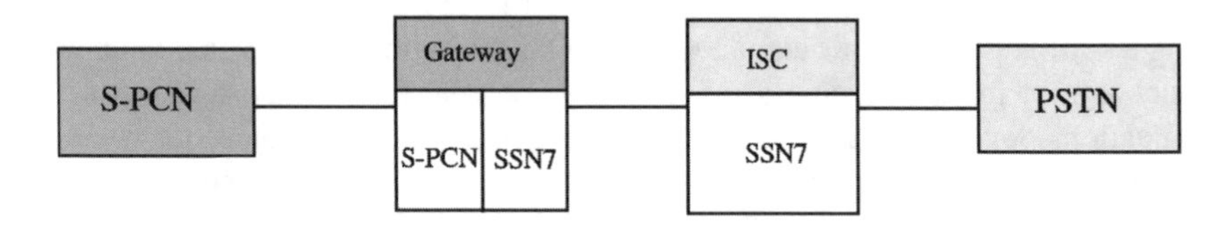

Figure 7.1 S-PCN-PSTN signalling connection.

7.2.2 Gateway Functions and Operations

Three functional components are required to support interworking functions at the gateway: the PSTN gateway switching centre (PGSC), the S-PCN gateway cell site switch (SGCSS) and the S-PCN database (SDB) as shown in Figure 7.2. Their functions are briefly described below.

- PGSC: the PGSC provides PSTN access to the S-PCN network. SSN7 can be used to

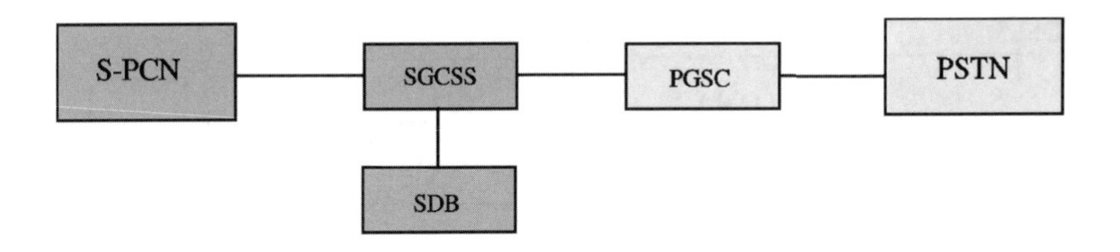

SGCSS S-PCN Gateway Cell Site Switch
SDB S-PCN Database
PGSC PSTN Gateway Switching Centre

Figure 7.2 S-PCN-PSTN gateway function.

perform any signalling exchange between the PGSC and the SGCSS. All calls destined for the S-PCN must be routed through this gateway.

- SGCSS: in analogy to PGSC, the SGCSS provides S-PCN access to PSTN. Signalling conversion between the S-PCN and SSN7 is also performed in this gateway. In addition, interworking functions such as voice encoding, decoding and bit rate adaptation are also supported in order to ensure end-to-end interoperability between the PSTN and the S-PCN gateway.
- SDB: this database contains information on the S-PCN mobile terminals such as the location, terminal characteristics, service profiles, authentication parameters and so on.

Figure 7.2 depicts a scenario whereby a PSTN interfaces with other external networks. It is also similar to that employed by GSM to interface with PSTNs. This approach has the advantage that little or no modification is required in the PSTN.

Upon receipt of a call request from a PSTN subscriber, the PGSC informs the SGCSS of such a request. The SGCSS then interrogates the SDB to check whether the called S-PCN subscriber is attached to the network. If the S-PCN subscriber is attached to the network, the called party's location will be identified. Signalling conversion operation is then performed by the SGCSS to translate the SSN7 call request from the PSTN to a S-PCN call request. The call is then routed to the appropriate FES. If the addressed S-PCN user is unavailable, the PSTN subscriber receives a notification, informing of the situation.

In the case when a call request is from an S-PCN subscriber, the SGCSS translates the request to an SSN7 call request and signals the PGSC of such a request. The PGSC then carries out relevant procedures to check if the called PSTN user is available. Appropriate signalling is translated and exchanged between the two networks to inform the S-PCN caller of the status of the call request through the two gateways.

7.2.3 Protocol Architecture of SSN7

7.2.3.1 Constituents

Before addressing the interworking features for signalling to interconnect the S-PCN and PSTN, the call set-up and release procedures between the two networks need to be analysed. Based on the assumption that the PSTN part will adopt the SSN7 signalling system, issues on how the message transfer part (MTP), the signalling connection control part (SCCP) and the

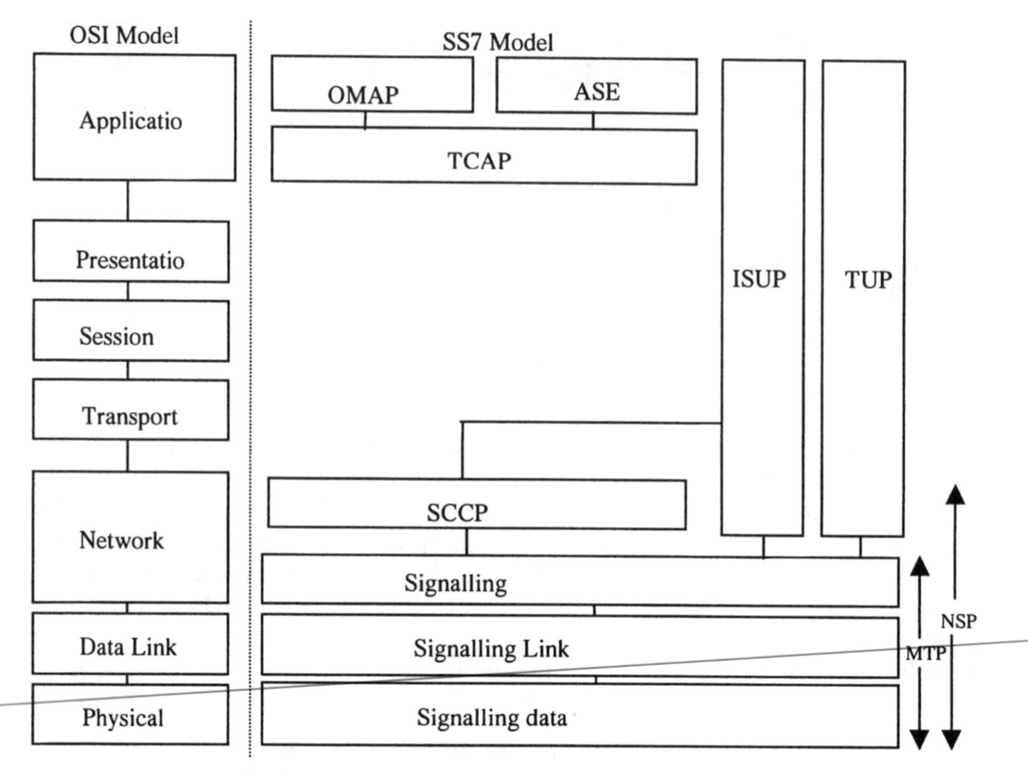

OMAP Operation, maintenance and administration Part
ASE Application Service Element
TCAP Transaction Capabilities Application Part
ISUP ISDN User Part
SCCP Signaling Connection Control Part
MTP Message Transfer Part
NSP Network Service Part

Figure 7.3 SSN7 signalling architecture.

telephone user part[1] (TUP) of this signalling can support the S-PCN call establishment procedures need to be investigated. As shown in Figure 7.3 [STA-95], the SSN7 signalling system has been specified in four functional levels.

- The MTP provides a reliable but connectionless service for routing messages through the SSN7 network. It is made up of the lowest three layers of the open systems interconnection (OSI) reference model. The lowest level, the signalling data link, corresponds to the physical level of the OSI model and is concerned with the physical and electrical characteristics of the signalling links. The signalling link level, a data link control protocol,

[1] The TUP layer may not appear in some reference books. It is used here for discussions on plain old telephone (POT) networks.

provides reliable sequenced delivery of data across the signalling data link. It corresponds to level 2 of the OSI model. The top level of the MTP is the signalling network level, which provides for routing data across multiple signalling points (SP) from control source to control destination. However, the MTP does not provide a complete set of functions specified in the OSI layers 1–3, in particular, the addressing function and connection-oriented service.
- The SCCP is developed to enhance the connectionless sequenced transmission service provided by the MTP by providing an addressing function and message transfer capabilities. The addressing function is supported by the destination point codes (DPC) and subsystem numbers (SSN). The addressing function is further enhanced by providing mapping facilities to translate global titles, such as the dialled digits, into the DPC + SSN format in order to route messages towards a signalling transfer point. The SCCP and the MTP together are referred to as the network service part (NSP). In the integrated scenario, the SCCP is required to be capable of routing messages through different networks based on different protocols in order to ensure interworking between the two different networks.
- Both the ISDN user part (ISUP) and the TUP form the user parts of the signalling system. They provide for control signalling required in an ISDN and telephone network to handle subscriber calls and related functions. They correspond to level 4 of the OSI structure.

The advantage of adopting signalling interworking using the SSN7 signalling architecture is that it intrinsically allows the definitions of interworking functions to ensure end-to-end interoperability between two different networks. This has also been the approach to interconnect PSTN with terrestrial mobile networks.

7.2.3.2 The MTP Layer

In order to provide interoperability with the SGCSS functional component in the S-PCN, the implementation of the MTP in the PGSC has to take into account the interworking requirements with the equivalent layers in the SGCSS. In plain old telephone (POT) networks, the MTP receives routing messages directly from the TUP, as shown in Figure 7.3, without having an intermediate layer such as the SCCP required by the ISUP. The message format of a routing label in the signalling information field of an SSN7 message signal unit consists of three parts, as shown in Figure 7.4.

- The signalling link selection (SLS): the value of the SLS is assigned by the user part in level 4. For a given source/destination pair, several alternate routes may be possible. The value of the SLS specifies the routing information. It is used to distribute traffic uniformly among all possible routes.

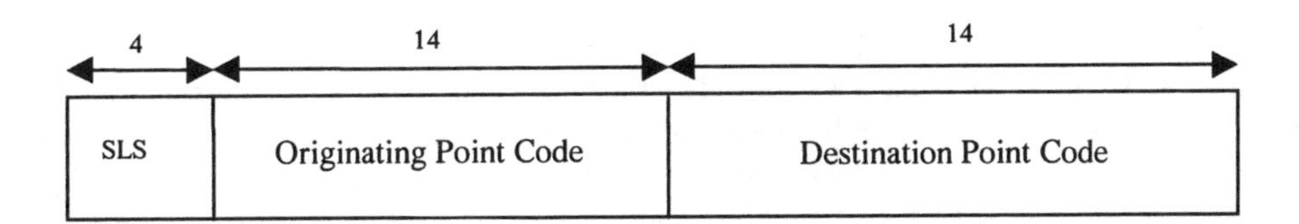

Figure 7.4 Routing label of SSN7.

- The originating point code (OPC): this field contains information relating to the source node such as the source address.
- The destination point code (DPC): this field contains information about the destination node such as the point code of the destination switch exchange. By means of the DPC, each intermediate switching exchange during the routing process can decide whether the message is destined for itself or whether it should be routed onward.

Since mobility is not supported in the PGSC, the mobile user's address cannot be translated into a recognisable DPC format by the PGSC during a call set-up procedure from the PSTN to the S-PCN direction. In this case, the PGSC will simply forward all the outgoing messages to the SGCSS, which will then query the SDB to obtain the mobile roaming data. Hence, in order to ensure a correct addressing operation, the following procedures are carried out.

1. Upon receipt of a call set-up message from the PGSC indicating a PSTN to S-PCN call establishment, the SGCSS associates the mobile user number with its roaming location and prepares the DPC accordingly.
2. The SGCSS will also translate the OPC to its own format so that within the satellite network itself, the OPC will refer to the SGCSS as the source address. In such a way, all subsequent switching centres within the S-PCN will be processed as if they were S-PCN messages.
3. The SGCSS will then translate the call set-up message into an initial address message (IAM) for setting up an end-to-end signalling path.

Similar procedures will be carried out in the PGSC with call set-up in a direction from the S-PCN to the PSTN.

7.2.3.3 SCCP Layer

For PSTNs that rely on the ISUP protocol, an SCCP layer is required to provide an intermediate addressing level between the user part and the MTP layer. The SCCP routing capabilities are supported by an international switching centre since they support ISUP protocols for layer 4 of the signalling system. However, addressing in SCCP is based on the global title, i.e. dialled digits, which is not supported by the SSN7 signalling network. The global title must then be translated by the MTP into an appropriate DPC. Furthermore, the SCCP layer implemented within the S-PCN will be tailored to its own requirements, which may not fit into the SSN7 signalling format. As a result, interworking functions between the PGSC and the SGCSS have to be carried out in order to translate the global address into the appropriate DPC. The MTP layer carries out the normal addressing and routing procedures as described in Section 7.2.3.2.

7.2.3.4 The TUP Layer

Although the TUP layer has been standardised by the ITU (Q.601–695), its implementation is still catered towards each country's switching network's requirements. The information transported by the TUP protocol which will have an impact on the interworking functionality with S-PCN includes the called party address, calling party address, circuit identity and the request of digital link (64 kps).

The *called party address* is identified by the called party's telephone number. The numbering scheme adopted in this address will impact on the call set-up procedure. Two types of numbering schemes can be implemented: A world-wide implementation with a common country code scheme or a nation-wide operation. In the former approach, an international switching centre (ISC) node has to be implemented to act as a gateway to the S-PCN network. Referring back to Figure 7.2, the PGSC can act as the ISC. In the latter approach – the nation-wide implementation – the numbering scheme is adopted as a specific PLMN identifier, which forms part of the whole telephone number. This approach is adopted in existing GSM networks within the same country. In this case, no ISC is required and all incoming traffic will be routed towards the nearest S-PCN gateway (the SGCSS) through the PGSC. This numbering scheme, however, requires at least one access point being accessible within each country.

The *calling party address* contains information on the calling party's telephone number. The PGSC provides this information upon receipt of a request from the SGCSS. The data supplied will depend on the implementation of the numbering schemes by the S-PCN operators. If a unique country code scheme is adopted, this information will also have to be supplied to the PSTN exchange.

The *circuit identity* specifies the characteristics of the signalling link established in the call set-up process. In an integrated environment, for an S-PCN to PSTN call set-up, the SGCSS will have a flag to indicate whether the communication is through the satellite or the terrestrial links so that appropriate actions can be taken.

The *request of digital link* requires that both the calling party's exchange and the called party's exchange must agree on the bit rate of the bi-directional digital circuit. This end-to-end requirement may be mapped onto the interworking functionality in the PGSC/SGCSS interface, in particular, the transcoding capabilities.

7.2.4 Access Functions

The access functions provide interconnection between an FES with an SCSS through the PSTN in order to exchange information during call establishment, user registration/location update and handover. Figure 7.5 shows the functional components for interworking access functions between the S-PCN and PSTN [CEC-95]. Three main functional components are involved:

- S-PCN cell site switch (SCSS): the SCSS is responsible for the switching, service and management functions between the S-PCN mobile terminals via the FES. It is normally coupled to the local exchange to provide PSTN interface functionality, which depends on

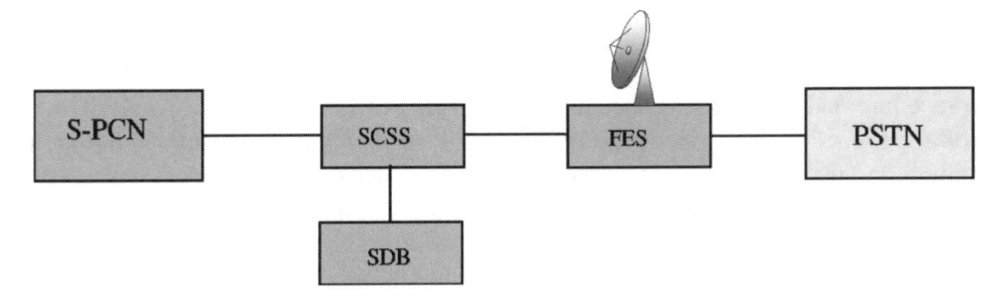

Figure 7.5 S-PCN-PSTN access function.

the connection type employed. There are two main types of connections: the dial-up type connection or a permanent connection. The on-demand type is similar to the dial-up connection, in which case the SCSS is coupled to the local exchange through one or more POT connections. For the permanent connection, a wide-band connection is established through the PSTN between the SCSS and the FES.

- S-PCN database (SDB): the SDB is similar to the SDB described in Section 7.2.2.
- Fixed Earth station (FES): the FES provides the interface between the S-PCN access network and the mobile terminal. Similar to the SCSS, the FES is connected to the local exchange through one or more POT connections or directly to the SCSS via a permanent leased line connection.

If dial-up connection is employed between the SCSS and the FES through the PSTN, connection between the SCSS and the FES needs to be established whenever information exchange takes place for different network procedures. This will introduce additional delay in exchanging information between the SCSS and the FES. The additional delay may prove significant in procedures such as registration, location update or handover. In addition, this type of connection is analogue in nature and exchange of information will have to make use of a dial-up modem, which may limit the data rates. On the other hand, if a permanent connection is deployed, there is no need to establish connection between the SCSS and the FES before any information exchange. Hence, no additional delay incurs. Furthermore, the connection is digital in nature which can offer higher data rates than those offered with the on-demand connection.

7.3 Integration with GSM

7.3.1 Introduction

In considering integration with GSM, the level of integration is the most important factor in determining the integrated network architecture. Two approaches can be drawn in designing such an integrated system: one is to modify the existing GSM infrastructure to suit the peculiarity of the satellite system; or to adapt the satellite system to existing GSM networks. As the development of GSM has already reached maturity, and future systems based on GSM, such as GPRS and HSCSD, are already standardised, an integrated system which would require significant modifications to existing GSM protocols or its infrastructure is seen as impractical. Hence, the discussion that follows assumes adaptation of and modifications to satellite network protocols for integration with the GSM network.

An integrated S-PCN and GSM system should aim to satisfy the user requirements as far as possible but at the same time should also be feasible from the network operators' perspective. From the users' perspective, the type of supported services and their quality, the authentication procedure and the charging mechanism form part of the user requirements. From the network operators' point of view, the requirements in interworking functions, the network procedures and management functions are important issues to be considered for a successful integration. Such requirements form the basis for the identification of possible integration scenarios.

Before going into detail on the integration requirements and scenarios, Table 7.1 summarises the similarities and differences between the S-PCN and GSM networks, as identified in Ref. [GMR-99]. They are based on the assumption that geostationary satellites are deployed in the S-PCN.

Table 7.1 Similarities and differences between a mobile-satellite network and a GSM network

Similarities	– The frequency re-use concept adopted in the GSM network can be applied to the S-PCN – A satellite spot-beam coverage is equivalent to a GSM cell coverage – Higher layer protocols of the GSM network may be adopted in the S-PCN with possible modifications
Differences	– Longer propagation delays in the S-PCN due to the long satellite-to-earth path and to the longer distance between the FES and the user terminal – Higher attenuation in the radio signal – Larger variations in conversational dynamics in voice communications – Increased echoes – Delay in double-hop connection for mobile-to-mobile call may become unacceptable – A more sophisticated timing synchronisation scheme is required – Higher attenuation in the radio signal in the satellite network due to the longer propagation delay – A spot-beam coverage is much larger than a terrestrial cellular coverage resulting in lower inter-spot-beam handover probability. However, the benefits of frequency re-use is diminished – A power control mechanism is required in a satellite network as the satellite power is shared by all the spot-beams over the entire coverage area This is different from a base station power in a terrestrial network, which is not a shared resource – Line-of-sight operation is required in a satellite network in order to compensate for the high attenuation in the radio signal in contrast to the use of multipath signals in terrestrial cellular mobile networks (see Chapter 4). An alerting procedure is required to inform users of any incoming calls when they are not in line-of-sight with the satellite – The satellite radio channel characteristics are described by the Rician channel as opposed to the Rayleigh channel in terrestrial networks – Adjacent cell interference in a terrestrial cellular network is a function of power and cellular radius; whereas adjacent spot-beam interference is a function of power and sidelobe characteristics of the satellite antenna array – User terminals can access and be accessed by any one of the gateways in satellite networks in contrast to terrestrial network access in which the user terminal in any given cell can only access an MSC associated with that cell – Optimum call routing is possible in satellite networks by routing the call to the nearest gateway to the called party. This is impossible in existing GSM networks – Döppler shift can be considerable in a satellite network especially during the initial period of operation

7.3.2 Integration Requirements

7.3.2.1 User Requirements

With GSM, the main services being provided are still very much voice-oriented. Future mobile users will require a wider range of telecommunication services, including those of messaging, voice communications, data retrieval and user controlled distribution services. Different services and applications may cater for different user groups, which can be divided into two main categories: business and private. It is expected that for business users, the services requested will require wider bandwidth, for example, data retrieval, high resolution image transmission, videoconferencing, etc. On the other hand, for private users, narrower bandwidth services are envisaged. From the users' perspective, the services provided have to be user friendly, ubiquitous, low cost, safe and of an acceptable quality. This has several implications:

1. User friendliness requires simple, easy-to-use and consistent human machine interface (HMI) design for user access regardless of the terminal type and the supported services. User interaction should be kept as low as possible. In the case of dual-mode terminals, the HMI should be consistent for access to both networks, i.e. the same access procedures for both networks and preferably the same user identity number for users to obtain access to both networks.
2. Ubiquitous service provision implies users can access services anywhere at anytime. This requirement has several impacts on the design of terminal equipment, the operating environment and the level of mobility being supported by the networks, hence their interworking functions in supporting mobility.
3. Terminal equipment can be broadly categorised into hand-held, portable and vehicular-mounted devices [CEC-97]. It is important that terminals should remain lightweight and compact for ease of carry and ease of storage. For vehicular-mounted terminals, this may include group public terminals mounted on buses, coaches, aeroplanes or sea-liners. The same HMI requirements as identified in point 1 above should apply to all terminal categories.
4. Ubiquitous coverage also implies that users can obtain access to services regardless of the type of environment they are in, including indoor and outdoor environments; urban, sub-urban and rural areas; motorways, railways, air and sea. This requires both networks to be complementary to each other. For example, the satellite network may fill in "holes" which are not covered by the terrestrial network.
5. Ubiquity also implies that users should be able to roam between different geographical areas, whether nation-wide or between countries. The ability to roam between the two networks in the case of dual-mode terminal (DMT) users should also be possible.
6. Cost is perhaps one of the most important factors that has an impact on the popularity of services. Needless to say, users require services that are good value for money. With this in mind, different charging options should be made available to cater for specific user needs. It is likely that there will be a significant difference in call charges and subscription fees between the terrestrial and satellite networks. Technical aspects such as inter-segment handover will require careful design and consideration in order to issue call charges when the user is switched from one segment to another.
7. A safe and secure service encompasses technical aspects, such as user authentication, terminal authentication, confidentiality of user profiles and location, ciphering and encryp-

tion, as well as health and safety issues, which may impose a limitation on the radiation power on the user terminal.

8. The QoS supported by both networks should be as consistent as possible. Every effort should be made to ensure the quality is compatible with that supported by the fixed network.

7.3.2.2 Network Operator Requirements

Network operators require the network to function in an efficient manner and at the same time to satisfy the customers' needs. Network operations involve three main control functions: call control, resource control and mobility control. In addition to these three control functions, network management functions are required to monitor the performance of the network operations as well as the maintenance of the network, i.e. the operation and maintenance (OAM) functions. In an integrated network environment, interworking functions between different networks are also required to ensure network inter-operability. The following gives a list of what is considered to be important for the integrated system from a network operator's point of view.

1. Modifications to existing standards or network infrastructure should be minimised. Thus, in an integrated GSM-S-PCN network, significant modifications to the GSM system are not expected. Hence, the definition of interworking functions and the implementation of such interworking procedures will mainly concern the S-PCN part.
2. Call control procedures adopted in the GSM network should preferably be re-used in the integrated network for compatibility. However, some of these procedures may not be suitable for the satellite counterpart due to the longer propagation delays. For instance, at least ten one-way [CEC-95] information exchanges between the BSS and the MT are required for call set-up in a GSM network. This will not be feasible in a geostationary satellite network in which a one-way transmission delay of about 270 ms will be incurred.
3. Resource control procedures are important in any radio network as the radio resource is scarce. This is especially so when satellite networks are considered. Depending on the level of integration, network resources in the terrestrial network and those in the satellite network can be shared so that overflow traffic can be redirected to the satellite network and *vice versa*. This may require special switching functionalities to be implemented in both the S-PCN and GSM networks in order to switch traffic between the two.
4. Mobility control procedures encompasses handover, location management, paging and call delivery. In an integrated network, the possibility of inter-segment handover should be taken into account. If inter-segment handover is considered, the call set-up procedure and the resource control mechanism will also need to take into account such handover mechanisms. The location management procedures in a satellite network will be different from that in the GSM network as discussed in Chapter 6. Hence, to ensure mobility between the two networks, an important interworking goal is to provide compatibility between network location registers. This implies that the numbering and addressing schemes need to be compatible between the two systems. This requires an examination of the feasibility of adopting existing GSM numbering and addressing schemes to the requirements of S-PCN operations. This may decide whether S-PCN would adopt a GSM-like numbering system

or whether a completely new scheme is required. Furthermore, the protocols used between location registers is also required to be examined. In GSM systems, MAP protocols are used to support mobility. Whether this or other protocols such as INAP will be adopted in S-PCN requires a close examination on the suitability of their use.

5. Signalling in all the above mentioned network control procedures should be minimised since this will consume both the frequency spectrum as well as the satellite power. Excessive signalling also increases signalling load and delay. This would affect the network efficiency as well as the QoS.

7.3.3 Integration Scenarios

7.3.3.1 Integration Levels

When considering the re-use of the GSM network architecture and structure, the equivalent counterparts of the S-PCN are indicated in Figure 7.6, where the use of 'G' in the acronyms refers to Gateway. Four levels of integration are depicted in the figure.

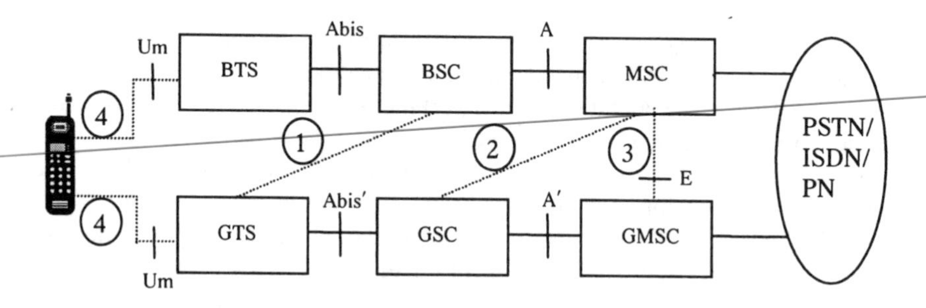

Figure 7.6 S-PCN-GSM integration scenarios.

1. Integration at the Abis-interface – separate BTS/GTS, common BSC and common MSC.
2. Integration at the A-interface – separate BTS/GTS and BSC/GSC, common MSC.
3. Integration at the E-interface – separate terrestrial and satellite networks but with inter-working functions between the MSCs.
4. Integration at the Um-interface – terminal integration.

In the above, the terminology adopted by ETSI is used to illustrate the satellite components of the network, that is GTS, GSC and GMSC.

7.3.3.2 Integration at the Abis-Interface

In this integration scenario, apart from the two different transceiver systems, the other network components are shared between the two networks as depicted in Figure 7.7. Inter-working functions between the two segments will have to be defined at both the Abis- and A-interfaces.

With different transceiver systems, the air interfaces in the respective segments can be optimised to suit the transmission characteristics of the individual propagation channels.

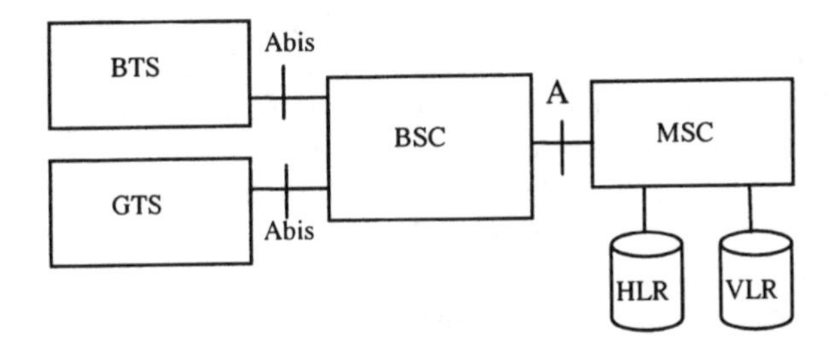

Figure 7.7 Integration at the Abis-interface – common BSC.

Since the BSC and the MSC are common to both networks, inter-network control functions such as inter-network mobility control, resource control and switching functions are easier to co-ordinate. Since the MSC is common to both networks, the HLR and VLR are available to both networks. This allows quick and complete access to all location registration information.

Due to the different transceiver systems, the QoS supported by one segment may not be consistent with the other segment. Furthermore, a common BSC and MSC implies that extra functions need to be implemented in these two network components in order to cater for the specific needs required by the satellite network, unless the same Abis-interface is implemented in both networks. This may require modifications to existing GSM standards, which are not considered likely. Although inter-network control functions are easier to co-ordinate, the design of the switching functions and control functions in the centralised BSC and MSC may not be optimum for each individual segment. For example, a fast switching between two BTSs within the same segment may not be possible. Since the BSC and MSC are responsible for the co-ordination of both segment specific and inter-segment network control functions, extra signalling functionalities are required to be carried out by the MSC which again would require modifications to the GSM standard.

7.3.3.3 Integration at the A-Interface

In this integration scenario, the satellite gateway provides functionalities similar to those of the BTS and the BSC of the GSM network. Figure 7.8 depicts this integration scenario.

Similar to the previous integration scenario, air interfaces can be optimised for each

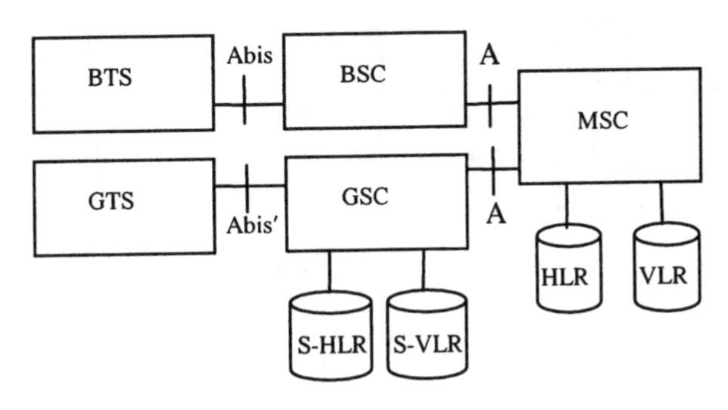

Figure 7.8 Integration at the A-interface – common MSC.

individual segment. Switching functionalities at the BSC and the GSC can be optimised for individual segments to enable quick switching between different BTSs or GTSs within the same segment. A single MSC can ease co-ordination of inter-network control functions, especially for inter-segment handover. Furthermore, as before, a common HLR and a common VLR implies full access to location registration information. This will ease signalling exchanges for subscriber data between different segments. The only drawback of this integration scenario is that modifications to the A-interface and the MSC are required in order that the MSC can establish signalling exchanges with both the GSM and the satellite access network. This implies that the functionality of the MSC may not be at an optimum for each individual segment.

7.3.3.4 Integration at the E-Interface

In this integration scenario, the GSM and S-PCN segments will have their own access networks, and each access segment will have a common network infrastructure as shown in Figure 7.9. This will enable calls to be routed transparently between the different access networks. With this level of integration, both networks can share a common HLR while maintaining a separate VLR. This common networking infrastructure allows transparent fixed user to mobile user connections since a unique calling number can be assigned to the mobile user regardless of the different segments. A single satellite gateway can be shared among different GSM networks with the satellite transponder capacity being dynamically shared among different traffic stations. In this respect, the satellite network operators may be considered as the provider of a common set of services to different service providers. Interworking functions between the two networks needs only be defined at the E-interface, providing switching and resource control functions between the two segments. The only modification to existing GSM standards is envisaged to be at the E-interface, if the functionalities of the GMSC and the MSC are not identical.

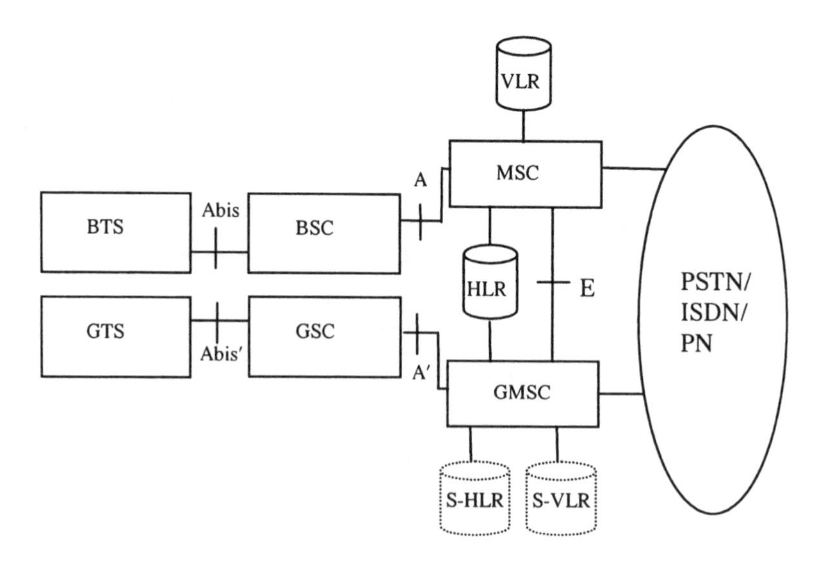

Figure 7.9 Integration at the E-interface – separate MSC.

7.3.3.5 Integration at the Um-Interface

This scenario has to be differentiated from other integration scenarios in which a dual-mode terminal is also required for dual-mode access. In this scenario, there is no integration between any network elements of the two networks. The only common point for the two mobile networks is at the dual-mode terminal as depicted in Figure 7.10.

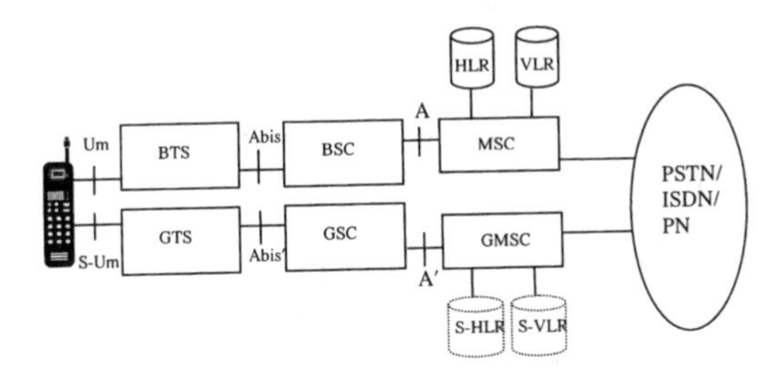

Figure 7.10 Integration at the Um-interface.

Segment selection will be performed during the call set-up stage. As there is no common network elements shared between the two networks, the dual-mode terminal merely serves the purpose of enabling users to obtain access to both segments. If roaming between the different networks is supported, a common HLR is required, which will then reduce the integration scenario to that of the split MSC integration option. Handover of calls between the two networks may only be possible when sufficient intelligence is implemented in the terminal such that handover of calls between the two networks will be controlled by the mobile, i.e. mobile controlled handover (MCHO). This will increase the complexity and the cost of the terminal.

7.3.4 Impact of Integration Scenarios on the Handover Procedure

In this section, only integration scenarios at the E- and A-interfaces are considered.

7.3.4.1 Integration at the E-interface Level

Referring back to Figure 7.9, which shows the integrated S-PCN/GSM network architecture at the E-interface, i.e. the S-PCN and GSM networks are stand-alone networks. In this figure, the HLR and VLR associated with the GSM network contain dual-mode user information, whereas the S-HLR and S-VLR contain information on subscribers registered to S-PCN only. Dual-mode users are managed by the GSM operators and are assigned with a GSM MSISDN number. Any calls from the fixed network to a dual-mode user will be routed through the GSM MSC by identification of the GSM MSISDN number. The impact of such an integration on handover, location management and call control functions are described below.

Impact on Handover From a GSM Cell to a Satellite Spot-Beam Using GSM Protocol GSM adopts mobile assisted handover (MAHO) as the handover controlling scheme, i.e. both the mobile station and the network monitors the received signal strength and the channel quality but the handover decision is made by the network. Furthermore, backward handover is adopted as the connection establishment scheme. Assuming that the GSM protocol is adopted as the baseline protocol for the integration scenario, Figure 7.11 shows the handover signalling flow. Due to the different network features and propagation delays, some parameters at the A-interface will have to be modified. In particular, a need for modification in the "Handover request" and "Handover request ack" messages have been identified. Furthermore the impact on handover interruption time due to the propagation delay difference between the GSM and satellite networks is also examined.

Modifications to Handover Request Message The "Handover request" message is a BSS-MAP message sent from the GMSC to the GWS. The information elements contained in the user data field of this message come mostly from the message "perform handover", which is sent by the GSM MSC to the GMSC. Some information elements are mandatory with fixed length while others are optional, as shown in Table 7.2 [CEC-95]. The effects of integration on three of these information elements are particularly discussed, namely, the *serving cell identifier*, the *target cell identifier* and the *radio channel identity*.

The *serving cell identifier* uniquely identifies the serving cell. Table 7.3 shows the format of this information field. In particular, the *cell identification discrimination* field indicates the whole or part of the *cell global identification (CGI),* which includes the *mobile country code* (MCC), *mobile network code* (MNC), *location area code* (LAC) and the *cell identity* (CI). For GSM, if handover occurs in different location areas, the LAC is used; whereas CI will be used if handover occurs in the same location area. The whole CGI is used when inter-PLMN handover occurs, which is not envisaged in the GSM handover process, although this option is provided. In an integrated S-PCN/GSM system, there may be more than one PLMN under the coverage area of a single satellite spot-beam. Hence, CGI will have to be used in order to uniquely identify a GSM cell. This implies that the GMSC will have to be responsible for translating the LAC, which was passed from the GSM-MSC to the GMSC in the "perform handover" message, into a CGI.

The target cell identifier identifies the spot-beam, the format of which is the same as that of the *serving cell identifier*. In order to uniquely identify the spot-beam, the satellite identifier number in the satellite constellation and the spot-beam number need to be identified. In addition, it also needs to distinguish the satellite network from the terrestrial network. Under this situation, it will be convenient to use the *location area identifier (LAI),* which includes the MCC, MNC and LAC. However, the meanings of the MCC and MNC will be different from those in the *serving cell identifier*. For example, the MCC can be used to identify that the target cell is in a satellite network, the MNC to identify the satellite number and the LAC to identify the spot-beam number.

The radio channel identity identifies the allocated radio channel. In GSM, this information element is optional and is only used for OAM purposes. The channel resource allocation function is performed by the base station. The MSC will not control such an allocation. In a satellite network, depending on the network architecture and functions, there could be the possibility that the MSC will be responsible for the allocation of channels for incoming calls and handover functions in order to improve resource utilisation efficiency. In this case, this information element will become mandatory in the integrated system. There are four fields in this information element: the *channel descriptor*, the *frequency channel sequence*, the *mobile allocation* and the *starting time*. The channel descriptor identifies the slot number and the frequency number. The slot number may remain unchanged if the GSM TDMA frame length

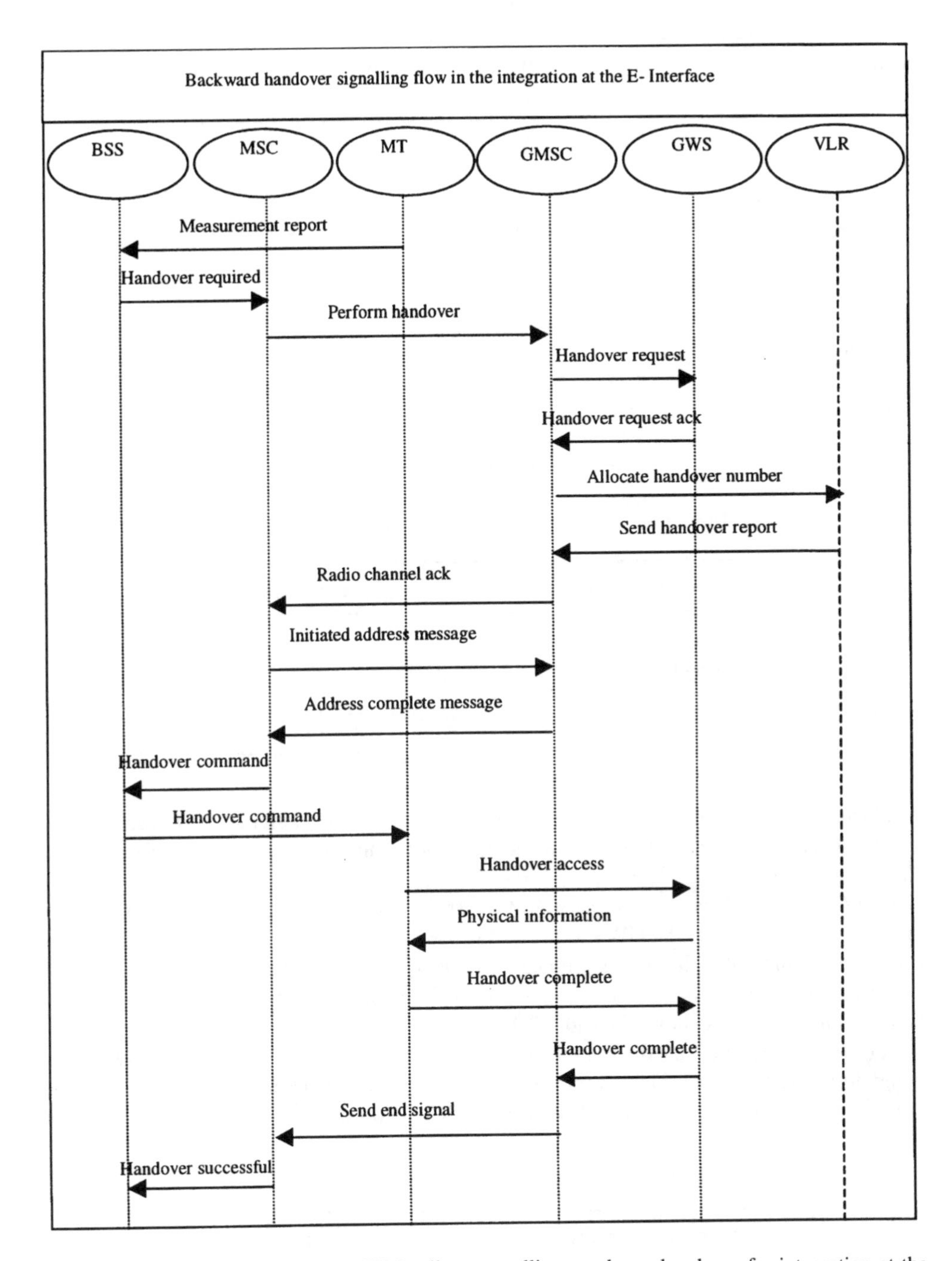

Figure 7.11 Signalling flow for a GSM cell to a satellite spot-beam handover for integration at the E-interface [CEC-95].

Table 7.2 Information elements in the "handover request" message in GSM

Information elements	Type (M: mandatory; O: optional)	Length (octet)
Message type	M	1
Channel type	M	5
Encryption information	M	3–20
Classmark information: 1 or 2	M	2 or 4
Serving cell identifier	M	5–10
Priority	O	3
Circuit identity code	O	3
Radio channel identity	O	5–38
Downlink DTX flag	O	2
Target cell identifier	O	3–10
Interference band	O	2

Table 7.3 Serving cell identifier field format

Information elements	Length (octet)
Element identifier	1
Length	2
Cell identification discrimination	3
Cell identification	4–10

is adopted in the satellite network. The maximum frequency number, however, may be subject to change, if the satellite network has more than 256 carriers, as opposed to the 124 carriers in GSM.

Modifications to Handover Request Ack Message This message is again a BSS-MAP message and is sent from the GWS to the GMSC to indicate that the handover request can be supported by the GWS. It will also indicate the radio channel to which the MT is assigned. The user data field in this message contains two information elements: the *message type* and *layer 3 information*, as shown in Table 7.4.

The *message type* is mandatory and is of length 1 octet. The *layer 3 information* field is also mandatory, the length of which varies from 9 to 56 octets. It contains the radio interface *handover command* message, which is then carried forward from the GMSC to the GSM-

Table 7.4 Handover request ack message format

Information elements	Type	Length (octet)
Message type	M	1
Layer 3 information	M	9–56

Table 7.5 Handover command message format

Information elements	Length (octet)
Protocol discriminator	2
Transaction identifier	2
Message type	2
Cell description	2
Channel description	3
•	
•	
•	

MSC. The GSM-MSC will forward this *handover command* message to the GSM-BSS transparently, as shown in Table 7.5. More than ten parameters are included in the *handover command* message. Among these parameters, two of which, namely, the *cell description* and the *channel description*, have to be modified in order to suit the satellite network.

The *cell description* field in GSM describes the target cell. In the satellite network, this field describes the target spot-beam. In GSM, this field has a length of two octets with four parameters, the format of which is shown in Table 7.6. The *BA-NO* is the frequency band identifier and is set to zero in GSM. The *NCC* and *BCC* together identify the target cell. For a satellite network, these two fields may be replaced by the LAI as described in the *target cell identifier* of the *handover request* message in order to identify the target spot-beam. Furthermore, the *BCCH carrier number* should identify the BCCH carrier number in the spot-beam. Depending on the number of carriers to be deployed in the satellite network, the *BCCH carrier number* field may span over two octets.

Table 7.6 Format of the cell description field

8	7	6	5	4	3	2	1
		Cell description IEI					
BA-NO			NCC			BCC	
		BCCH carrier number					

The *channel description* has several parameters, however, only two of them require modifications to suit the satellite network, namely, the *time slot number (TN)* and the *RF channel number (ARFCN)*. The *time slot number* takes 3 bits in length while the *RF channel number* takes 7 bits. For the same reason as indicated previously, the *RF channel number* may be modified according to the maximum channel number in the satellite network.

Handover Interruption Duration When handover occurs in a GSM system, the MT has to switch to the allocated new traffic channel from the old traffic channel. This switching process results in a short interruption to the service. This interruption mainly arises from the signal processing delay, transcoding delay, propagation delay and packet transmission delay.

Synchronisation between the MT and the base station has to be performed. Although the interruption duration may be up to 150 ms, it is distributed over a long time interval such that the interruption is hardly noticeable by the users.

In an integrated GSM/S-PCN system, the interruption will be much higher due to the longer propagation delay in the satellite to Earth path. If the GSM protocol is used for handover between the terrestrial and satellite networks, the MT needs to synchronise with the target satellite spot-beam after receiving the *handover command* from the GSM-BSS, as depicted in Figure 7.11. The MT sends a *handover access* burst to the GWS. The GWS calculates the single trip delay for signal transmission between the MT and itself. A *timing advance (TA)* will then be allocated to the MT through the *physical information* message. Upon receiving this message, the MT sends a *handover complete* to GWS and begins transmission on the allocated traffic channel based on the assigned TA value. As a result, three single trip delays are encountered from the moment the MT receives the *handover command* to *handover complete*. Depending on the satellite constellation type, the interruption duration can range from 216 to 965 ms, which can degrade the QoS significantly. One solution is to modify the GSM protocol so that a TA can be provided to the MT by the GSM-BSS through the *handover command* message. This can reduce the number of single trip delays to one. However, this would imply modifications to GSM protocols.

Impact On Handover from an Satellite Spot-Beam to a GSM Cell Using GSM Protocol The signalling flow for this type of handover can re-use that described by Figure 7.11, with the BSS and the MSC replaced by the GWS and GMSC, respectively, and *vice versa*. An important issue in this type of handover is that the satellite network needs to know the exact position of the mobile terminal in order to handover to the correct GSM cell. The MT can send its positional information to its GWS via the *measurement report* message periodically. In addition, the positional information should uniquely define the GSM cell in relation to the satellite spot-beam. This requires the use of CGI as described before. Therefore, some signalling parameters in the A-interface have to be modified. Referring to Figure 7.11 and considering it for the use of handover from a satellite spot-beam to a GSM cell, two messages, namely *handover required* and *handover command*, require parameter modifications. Furthermore, due to the difference in propagation delay between the satellite and terrestrial networks, packets may overlap during handover. These issues, together with the identification of the MT's position in terms of a GSM cell and spot-beam are discussed below.

Modifications to Handover Required Message This is a BSSMAP message sent by the GWS to the GMSC to indicate that a given MT requires a handover. The user data field of this message is shown in Table 7.7.

Among the parameters indicated in Table 7.7, *cause, cell identifier list* and *environment of BSS "n"* will have to be modified to suit the satellite network.

The *cause* element indicates the reason for handover initiation. The causes of handover initiation in a pure GSM network will have to be extended to cover the characteristics of an integrated S-PCN/GSM network and to include the complementary role that the satellite network plays in this integrated S-PCN/GSM environment. For example, a handover from the satellite to the terrestrial network may be due to economic reasons or to traffic overflow in the terrestrial network.

The *cell identifier list* provides a list of GSM cell identifiers in a pure GSM system. Each

Table 7.7 User data field in *handover required* message

Information elements	Type	Length (octet)
Message type	M	1
Cause	M	3–4
Response request	O	1
Cell identifier list	O	$7n + 3$
Current radio environment	O	15–n
Environment of BSS "n"	O	7–n

cell identifier uniquely identifies a cell within a BSS by its LAC or its CI during handover. However, due to reasons mentioned before, a CGI should be used in order that a GSM cell can be identified within a satellite spot-beam. In addition, the list will also have to be extended in order to include a list of satellite spot-beams using the LAI so that inter-spot-beam handover can be taken into account.

The *environment of BSS* "n" gives the average power levels of the corresponding cells in the *cell identifier list*, together with their BSIC and BCCH frequency numbers in a pure GSM system. Modifications have to be made to this element so that CGI is used instead of BSIC for reasons stated before. Furthermore, measurements of average power levels of the corresponding spot-beams included in the extended *cell identifier list* should also be included in this element.

Modifications to Handover Command Message The *handover command* is a direct transfer application sub-part (DTAP) message sent from the GMSC and GWS. The DTAP is a sub-part of the base station system mobile part (BSSMP) and is used to transfer call control and mobility management messages to and from the mobile station. The information field of this message comes from the GSM-BSS and will be transmitted to the radio interface transparently. The format of the *handover command* is the same as that shown in Table 7.7. As before, two elements need to be changed in order to accommodate the satellite network: the *channel description* and the *cell description*. The modifications should be according to those described before. Since this message comes from the GSM network, any modifications to the GSM protocol would be unavoidable.

Packet Overlapping During Handover Due to the propagation delay difference between the satellite and the terrestrial network, packets transmitted during handover over the S-PCN may overlap with those transmitted over the GSM network at the GSM-MSC in the uplink and at the mobile station in the downlink. There are two possible solutions: the first is to buffer those packets from the GSM network in the GSM-MSC until packets transmitted from the satellite network to the GSM-MSC have finished; the second solution is to discard overlapping packets. The first solution results in significant delay for packet transmission in the GSM network, while the second solution results in packet loss. Hence, a degradation in the QoS is inevitable.

Identification of Mobile Station Position In order to perform handover between the S-PCN and GSM network, the exact position of the mobile station recognisable to both networks is essential for two reasons: (i) so that the mobile station can receive the BCCH allocation list from its serving GWS to measure the power level of its surrounding BSSs (note that MAHO is assumed); and (ii) so that the serving GWS can contact the specific GSM-MSC towards which the mobile is moving. In GSM, the mobile station knows its surrounding BSSs through the use of the BSIC and BCCH frequency in the measurement report. In an integrated S-PCN and GSM network, a single satellite spot-beam can cover several GSM cells. Different cells belonging to different countries may have the same BSIC and BCCH numbers. This makes it difficult for the mobile station to identify the serving BSS and its surrounding BSSs.

A possible solution is to have a dual-mode terminal with two transceivers capable of listening to two channels at the same time, one for the GSM network and another for the satellite network. In this case, the mobile station is able to obtain its exact position through the GSM BCCH channel while it communicates with the satellite network. Two parameters are included in the message transmitted by the GSM BCCH channel: the location area identification (LAI) and the cell identity (CI). These two parameters enable the satellite network to identify the mobile station's exact position and hence the GSM BSS towards which the mobile is moving. This requires that the mobile station reports these parameters to its serving GWS soon after the signalling links are established between them during the call set-up stage.

7.3.4.2 Integration at the A-Interface

With this integration scenario, a layered cellular architecture by overlapping the terrestrial and satellite coverage can be obtained and the number of network elements for the integrated network to provide a global network infrastructure can be reduced. However, a closer form of collaboration between satellite and terrestrial network operators is required in order to define inter-network functions and the sharing of common network elements.

Several advantages can be offered with this integration scenario. First, a more flexible and optimised radio resource allocation strategy can be implemented by exploiting the network, which can offer better features or by shifting the network load from one network to another. For an S-PCN/GSM integrated network, it is expected that the satellite part will play a complementary role to the terrestrial part in order to extend the terrestrial coverage to form a truly global network. With this in mind, the terrestrial network will carry most of the user information and non-dedicated mode signalling traffic. The satellite network can off-load the terrestrial traffic during peak load periods.

With the single-MSC integration scenario, the location management strategy has an important and close relationship with inter-segment handover. Two location management scenarios can be identified. The first scenario assumes that the satellite coverage is part of each MSC/VLR area. This implies that the GWS is connected with multiple MSCs through the A-interface and a single LAC representing the satellite coverage is allocated. Since the satellite coverage will be treated as another cell in the same MSC/VLR area, inter-segment handover will also be treated as an intra-MSC handover in a pure GSM network. The second scenario assumes that satellite coverage belongs to only one MSC/VLR such that the GWS is connected with only one MSC. The dual-mode user terminal data is kept in the VLR of this MSC/VLR area. Hence, the GWS always communicates with the same MSC. As a result, both intra- and inter-MSC exists in this case.

A possible strategy for location registration/update in this integration scenario is to execute such procedures via the terrestrial network whenever a terrestrial network is reachable and accessible by the user. This reduces the amount of non-dedicated mode signalling in the satellite network during both the location update and paging procedures, hence, saving the satellite bandwidth. It is also possible that location data required by the satellite network can be made accessible through the terrestrial network. However, an additional procedure may be required in a dual-mode terminal in order to complete such a location strategy. It may also result in heavy signalling flow between the terrestrial and satellite networks and changes in the GSM signalling protocols may be required.

Referring to Figure 7.8, a single MSC is shared between the terrestrial and satellite radio sub-systems. The satellite GSC is connected with a satellite location register (SLR: the S-HLR and S-VLR), which is a database strictly restricted to the satellite access network. It contains position information of MTs roaming in an area which is controlled by the FES (or the equivalence of GWS in ETSI terms) as well as any information useful for location updating and paging procedures. Information exchanged between the SLR and the FES is supported by an interface similar to the B-interface between the MSC and the VLR in GSM networks. It has to be noted that the SLR is solely managed by the FES and that it does not have direct interface with other entities in the GSM network. Message exchanges between the MSC and the FES are supported by the A-interface, which is also used for signalling connections between the MSC and the BSS in GSM networks. Thus, in this integration scenario, the A-interface carries information for BSS/FES management, call handling as well as for location management. Furthermore, in a pure GSM network, mobility management operations such as location update are supported by the MSC and these operations are transparent to the BSS. However, in satellite networks, the FES is responsible for mobility management functions. This results in extra mobility management functionalities being performed outside of the MSC.

The following sections describe the impact of the integration scenario on the GSM protocols in the call set-up, location management and handover procedures.

Impact on Handover from GSM to Satellite using GSM Protocol Two handover scenarios can be considered: intra-MSC handover and inter-MSC handover. Intra-MSC handover occurs when the target FES and the serving BS of the GSM network are connected to a common MSC; whereas inter-MSC handover occurs when the serving BS is connected to an MSC which is different from that of the target FES.

Intra-MSC Handover Intra-MSC handover in the integrated S-PCN/GSM network is similar to inter-BSS handover in a pure GSM network. For this type of handover, only the A-interface is used. Signalling on the A-interface uses Signalling System No. 7 with MTP, SCCP, BSSMAP and DTAP, with a majority of signalling to and from an MT carried by DTAP. As a result, for an intra-MSC handover, the modified signalling protocol should be DTAP and BSSMAP. The signalling exchange for this type of handover in a GSM network is shown in Figure 7.12. However, the suitability of such a protocol in an integrated S-PCN/GSM network requires further analysis since the traffic resource management in the satellite network is different from that of the GSM network. In GSM, the same BSS is responsible for the management of traffic resources and the call routing functions. However, in a satellite network these two functions may be performed by two different FESs, which may not be in the same physical location. For an FES which performs both functions, it has the combined functionalities of both the GWS and the GMSC; whereas for an FES which does not perform any traffic resource management functions, it is equivalent to a GWS in ETSI terms. For simplicity, the former will be termed as “FES/c” and the latter “FES” hereafter.

In adapting the GSM inter-BSS protocols to intra-MSC handover between the terrestrial

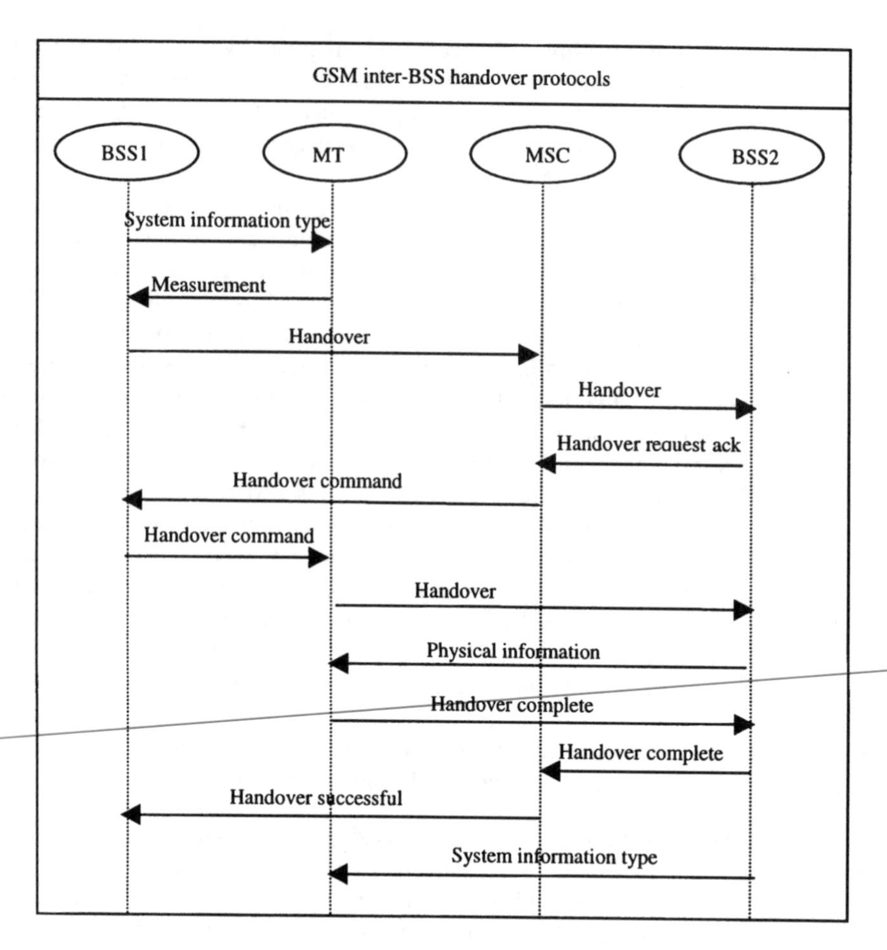

Figure 7.12 GSM Inter-BSS handover protocol.

and satellite networks, two possible options can be considered, assuming that the FES responsible for traffic resource allocation (FES/c) is not co-located with the FES that the call is to handover to. The first option assumes that the MSC is responsible for locating the most suitable FES for handover and signalling exchange goes directly between the MSC and the FES for handover request. Upon receiving the handover request message, the FES exchanges signals with the FES/c for traffic channel allocation. The signalling flow for this option is shown in Figure 7.13. The second option assumes that the FES/c is responsible for locating the most suitable FES for handover. Hence, the MSC sends a handover request message to the FES/c directly and signalling exchange is then carried out between the FES and FES/c for traffic channel allocation. The signalling flow for this option is shown in Figure 7.14.

Inter-MSC Handover Inter-MSC handover in a pure GSM network is concerned with handover of calls between BSSs belonging to different MSCs. The interfaces involved in this type of handover are A, B and E. Signalling on the B- and E-interfaces uses SSN7 with MTP, SCCP, TCAP and MAP. Modifications to the signalling protocol will be on the MAP and

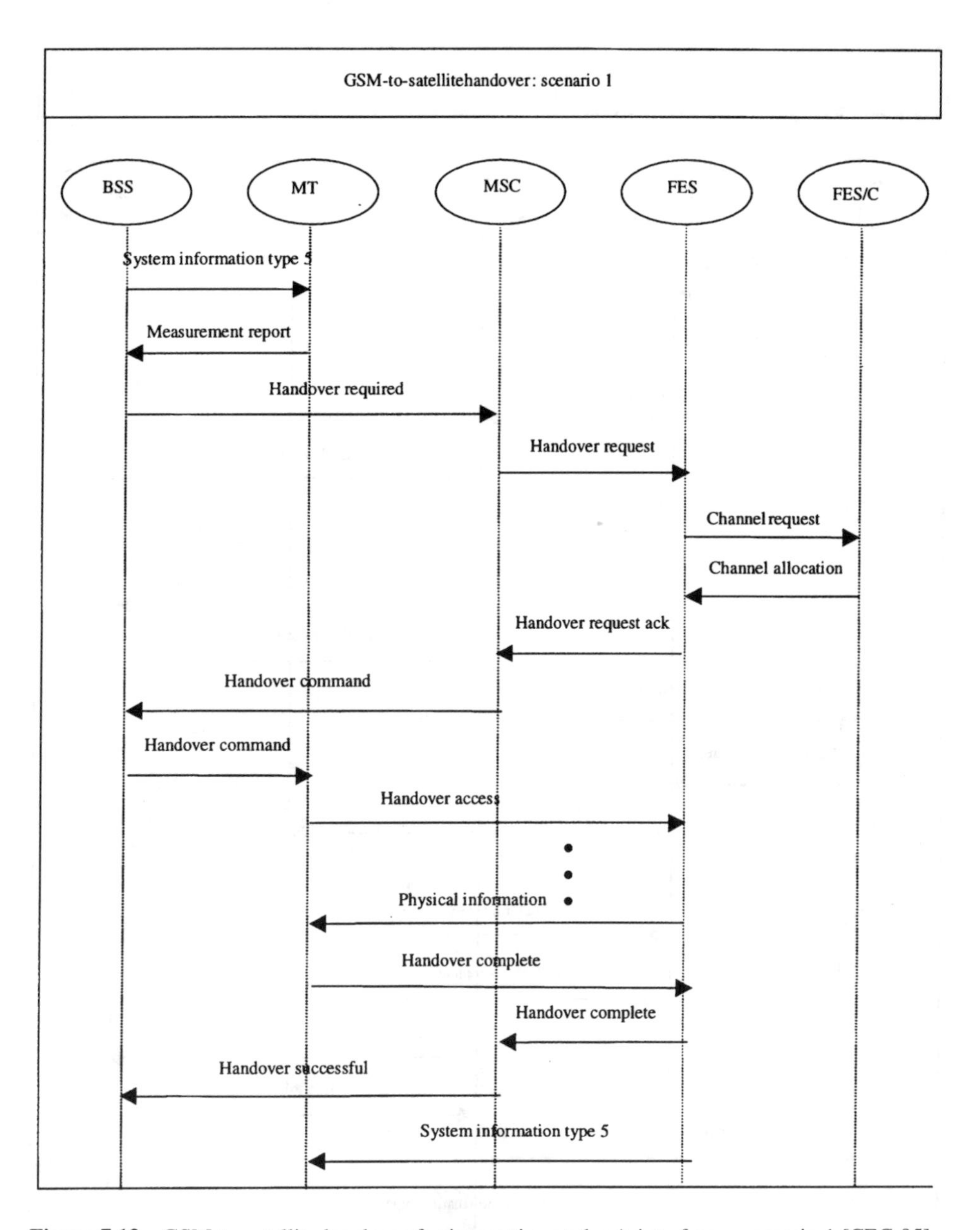

Figure 7.13 GSM-to-satellite handover for integration at the A-interface – scenario 1 [CEC-95].

TCAP. The signalling flows for this type of handover in a pure GSM network are already shown in Figure 7.9. However, the application of the signalling protocols shown in Figure 7.9 needs to take into account the fact that the S-PCN and GSM network share a common MSC. This has a direct impact on the *measurement report* message, which contains information on the signal strength received from neighbouring cells. Furthermore, modifications in the

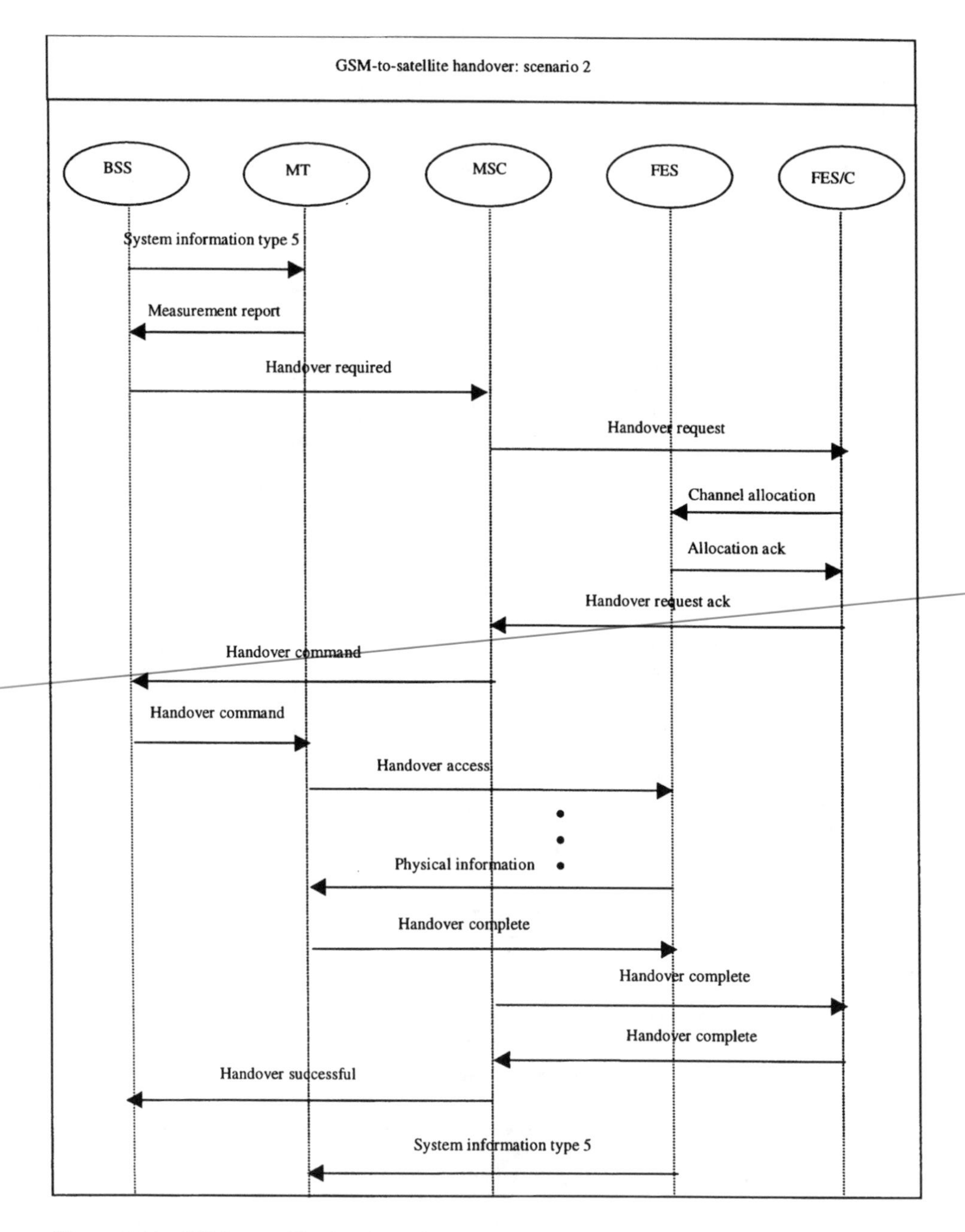

Figure 7.14 GSM-to-satellite handover for integration at the A-interface – scenario 2 [CEC-95].

handover required and *handover request* messages are also required. The following describes the necessary modifications to the different messages.

Modifications to Measurement Report Message The *measurement report* message is sent by the MT to the serving BSS. It contains information on the received signal quality from the serving BSS as well as information from the neighbouring cells. The format of this message is shown in Table 7.8.

Table 7.8 Measurement report message format

Information elements	Length (octet)
Protocol discriminator	2
Transaction identifier	2
Message type	2
Measurement results	16

In a pure GSM network, the "measurement results" field contains information for each neighbouring cell concerning the base station identity code (BSIC), the signal strength and the quality indicator. The candidate target cell is identified by the serving BSS by means of the BCCH carrier and the use of BSIC as a means of ambiguity resolution, since multiple cells might be received by the MT with the same BCCH frequency. The BSIC is made up of 3 bits of the PLMN colour code and 3 bits of the base station colour code (BCC). In the single MSC integration scenario, the measurement report should also contain information on the address of the target FES together with information on the satellite number and the spot-beam number. As a result, a specific BSIC has to be defined for the satellite network. This may be replaced by the LAI as described in Section 7.3.4.1. Furthermore, the *BCCH carrier number* should identify the BCCH carrier number in the spot-beam. Depending on the number of carriers to be allocated in the satellite network, the BCCH field may have to be extended.

Modifications to Handover Required Message Upon receipt of the *measurement report* message from the MT, the BSS initiates the handover process by sending a *handover required* message to the serving MSC. The format of this message is already described in Section 7.3.4.1. With the single MSC integration scenario, the "Cell identifier list" field in the message has to include identifiers for the satellite network. The same modifications described in Section 7.3.4.1 for the *handover required* message in the case of handover from a satellite spot-beam to a GSM cell can be applied.

Modifications to Handover Request Message The *handover request* message is a message sent by the new MSC to the FES. The format and modifications to this message to suit the satellite network have already been described in Section 7.3.4.1.

Modifications to Handover Request Ack Message Modifications to the *handover request ack* message to take into account the satellite network have already been dealt with in Section 7.3.4.1.

Impact on Handover from a Satellite Spot-Beam to a GSM Cell using GSM Protocol In this type of handover, the serving BSS is the FES. Again, two scenarios can be considered for this type of handover: intra- and inter-MSC handovers.

Intra-MSC Handover The treatment of this type of handover is the same as that for handover from a GSM cell to a satellite spot-beam. Two signalling options can be considered: the first option assumes the common MSC will choose a suitable FES for handover. The second option assumes that the common MSC communicates directly with the FES/c for a suitable FES for handover. Figures 7.15 and 7.16 show the signalling flow for the two options.

Inter-MSC Handover The signalling flow for this type of handover is similar to that shown in Figure 7.9. The FES must be able to define the CGI of the target GSM cell from the measurement message. Furthermore, modifications to the message format will be concentrated on the terrestrial cell identifiers which can be recognisable to the satellite network and *vice versa*. In previous sections, such modifications to take into account the

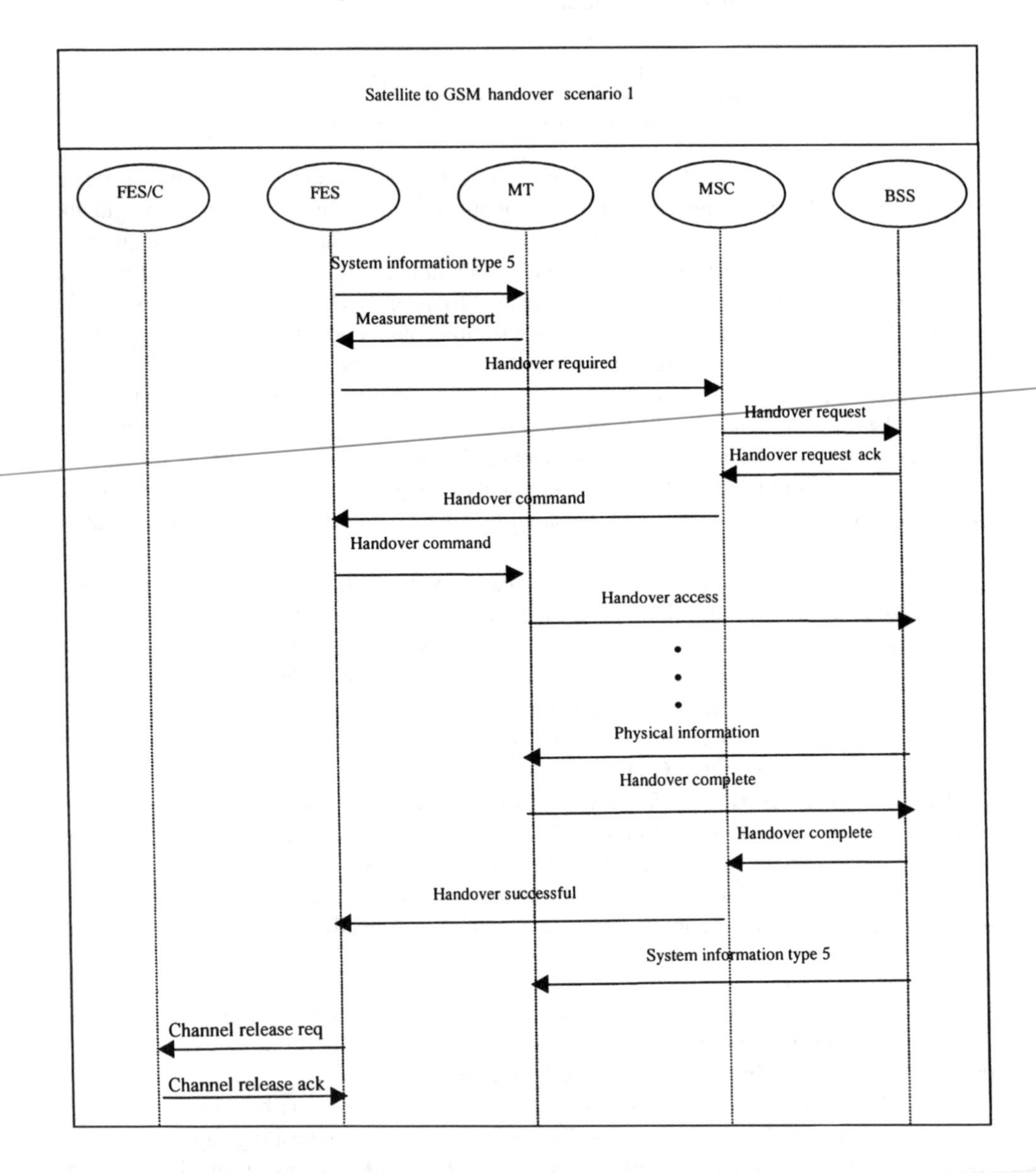

Figure 7.15 Satellite to GSM handover – scenario 1 [CEC-95].

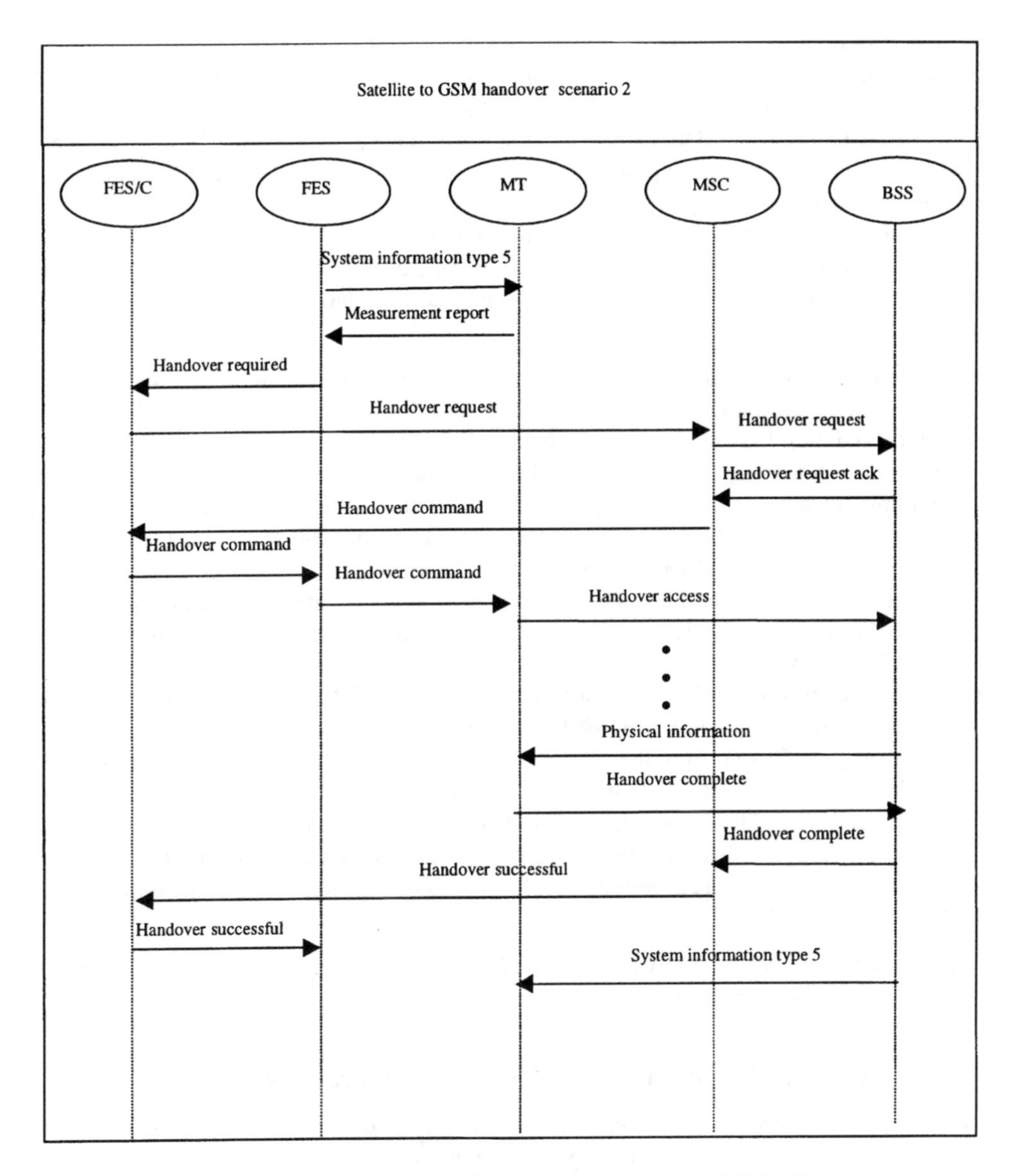

Figure 7.16 Satellite to GSM handover – scenario 2 [CEC-95].

satellite network have already been described. The interested reader should refer to previous sections for a description.

7.3.5 Impact of Integration Scenarios on the Location Management Procedure

7.3.5.1 Introduction

Location management mainly consists of two procedures: location update and call delivery. In Chapter 6, the approaches for location update in S-PCN have been described. In an integrated satellite and terrestrial network, the procedures used in the terrestrial network require some modifications in order to suit the needs of the satellite network. In the terrestrial network, location update is mainly based on the signal strength quality, although other parameters will also be used for management purposes such as load sharing between adjacent cells. Once the location update procedure starts, selection of an appropriate BSS and consequently the location area is performed.

In an integrated S-PCN/GSM network, different approaches with regard to the location area definition to that adopted in the GSM network can be used. In Chapter 6, two basic approaches for defining a location area in a satellite network have been discussed, which are summarised as follows:

- The guaranteed coverage area (GCA) approach: this approach defines the area served by a given FES at all times. Within this approach, two methods can be considered. The first method assumes that the FES area can be discriminated by monitoring the BCCHs broadcast by the spot-beams. This method is similar to that adopted in GSM and would require minor modifications to the GSM protocols in which the LAC is substituted by the FES area code. The second method requires that the network has the ability to compute the terminal's position at the FES by measuring the Döppler shift and the propagation delay between the MT and the FES. The latter method assumes that two satellites are visible to the MT at all times. Additionally, the latter method would require major modifications to the GSM protocols.
- The terminal position (TP) based approach assumes that an MT has the ability to compute its position using signals transmitted by satellites of the constellation. An uncertainty radius (Ru) originating from the estimated MT position is established for each MT. The location area is then defined as the circle formed by the locus of the uncertainty radius, within which the MT is free to roam. The MT measures its position periodically and if it detects the difference between the current position and the last measured position is greater than *Ru*, a location update is performed. This method would require major modifications to the GSM protocols.

The second method, the GCA approach and the TP based approach require major modifications to the GSM protocol.

7.3.5.2 Signalling Flow for the Location Management Procedure

Before proceeding directly to the signalling flow for the location management procedure in an integrated S-PCN/GSM network, the signalling flow for location management procedures in a GSM network and in an S-PCN will first be described. The signalling flow for the integrated network, in particular the single MSC integration scenario will be discussed based on these signalling exchanges.

Signalling Flow for Location Update in GSM In GSM, location update procedures mainly involve the MSC and VLR; the involvement of HLR is minimal. Figure 7.17 shows the location update signalling exchanges when the old and new location areas do not belong to the same MSC/VLR. Upon receiving from the MT the *location update message*, which contains the old TMSI and the LAI of the MT, the BS converts the message into a DTAP protocol format and forwards it to the serving MSC. This allows the

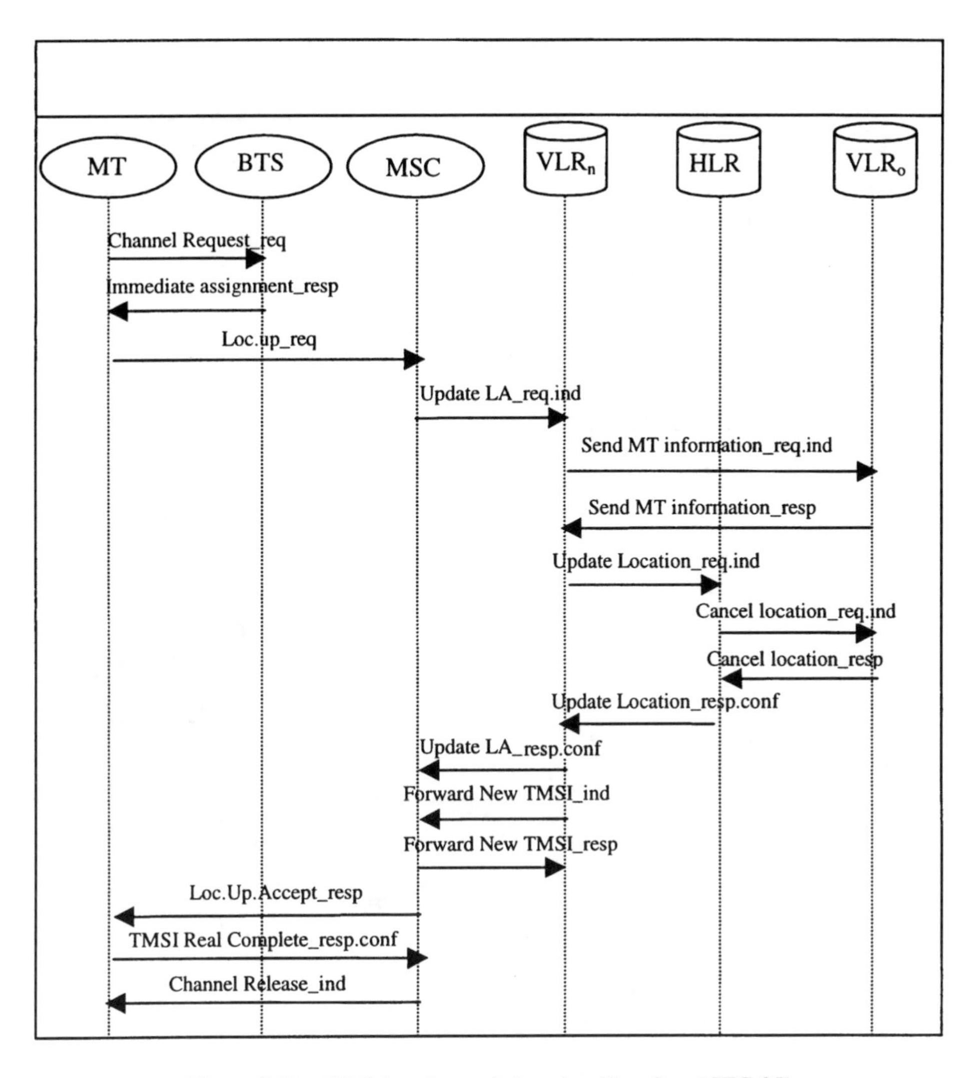

Figure 7.17 GSM location updating signalling flow [CEC-95].

MSC to identify the new LA of the MT. Since the MSC is able to identify the new LA and receives the old LAI, location update in the network is then possible. Upon receiving the *update LA* message from the MSC, the new VLR records the new LAI of the MT and contacts the old VLR using the old TMSI to retrieve the authentication and ciphering setting which are stored in the old VLR. The new VLR will also contact the HLR to update the VLR number. The HLR in turn informs the old VLR to delete all the data related to the MT. A new TMSI is then allocated to the MT.

Signalling Flow for Location Update in S-PCN As there are two approaches for location update in an S-PCN, Figures 7.18 and 7.19 show the signalling flows for location update procedures in the GCA and TP approaches, respectively.

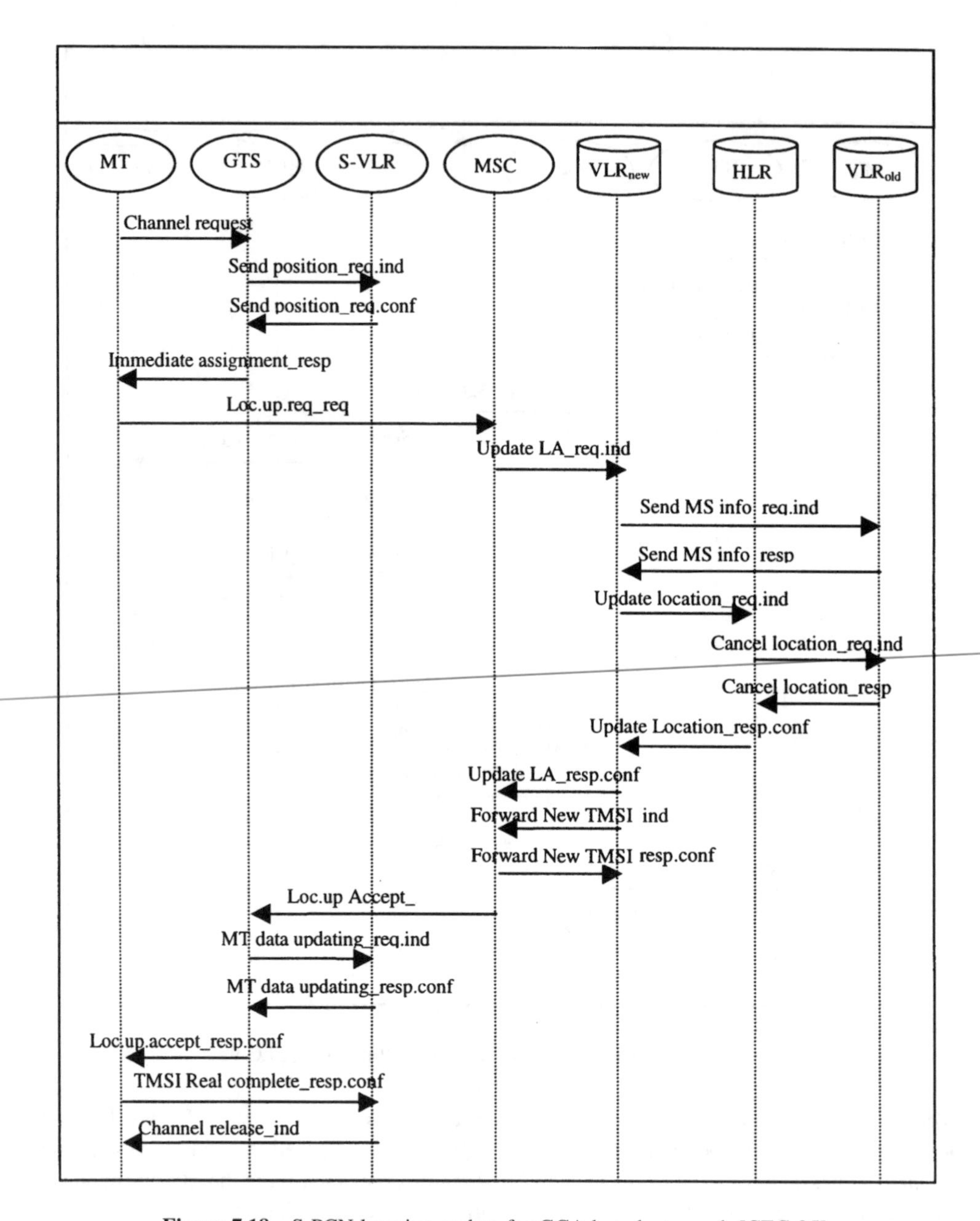

Figure 7.18 S-PCN location update for GCA based approach [CEC-95].

In the GCA approach, the MT periodically sends a request to the network via the *channel request* message for its positional information. This message can adopt the format in GSM, although a different format can also be used. Since the FES has the capability of measuring the MT's position, it can decide whether the MT is roaming inside of its controlled location area. If the MT is still inside the FES SLA that it is registered with, a location update is not

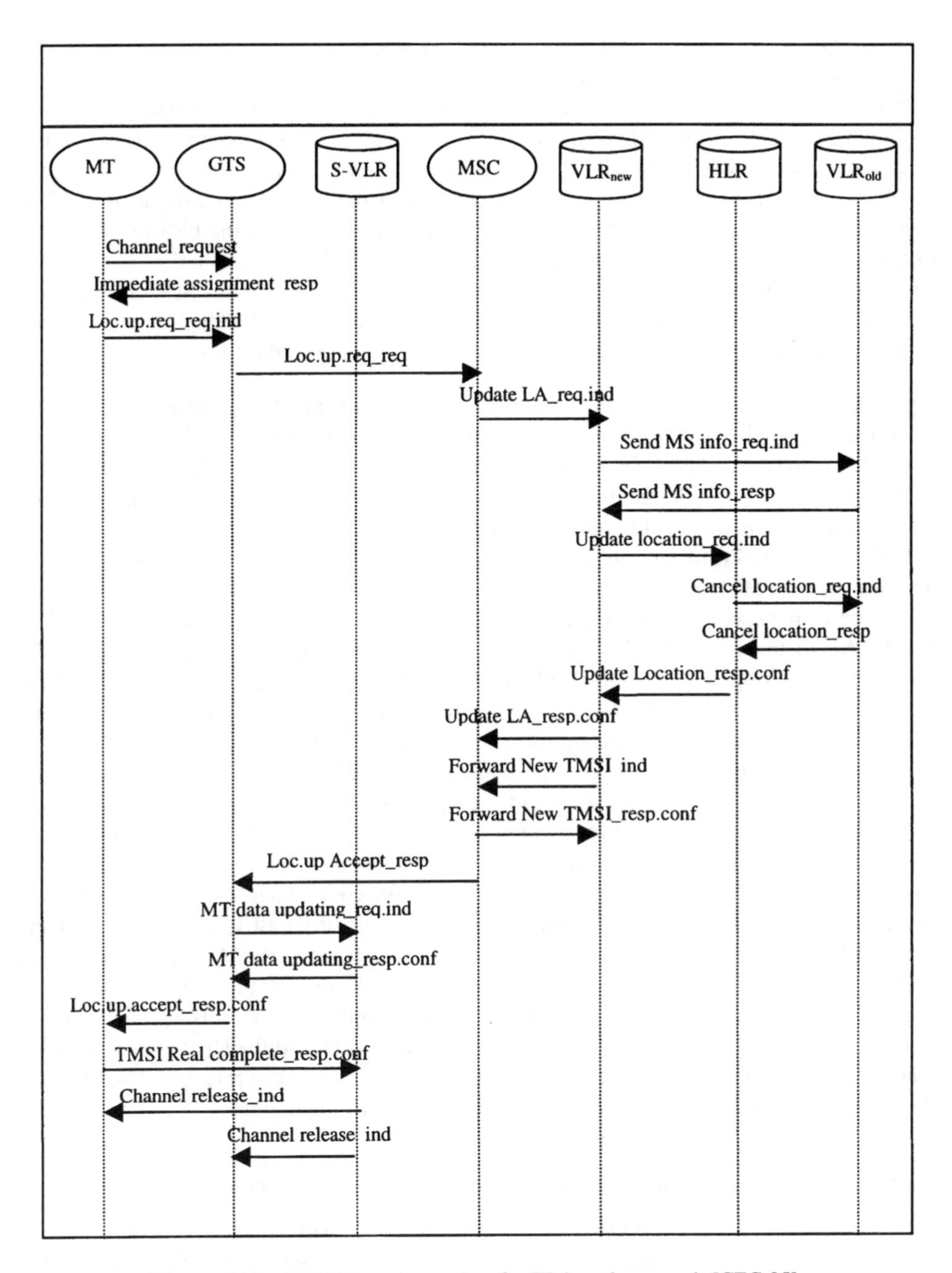

Figure 7.19 S-PCN location update for TP based approach [CEC-95].

required and the FES sends an *acknowledge response* message to the MT. If the FES that the MT contacted decides that a location update is necessary, it sends an *immediate assignment* message to the MT to inform the MT to start a location update procedure. The MT then sends a *Loc.Up.Req* message to the MSC that is associated with the contacted FES. The remainder of the location update procedures then follows.

In the TP based approach, the MT itself measures its position and compares its current position with the last measured one. When the MT detects that a location update is necessary, it requests a dedicated control channel by sending a *channel request* message to the most suitable FES. The FES allocates a control channel to the MT by sending an *immediate assignment* message. The MT then sends a *Loc.Up.Req.* message to the FES to inform the FES of its position and of the FES code. The FES code can be considered as the satellite counterpart of the GSM LAC. The FES stores the MT positional data in the SLR and forwards the request message containing only the FES code to its associated MSC. In this approach, the start of the location update procedure is no longer transparent to the FES.

Signalling Flow for Location Update Procedures in an Integrated S-PCN/GSM Network The location update procedures in an integrated S-PCN/GSM network can be associated with GCA and the TP based approaches for defining the SLA. Slight modifications to Figures 7.18 and 7.19 are required to take into account the different location area information used in the satellite and terrestrial networks, respectively.

In the case of GSM to satellite location updating, if the necessity of a location update is detected and the MT decides to select a satellite resource, it will start a location updating procedure with the satellite network by sending the *Loc.Up.Req* message, which contains its GSM LAI, to the FES. The FES then contacts its associated MSC and forwards it the *Loc.Up.Req.* message. If the contacted MSC is different from that controlling the GSM location area, the new VLR retrieves information related to the MT from the old VLR. The HLR updates the mobile location and requests the old VLR to delete data related to the MT. However, if the new MSC is the same as the old MSC (i.e. intra-MSC), there is no need to update the HLR. The remainder of the signalling flows is implicit.

In the case of satellite to GSM location updating, the MT is registered with both the terrestrial database (the HLR and the VLR) and the satellite database (the SLR). In this case, the SIM card and the visited VLR store the FES code under which the MT roams. Once a GSM resource is available, the MT selects a terrestrial BCCH and compares the received LAC with the stored FES code. Should a mismatch occur, the MT initiates a location update procedure by sending a *Loc.Up.Req* message, which contains the FES code to identify the old VLR. The GSM base station then establishes a connection with its associated MSC. The old VLR will delete all the information related to the MT and the HLR is then updated. Once a new TMSI is allocated to the MT, the satellite channel is then released. The FES will then delete the MT data contained in the SLR when a set time expires.

7.3.6 Impact of Integration Scenarios on the Call Set-up Procedure

Again, it is assumed that the GSM call set-up procedure will be used as a basis for the integrated S-PCN/GSM system. Assuming that the MT is registered with the satellite network, a call to the MT requires the network to identify the location of the MT and its routing address and the mobile terminal roaming number (MTRN), which can be provided by the HLR. In order to do this, the call will be directed to a gateway MSC (GMSC). Upon receiving the incoming call notification, the GMSC interrogates the HLR for the MTRN of the called MT. The HLR in turn interrogates the VLR with which the

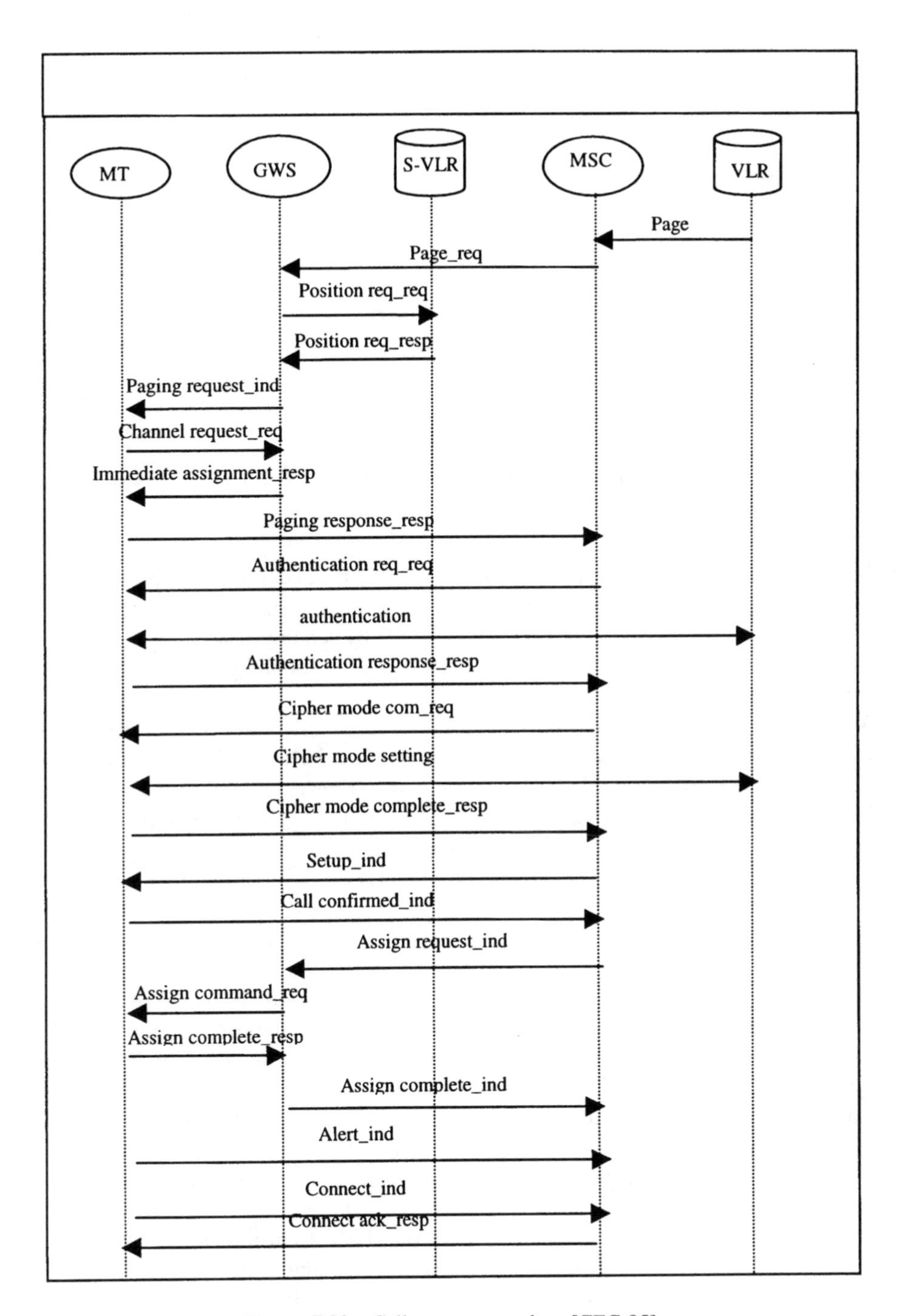

Figure 7.20 Call set-up procedure [CEC-95].

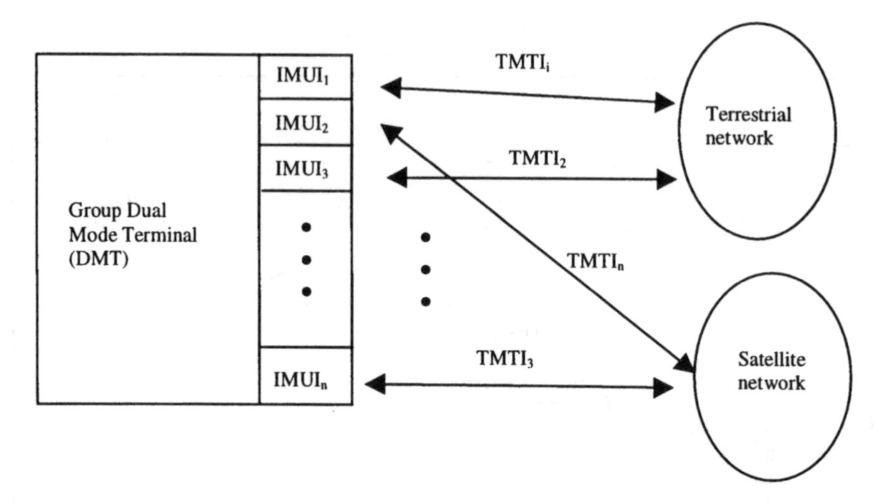

Figure 7.21 Mapping of IMUIs and TMTIs onto different network segments for a group DMT.

mobile terminal is currently registered for the MTRN. By using the MTRN, the GMSC is able to route the call to the visited MSC, which then requests the VLR for information related to the MT. The VLR acknowledges the MSC by sending related information of the MT and invokes the paging service. The signalling flow for the call set-up procedure is shown in Figure 7.20.

When setting up a paging service, the MSC has to identify the correct location area in which the mobile currently roams. If the MT is initially registered with the satellite network, the location area identifier would contain the FES code. Hence the MSC will contact the FES and forward it the paging message. The FES will then interrogate the S-VLR for the MT's positional data in order to page the mobile terminal.

In the terminal position (TP) approach, the MT may be covered by more than one spot-beam belonging to different satellites and under the control of different FESs. Hence, the FES contacted by the MSC may need to contact the other FESs. This requires some switching functionalities to be implemented in the FESs.

The *paging request* message broadcast by the FES will be followed by the *channel request* and *channel assignment* messages. A control channel is then established to link the MT and the network. The MT then sends a *paging response* message to the MSC and the call set-up procedure follows that of the GSM call set-up procedure.

The GSM call routing procedure requires that the MSC/VLR should only point to one FES for paging the terminal. This requirement can be easily fulfilled when the GCA approach for location management is adopted, though the FES distribution has to be carefully planned. For the TP based approach, this requirement is difficult to fulfil since the terminal's position is not bounded by only one FES. Hence, some of the procedures in GSM would have to be modified. Another issue in defining the call set-up procedure is the adaptation of the time-out runners according to different satellite orbits, which have an effect on the propagation delay. As shown in Figure 7.20, for almost every signalling message required for the call set-up phase, an acknowledgement message is assumed. This implies that the total call set-up delay is equal to the product of a double-hop time delay (in the case of transparent satellites) and the total number of signalling messages. Hence, all the time-out runners related to the call control procedure will have to be dimensioned according to the adopted protocols and to the type of satellite constellation.

7.3.7 The Role of Dual-mode Terminal in Terrestrial/S-PCN Integration

7.3.7.1 Basic Requirements

A dual-mode terminal provides further interworking capabilities between the terrestrial/S-PCN networks in an integrated environment, allowing each segment to have its own distinct radio interfaces and radio controllers. From the service point of view, a single mobile subscriber number should be used for both segments for ease of use. However, this requires close agreement and collaboration between service providers and network operators alike. Thus, as a basic requirement, network operators should support users with dual-mode terminals which are able to access both the terrestrial and satellite networks, as well as users with single-mode terminals dedicated to specific systems. The discussion that follows assumes a single MSC scenario.

The operations of a dual-mode terminal (DMT) can be categorised into two states: *idle* and *active*. In idle mode operation, the session set-up and registration procedures are used to establish a connection between a mobile terminal and the service provider. In the active mode, the DMT exchanges signals with the MSC for mobile originating and mobile terminating call set-up procedures. The DMT will automatically enter into an active mode upon receiving a paging message from the MSC for a mobile terminating call set-up or upon a mobile originating call set-up procedure (i.e. the mobile subscriber initiates the call set-up procedure).

7.3.7.2 Session Set-up

Session set-up is initiated once the terminal is switched on so that a location area identifier and a TMTI can be obtained. It also ensures that the appropriate registration process can be carried out. This is a procedure that is independent of the service type. Session set-up consists of several logical steps as outlined in Chapter 6, namely, select user role, access control, access data, session data and authentication.

Select User Role The *select user role* allows the user to be associated with different roles every time the terminal is switched on. These different roles are distinguished through a subscriber identity device (SID) which contains a set of international mobile user identifiers (IMUI). At power on, the user chooses a preferred IMUI and hence, the service provider that will support the requested service. This procedure involves only the user and the terminal, i.e. no network interaction is required.

Access Control Access control assigns a signalling connection between the DMT and the service provider chosen by the user. At this stage, the user's preferred segment has to be determined. Segment selection can either be performed manually or automatically. Three alternatives can be provided for segment selection: (a) terrestrial mode; (b) satellite mode; (c) automatic selection. The terrestrial mode and the satellite mode can be regarded as manual operations, which allows the user to select segments according to the service requested and to *a priori* knowledge that the user has of the system, in terms of service availability within the operating area. For example, if the user is in an aeronautical/maritime/rural environment, it is likely that the satellite segment would be chosen for the provision of service since the terrestrial network may not be available. This eliminates the time that otherwise would be spent on the automatic selection procedure to search for a suitable and available segment.

In the automatic selection mode, the DMT will select a segment according to a user preference list, which is previously defined by the user. The DMT makes use of the preference list to prepare a priority list. However, for economic reasons, it is logical that the terrestrial segment will be selected as the priority segment. Furthermore, the selection procedure will also take into account the following factors:

- The preferred segment availability following the switching on of the terminal;
- The user access rights to the selected segment.

Since the preference list is predefined by the user through the terminal user interface or during subscription, no extra signalling overhead is incurred in the network.

Access Data Access data is responsible for assigning a location area identifier to the DMT. The approaches used to define a location area in a satellite environment have already been discussed in Chapter 6 and in previous sections. In assigning the location area identifier, signalling exchanges are required between the DMT and the MSC. The MSC is responsible for the LA mapping and comparison functions. Once the access data functions are completed, the MSC will then associate an LA and a TMTI for the selected segment to the subscriber's IMUI. Thus, the set (IMUI, LA, TMTI) uniquely identifies the user and the environment to which the MSC can reach the mobile terminal. Depending on the type of terminal, different users may be able to register with different segments using the same terminal, in which case the MSC will assign different TMTIs to the corresponding IMUIs.

It has to be noted that a mobile subscriber can register for different services with different segments on different terminals. Consequently, the MSC will be able to assign different TMTIs to the same IMUI. However, subscribers cannot register for the same services on two different terminals since the second registration will force the MSC to release any connection previously established for the first terminal. Furthermore, the subscriber will not be able to access both the terrestrial and satellite domains using the same terminal as this implies that an IMUI is associated with two distinct TMTIs, both pointing to the same DMT.

Session Data The session data is responsible for managing data which are required for maintaining the session establishment between the user and the service provider. The data include the IMUI, TMTI, the selected service provider, and the user keys for security after session set-up. The MSC is responsible for associating the IMUI with the other data during session set-up. This set of data will be used in the call set-up process.

7.3.7.3 Registration

Registration is invoked after the successful completion of session set-up. It is a process which allows the user to select an environment to receive and initiate a service. Registration can force a DMT to register with a segment other than that selected at the session set-up stage. The functions of the MSC are listed below, these being dependent upon the segment selected by the DMT and the user, and the capability of the network to support the required service.

- In the case where the segment selected by the user is different from that selected by the DMT at session set-up, the MSC can either force the user to re-run the session set-up or to perform an automatic session set-up with the DMT without user intervention.
- In the case where the requested service cannot be supported by the selected segment at session set-up, the MSC can either check if another segment is available and request a session set-up re-run with the DMT or stop the registration and inform the user that the service is unavailable.
- In the case where the above two events run successfully, the MSC performs a user capability check to ensure the user has access rights to the requested service. The MSC will then interrogate the user's home network or the visiting network, depending on where the registration process takes place, for relevant subscription data.

The MSC completes the registration process by updating the current user profile located in the visiting network and associates the TMTI with the assigned service. Hence, after the registration process, the user profile data will be updated to contain the following data set: (IMUI, LA, TMTI, service).

7.3.7.4 Call Handling

For a mobile-originated call, the call set-up procedure starts with the registration process. Upon completion of the registration, the MSC will then proceed with the call establishment procedure with the called party. The call establishment procedure that follows depends on whether the satellite or the terrestrial segment is selected and the MSC should be able to handle both terrestrial and satellite call set-up procedures.

The MSC is responsible for locating the called mobile in the case of a mobile terminated call. Upon receiving the call set-up message containing the IMUI of the called mobile from a remote gateway MSC (GMSC), a series of signalling exchanges takes place between the GMSC and the HLR in order to identify the visiting network in which the called mobile is currently located. This is then followed by another series of message exchanges between the VLR, HLR and the visiting MSC in order that the following data can be supplied:

- The location area identifier (LAI)
- The base station identifier (either the satellite or terrestrial)
- The TMTI

It is expected that the signalling exchanges will be similar to that for mobile-terminated call set-up in the GSM network. The GWS will then map the location data set with a list of spot-beams, if the mobile is registered with the satellite network, in order to page the mobile. However, if the terrestrial network is used, the MSC will prepare a list of candidate terrestrial base stations.

Recall in the session set-up stage that different mobile subscribers can use the DMT to access different services through different segments as long as there is a one-to-one correspondence between the IMUI and the TMTI (Figure 7.21).

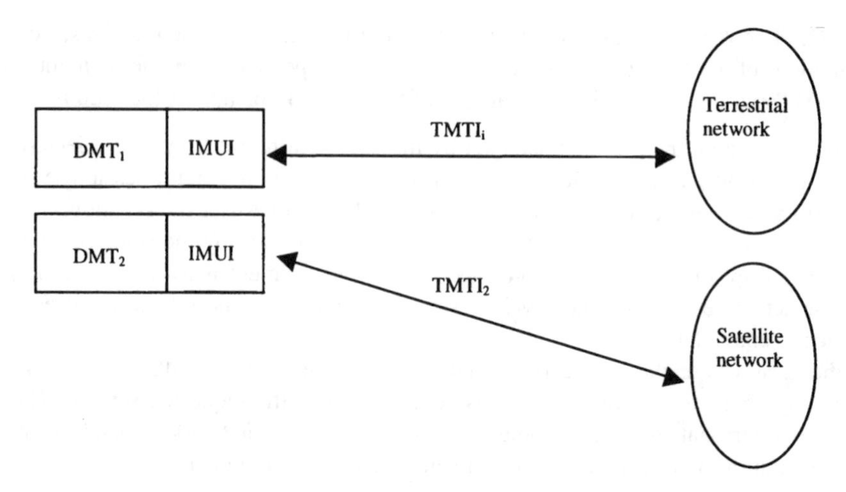

Figure 7.22 Mapping of the same IMUI onto different TMTIs with different network segments through two different DMTs.

Or in another case as shown in Figure 7.22, a mobile subscriber can use different terminals to access different services through different segments, in which case a single IMUI is mapped onto two different TMTI but pointing to different terminals. In both cases, the MSC will use the service type to identify the most suitable segment and the corresponding TMTI for paging the mobile terminal.

However, in the case of a DMT being associated to two different TMTIs but a single IMUI, i.e. a mobile subscriber intends to use the same terminal to access both segments at the same time for two different services, extra functions will have to be performed by the MSC for the following two scenarios.

1. The called mobile is registered with either one of the two segments for a particular service at the time when its associated MSC receives a call set-up message. If the incoming call is from the same segment as the called mobile is registered with but is associated with a service other than the one which is currently registered by the called mobile, the MSC may

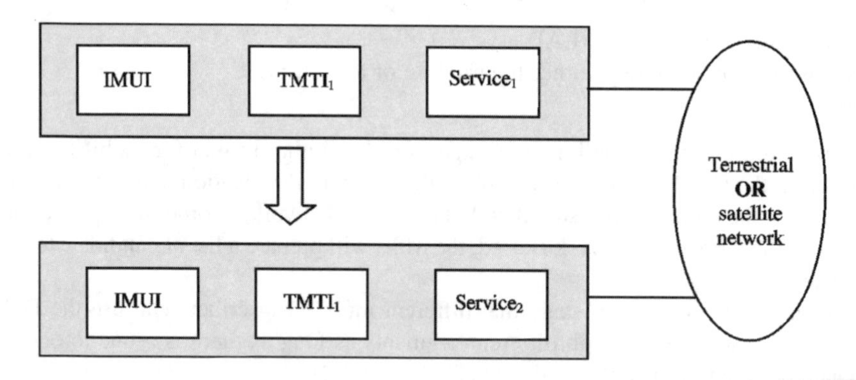

Figure 7.23 Re-registration mechanism for changing service association – scenario 1.

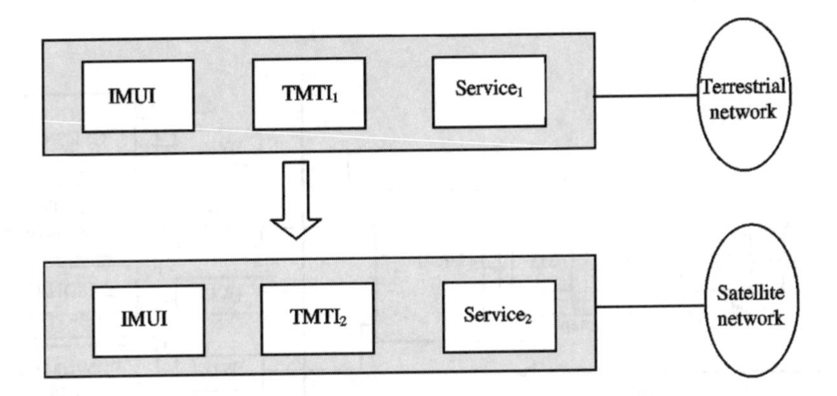

Figure 7.24 Re-registration mechanism for changing service association – scenario 2.

force the called mobile to re-register with the service associated with the incoming call. The mechanism is shown in Figure 7.23.

2. An incoming call associated with a different segment and a different service to the called mobile is received by the MSC. If the segment that the called mobile currently registered on is incapable of supporting the service associated with the incoming call, the MSC may force the called mobile to re-register on the other segment. The changes in the registration mechanism are as shown in Figure 7.24.

7.4 Integration with Third Generation (3G) Networks

7.4.1 Concept of Interworking Units

As we have seen, the universal mobile telecommunications system (UMTS) is the 3G mobile system for Europe and will deliver a convergent network incorporating cellular, cordless, wireless local loop and satellite technologies. Its network architecture and concepts have already been outlined in Chapter 1. The use of satellites to complement available terrestrial-UMTS networks will play an important part in establishing UMTS in Europe and beyond. In addition to the development of 3G terrestrial and satellite components, UMTS will provide backward compatibility with second-generation (2G) mobile networks. Unlike UMTS, 2G networks have been designed largely independently of each other, resulting in a number of different radio interfaces and subsequent incompatibility between systems. An important aspect in designing the UMTS architecture is its inter-operability with existing networks.

The underlying approach in the initial development of the UMTS network architecture is the use of interworking units (IWU) for inter-operability with other core networks. Figure 7.25 shows the initial concept of the UMTS network architecture as defined by Ref. [ETS-98] in the implementation of the mobile equipment domain and the various core network domains.

In Figure 7.25, the mobile equipment is divided into three sub-components: the terminal equipment (TE), the terminal adapter (TA) and a mobile terminal (MT). The UMTS radio access network (URAN) accommodates a base transceiver system with UMTS specific radio

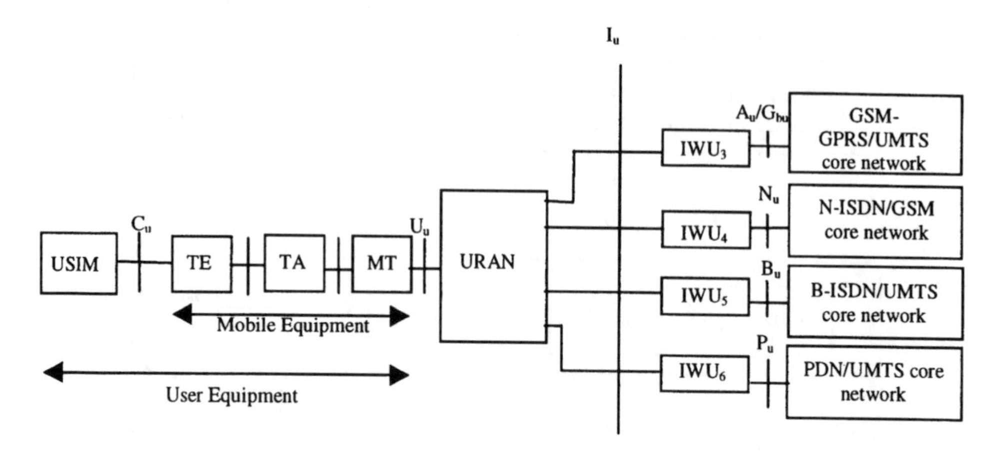

Figure 7.25 Initial UMTS mobile equipment domain and core networks concept [ETS-98].

interfaces. The IWUs guarantee the networks' interoperability while decoupling their procedures and protocols [DEL-99]. Using this approach, no common protocols are required between different networks. Furthermore, modifications to existing networks' infrastructure and protocols can be avoided.

7.4.2 The Radio-Dependent and Radio-Independent Concept

The second step in the UMTS development is to enable the UMTS radio access network to encompass different radio access networks. To enable such a realisation, higher layer protocols need to be independent of the radio access mechanisms. The European Fourth Framework ACTS RAINBOW project has introduced the concept of separating the radio-dependent and radio-independent parts of the UMTS network. The boundary between the radio-dependent part and the radio-independent part lies on the interface between the two sublayers of the

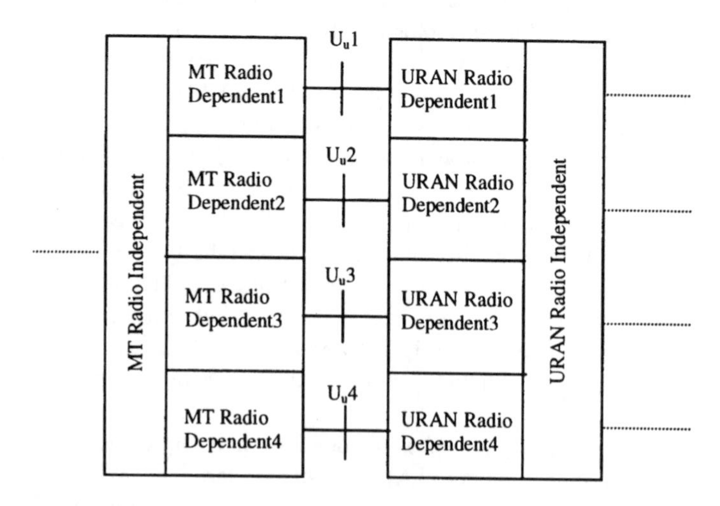

Figure 7.26 Radio-dependent and radio-independent concept.

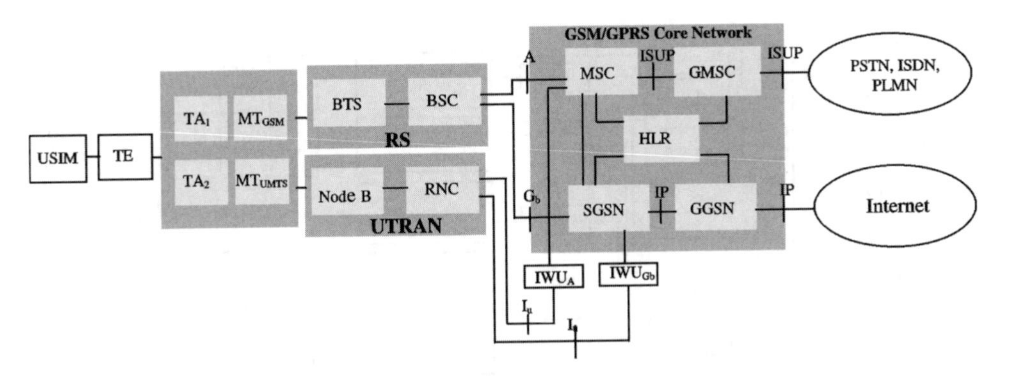

Figure 7.27 First phase evolution: from GSM/GPRS to UMTS.

data link layer: the link access control (LAC) and the medium access control (MAC). The former is radio independent while the latter is radio-dependent. A generic interface between the LAC and the MAC has been designed in the RAINBOW project to work with any radio access technique [VRI-97]. With this generic interface, higher layer protocols of the UMTS can be used for all UMTS radio access. This allows different radio access networks, including those of satellite systems, to co-exist within UMTS. Figure 7.26 shows the concept in the separation of radio-dependent and radio-independent functions [DEL-99] in the mobile terminal and the URAN.

7.4.3 Satellite Integration with UMTS – a UTRAN Approach

Specifications for the future mobile core networks is carried out by the 3G Partnership Project (3GPP) [3GP-99a, 3GP-99b]. It is envisaged that the Phase I deployment of UMTS will be based on an existing GSM/GPRS core network architecture to support both circuit-switched and packet-switched services. Figure 7.27 shows the Phase I evolution concept from GSM/GPRS to UMTS.

In Figure 7.27, a better defined UMTS radio access network called the UMTS Terrestrial Radio Access Network (UTRAN) was developed (Chapter 1). The operational frequency range of UTRAN is in the region of 2 GHz and achieves bit rates of up to 2 Mbps. The first step evolution from GSM/GPRS to UMTS relies on two IWUs placed between the UTRAN and the SGSN and GGSN of the GSM/GPRS core network. In Figure 7.27, the GSM/GPRS and the UMTS radio access parts are still seen as two separate systems, although the hosting core network is based on GSM/GPRS. Continuing with the network evolution concept, the UTRAN can eventually be connected directly to the SGSN and GGSN without relying on the IWUs. In such a way, the GSM/GPRS core network will act as the UMTS core network with possible upgrade of functionalities in the SGSN, the so-called 3G-SGSN as opposed to the 2G-SGSN for the GSM/GPRS core network.

By adopting the RAINBOW project approach and network architecture of the EU ACTS SINUS [EFT-97] and INSURED [BRA-99] projects, [DEL-99] derived the following network architecture for the integration of 2G GSM-based satellite systems into the UMTS network, as shown in Figure 7.28. This follows ETSI's migration approach for the evolution of GSM/GPRS to UMTS.

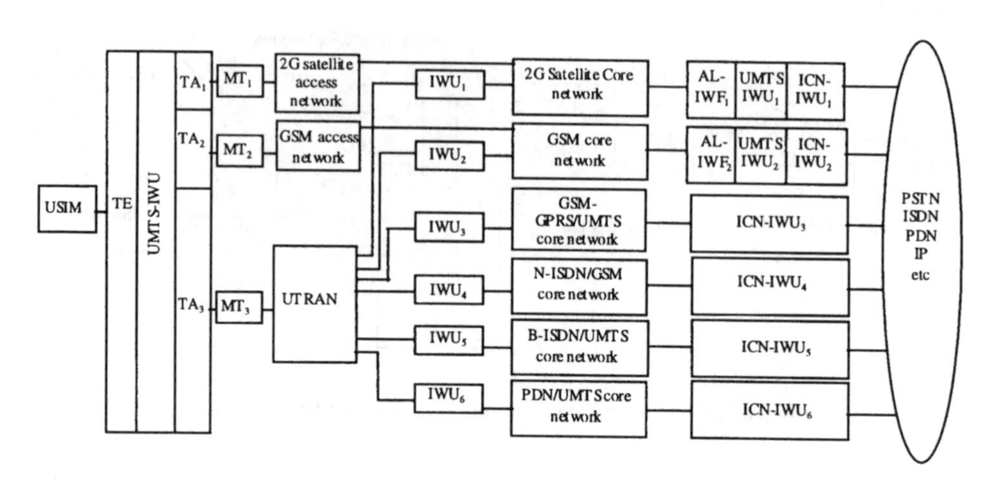

Figure 7.28 First phase integration of 2G systems into UMTS[DEL-99].

In Figure 7.28, the intercore networks IWUs (ICN-IWUs) allow interoperability between 2G terrestrial and satellite GSM networks, the 3G hosting core networks as well as the PSTN. The IWUs connected to the UTRAN have the same functions as the IWUs in Figure 7.27. Furthermore, the mobile equipment domain is further subdivided into three components: the terminal equipment (TE), the terminal adapter (TA) and the mobile termination (MT). In here, the user service identity module (USIM) is normally a stand-alone smart card which contains data and procedures necessary for identification and authentication purposes. The TE contains only the application stratum function. It is connected to a TA to provide adaptation functionality between the application and the transport strata. The MT itself contains no application stratum. In this arrangement, the application stratum can be differentiated from the other strata. The configuration shown in Figure 7.28 allows first phase UMTS development for interoperation with 2G mobile (both terrestrial and satellite) networks.

As the networks gradually evolve into UMTS, the IWUs will disappear and the UTRAN can be connected directly to a central UMTS core network. For integration of a 3G satellite network into the UMTS, modifications and upgrading, including both software and hardware, of various satellite network elements such as the MS, GTS and GSC, are required to take into account the radio access scheme and for the introduction of new network nodes, the Node B and RNC, as specified in the UTRAN.

7.4.4 *Satellite Integration with GSM/EDGE – a GERAN Approach*

Instead of adopting UTRAN as the UMTS radio access network, the GSM/EDGE radio access network (GERAN) technical specification group within 3GPP defined another evolution path for GSM to UMTS. The underlying core network of EDGE is still based on the GSM/GPRS core network. However, the radio access network, i.e. the GERAN, still adopts the GSM architecture. One of the major changes in GERAN is the inclusion of an inter-NSS interface, the I_{ur-g}-interface. Figure 7.29 shows a GERAN-based UMTS network architecture.

Higher data rates are supported by using a different modulation scheme and different error correcting codes. With this evolutionary approach, modifications to the existing GSM/GPRS

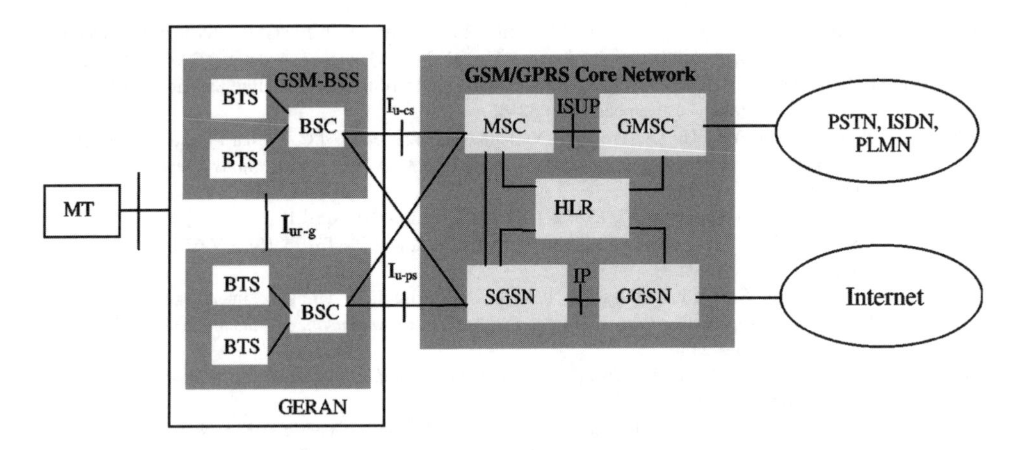

Figure 7.29 A GERAN-based UMTS network architecture.

network architecture are limited to software upgrades in the GSM-BSS and the GSM/GPRS core network as opposed to both software and hardware upgrades required for the introduction of new network elements in the UTRAN approach. If a satellite network adopts the GMR specifications, the same requirements on the software upgrades as those for GSM are envisaged.

7.4.5 Conclusion

3GPP is now investigating the all-IP solution for UMTS. Regardless of this and of the integration approach, important issues on mobility management, location management, call set-up (or PDP context in IP terms) procedures need to be defined in order that different networks can inter-operate with each other. Several EU projects, such as the ACTS INSURED [BRA-99] and ACCORD [CON-99] projects, the IST SUITED [CON-00] and the VIRTUOUS [OBR-00] projects, etc. are now on the way to defining such procedures. Working groups within the IETF and 3GPP have been set-up to define the mobility management issues for the UMTS network. The outcome of those working groups and projects will contribute to the standardisation process for satellite integration into UMTS.

References

[3GP-99a] "General UMTS Architecture", *3GPP 3GTS 23.101 version 3.0.1*, 1999.

[3GP-99b] "Evolution of GSM platform towards UMTS", *3GPP 3G TR 23.920 version 3.1.0*, 1999.

[BRA-99] H. Brand, A. Vesely, F. Delli Prisoli, A. Giralda, "Intersegment Handover Results in the INSURED Project", *Proceedings of 4th Mobile Communications Summit*, Sorrento, 8–11 June; 145–150.

[CEC-95] EC RACE Project SAINT, "Interworking with Other Networks", *CEC deliverable SRU/CSR/DR/P216-B1*, November 1995.

[CEC-97] EC RACE Project SAINT, "System Requirement Report", *CEC deliverable AC004/NTS/IIS/DS/P020-A1, February 1997.*

[CON-99] P. Conforto, F. Delli Priscoli, V. Marziale, "Architecture and Protocols for a Mobile Broadband System", *Space Communication*, 15(4), 1998; 173–188.

[CON-00] P. Conforto, G. Losquadro, C. Tocci, M. Luglio, R.E. Sheriff, "SUITED/GMBS System Architecture", *Proceedings of IST Mobile Communications Summit 2000*, Galway, 1–4 October 2000; 115–122.

[DEL-99] F. Delli Priscoli, "UMTS Architecture for Integrating Terrestrial and Satellite Systems", *IEEE Multimedia*, 6(4), October – December 1999; 38–45.

[EFT-97] N. Efthymiou, Y.F. Hu, A. Properzi, V. Faineant, H. Medici, J.C. Bernard-Dende, "The SINUS Network Architecture", *Proceedings of the 2nd ACTS Mobile Communications Summit*, Aalborg, 7–10 October 1997; 168–174.

[ETS-98] "General UMTS Architecture", *ETSI UMTS 23.01 version 0.6*, April 1998.

[GMR-99] GMR-1 01.202, "GEO Mobile Radio Interface Specifications", *ETSI TS 101 377-01-03 v0.0.08*, October 1999.

[OBR-00] V. Obradovic, S.-H. Oh, F. Grassl, L. Falo, F. Delli Priscoli, A. Giralda, "Inter-Segment Roaming – VIRTUOUS Approach", *Proceedings of IST Mobile Communications Summit 2000*, Galway, 1–4 October 2000; 209–214.

[STA-95] W. Stallings, *ISDN and Broadband ISDN with Frame Relay and ATM*, 3rd Edition, Prentice-Hall, Englewood Cliffs, NJ, 1995.

[VRI-97] J. De Vriendt, P. Lucas, E. Berruto, P. Diaz, "RAINBOW: A Generic UMTS Network Architecture Demonstrator", *Proceedings of 2nd ACTS Mobile Communications Summit*, Aalborg, Denmark, 7–10 October 1997; 285–290.

8

Market Analysis

8.1 Introduction

As illustrated in Chapter 2, the start of the last decade resulted in a number of new satellite proposals to address the perceived needs of the mobile communications market. Significantly, at this time, second-generation mobile phones were just starting to appear on the high street and market predictions for cellular communications suggested a relatively modest take-up of the technology, based on regionally deployed networks. In such an environment, there was clearly an opportunity for a number of satellite networks to enter the global mobile communications market. A decade later, the reality of the situation is quite different. The growth in terrestrial cellular services has been nothing short of spectacular. Figure 8.1 illustrates the world-wide availability of the GSM network at the turn of the century. Of course, this figure may prove to be slightly misleading, in that blanket coverage of the network in certain parts of the world does not exist, as yet. However, irrespective of the degree of coverage in each country, the globalisation of the standard is clear to see.

The adoption of GSM as a virtual global standard will have a major impact on the perceived market for mobile satellite communications. For example, when S-PCN systems were first proposed at the start of the 1990s, there were only 10 million cellular subscribers and year 2000 predictions were in the region of 50–100 million world-wide, a figure less than the total number of subscribers within the EU! This clearly demonstrates the difficulty that satellite operators face when trying to predict long-term trends.

GSM is not unique in its global success. The global mapping of the availability of cdmaOne, for example, would result in a similar picture, complementing and in places co-existing with the GSM coverage pattern.

The previous chapters have illustrated the technologies behind present and near-future satellite systems. Although systems such as NEW ICO and GLOBALSTAR may appear to require development of relatively sophisticated, new technology, it is important that such technological development reflects user requirements and market demand. The key issues of user and service requirements, service costs and the potential number of users have to be addressed at an early stage of the design process and may continue to need re-assessment throughout the development of the system. This can be seen with developments in NEW ICO, where the delay in service launch allowed the importance of mobile Internet access to be incorporated into the design.

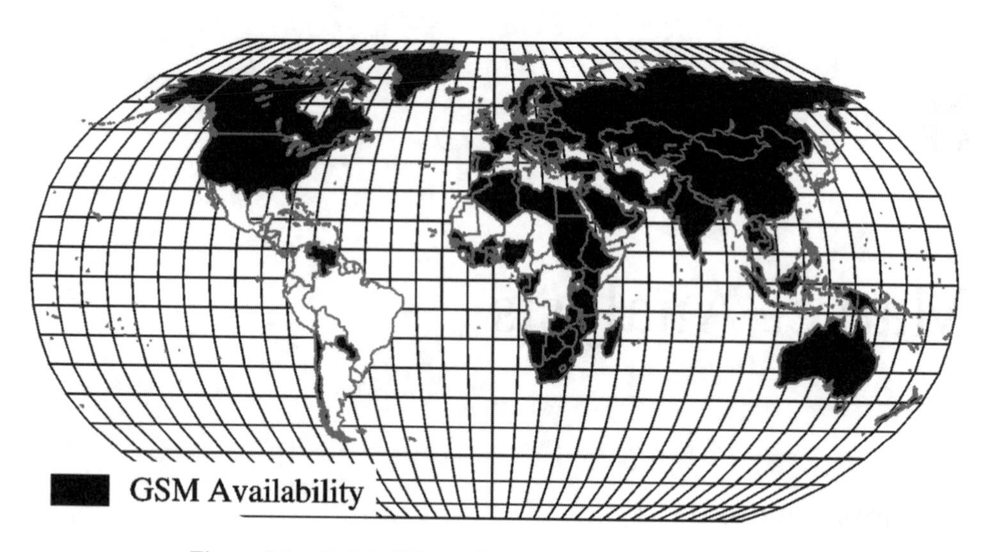

Figure 8.1 Global GSM availability at the turn of the century.

In addition to putting together a business case for a prospective new satellite system, market analysis information is also used to derive potential satellite traffic characteristics, which are used during the system design phase. For example, system engineers make use of traffic prediction models to dimension required satellite beam capacity, in terms of the number of available channels; and, subsequently, satellite power requirements, given knowledge of the required EIRP/channel. In the case of non-geostationary satellites, the coverage area of each satellite will constantly change relative to the Earth, hence the traffic load seen by a satellite will change continuously as it passes over areas of little or no traffic, e.g. the sea, to regions of high traffic density.

Regulatory bodies, such as the ITU, make use of market prediction studies to determine the spectral needs that will be required to sustain demand for a particular category of service. This can be achieved by sectorising the market into particular terminal/user types from which services and associated bit rates can be applied. An example of how this methodology was applied by the UMTS Forum to estimate the spectral requirements for UMTS can be found in [UMT-98].

Of course, the difficulty with satellite-personal communication networks is that they are highly dependent on the success or failure of the terrestrial mobile communications industry. For a number of reasons, not least of all cost, it is not feasible for satellites to compete with their terrestrial counterparts, hence satellites play complementary roles by essentially filling in the gaps in coverage. The size of this complementary role is determined by how well the terrestrial mobile networks are established. The longer a satellite system takes to move from initial design to reality, the more established the terrestrial networks become. Hence, there is a need for accurate long-term market prediction analysis combined with a satellite implementation schedule that is able to meet the markets identified and at a cost which will enable a profitable service to be delivered.

Satellites come into their own when used to provide services to areas unreachable by terrestrial means. The success of Inmarsat demonstrates that satellites can be used to provide mobile services to specialist, niche markets. For many years, the maritime sector has been reliant on Inmarsat for its communication facilities, while satellite delivered aeronautical services are becoming an increasingly important market sector.

The introduction of third-generation (3G) mobile systems will provide the next major opportunity for satellite service providers and terminal manufacturers to enter into the mobile market. In UMTS/IMT-2000, the satellite component is foreseen to provide mobile multimedia services at rates of up to 144 kbit/s. At such rates, the possibility to provide video facilities to lap-top-like terminals is certainly feasible.

8.2 Historical Trends in Mobile Communications

At the end of the 20th Century, the total number of cellular subscribers in Europe was just under 180 million, with 82% of these belonging to EU-15 member states (that is Austria, Belgium, Denmark, Finland, France, Germany, Greece, Ireland, Italy, Luxembourg, The Netherlands, Portugal, Spain, Sweden, UK) [MCI-00]. Roughly 8 years after the introduction of the first GSM services into Europe, the average market penetration for EU-15 countries for cellular services was in the region of 45%, with Scandinavian countries achieving over 60%.

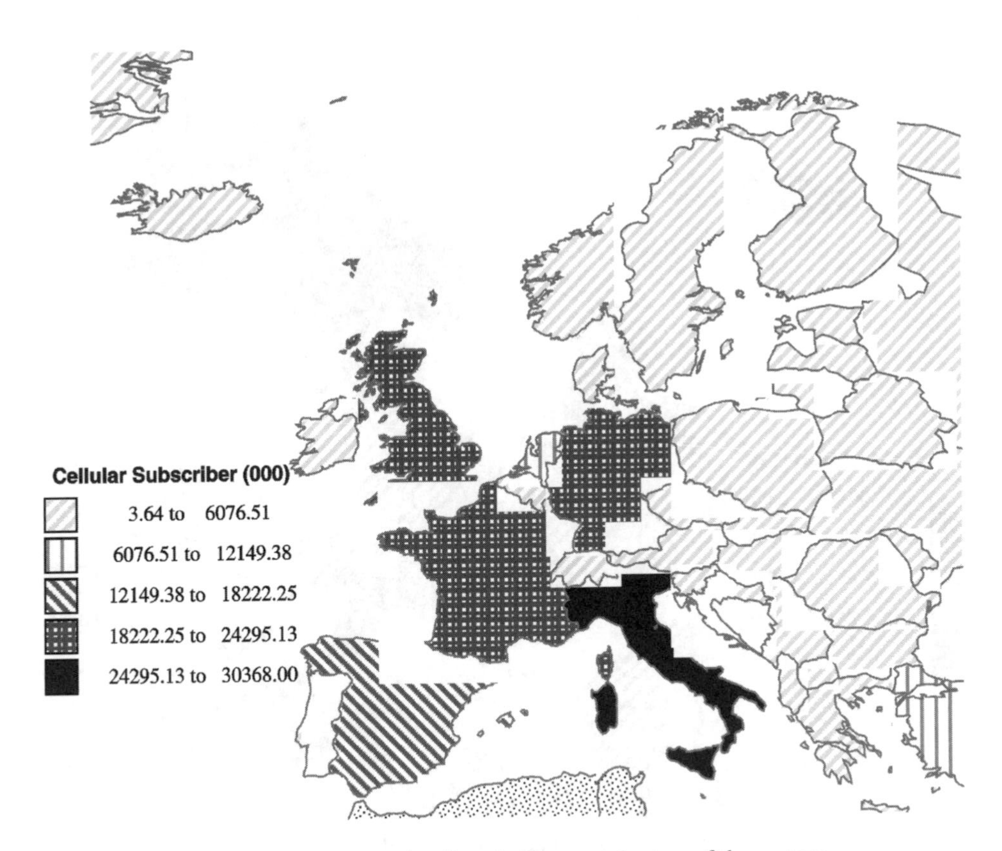

Figure 8.2 Cellular subscribers in Europe at the turn of the century.

The number of cellular subscribers in Europe and the percentage of market penetration at the end of the 20th Century are shown in Figures 8.2 and 8.3, respectively.

For non-EU countries, with the notable exception of the Western European countries Norway, Switzerland and Iceland, the market penetration for cellular services is less spectacular. This is particularly noticeable in the former USSR states, where market penetration achieved only 13% of the population at the turn of the century. This may be due to several factors including economic instability and the delay in introducing such services to Eastern Europe.

At the start of the 21st Century, there were in the region of 400–450 million cellular subscribers world-wide. Without doubt, the market for mobile-satellite services is seriously affected by this world-wide take-up of cellular services.

As terrestrial cellular services converge on a global scale, the appearance of multi-mode, multi-band terminals on the market, aimed very much at the international business traveller, is likely to have a further impact on the mobile-satellite market [SAT-98]. Conversely, as the populace becomes more mobile aware, there will be an expectancy for mobile access in all environments. The ability to communicate in an aircraft with the same ease and facilities as on a train, for example, appears a reasonable expectation for both business and leisure travellers.

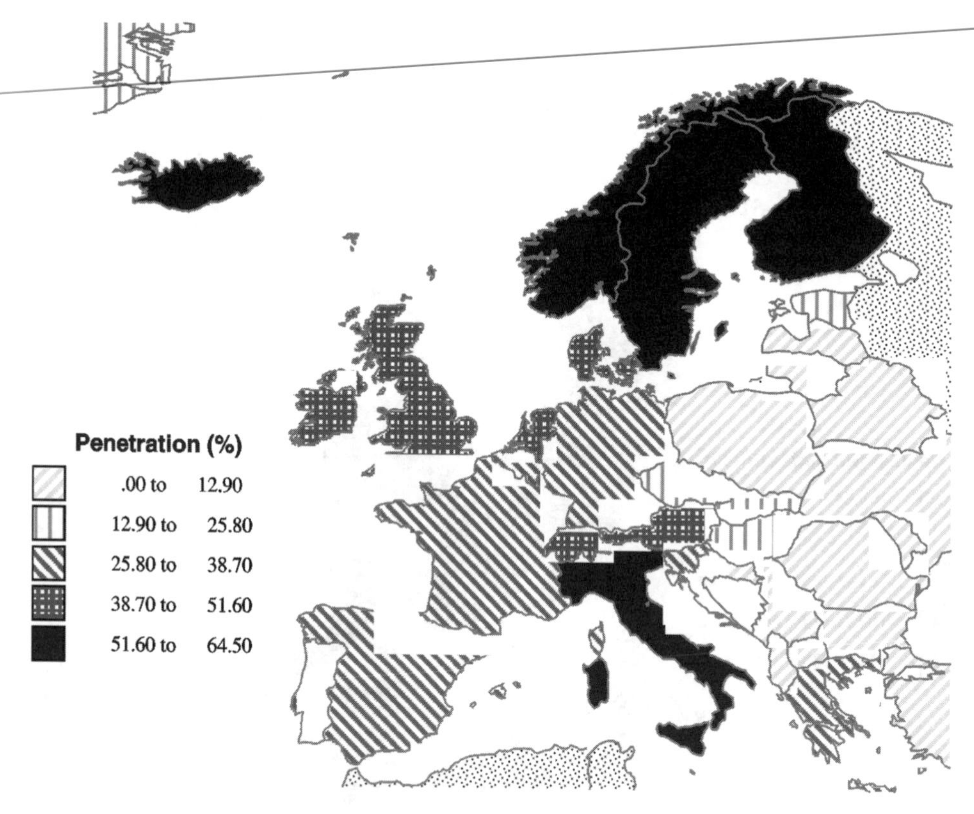

Figure 8.3 Cellular penetration in Europe at the turn of the century.

8.3 Prospective Satellite Markets

8.3.1 *Objectives*

Before the role, and hence the prospective markets that can be addressed by the satellite component of a future integrated mobile network, can be established, initially the general objectives of the satellite component must be defined. Both ETSI and the RACE II project SAINT [SAI-94] have approached this task from a satellite-UMTS perspective, the collective results of which are summarised below:

- To provide access to UMTS satellite services throughout the European region and to extend this facility world-wide;
- To guarantee coverage within large contiguous areas which are also covered by the terrestrial network infrastructure, irrespective of regional demographics;
- To provide complementary service coverage to the terrestrial component;
- To provide seamless quality of service availability within the coverage area;
- To provide paging/SMS capacity in areas of poor or non-existent terrestrial coverage;
- To augment the development of telecommunication services in developing countries;
- To provide transparent access to the fixed network with a quality of service (QoS) comparable to ISDN and commensurate with affordable cost;
- To provide highly reliable emergency services across a wide area irrespective of geographic, economic or demographic considerations;
- To provide rapid and cost effective deployment of UMTS services;
- To support the use of small pocket-size terminals in addition to other mobile and fixed terminal types;
- To operate within recommended health limits.

The above can be used to establish the roles that a satellite can play in an integrated network such as UMTS or IMT-2000.

8.3.2 *The Role of Satellites*

Taking the S-UMTS objectives into account, the following roles can be identified for the satellite component:

- *Coverage completion*: where a terrestrial mobile service is well established, such as in Western Europe, it is unrealistic to think of a competitive satellite service, it is more likely that the role of the satellite will be to provide a complementary back-up service. The implementation of terrestrial cellular systems is a continuous, gradual, process; initially mobile services are deployed in densely populated urban areas, followed by extension to suburban and semi-rural areas, airports, motorways, etc., and finally rural areas. The economics and demographics associated with a region largely control progression down this implementation path. Due to the large coverage area offered by a satellite beam, which provides equal priority coverage for all areas of population density, the satellite component can be used to complete the coverage of the terrestrial network. Furthermore, aeronautical and maritime users may be solely dependent on the satellite component for the provision of services.

- *Coverage extension*: in this instance, the satellite system can be used to extend coverage boundaries of the terrestrial network. For example, GSM is now well established in Western Europe, however, the degree of penetration into Eastern Europe is still relatively low. A satellite system capable of illuminating all of Europe can be used to rapidly extend coverage into these regions.
- *Disaster proof availability*: satellite systems can provide a back-up service if some form of natural or man-made disaster reduces the effectiveness of the terrestrial network. Recent years have shown how the catastrophic effect of earthquakes can nullify terrestrial communication facilities at a time when they are needed most. Satellites have played an important role not only in relaying imagery of such disasters but also in providing the necessary communication facilities for the co-ordination of disaster relief efforts.
- *Rapid deployment*: satellite systems can be used to rapidly extend the coverage of the terrestrial network where deployment of the terrestrial network has fallen behind schedule. In the first few years of UMTS/IMT-2000 deployment, when terrestrial coverage may not be so prevalent, satellites could play an important part in the roll-out of the service.
- *Global roaming*: satellite systems can provide global roaming for users of UMTS/IMT-2000 terminals in support of the Virtual Home Environment concept.
- *Dynamic traffic management*: the satellite resource can be used to off-load some of the traffic from the terrestrial network. For example, a mobile moving from one terrestrial cell to another where no channels are available due to user demand could be re-routed over the satellite.

All, bar the last, of the above roles can be considered as being complementary to the terrestrial service. The final role is supportive, which can lead to a decrease in the blocking probability of the terrestrial network, or alternatively an increase in the network capacity for the same grade of service. The effectiveness of the satellite's supportive role will largely be determined by the resource assignment strategy adopted by the network.

8.3.3 Satellite Markets

The UMTS Forum Report "A Regulatory Framework for UMTS" published in June 1997 predicted that the annual market revenue in Europe for mobile multimedia services will be at least 34 billion ECU (made up of 24 billion from services and 10 billion from terminals) with at least 32 million mobile multimedia services users [UMT-97]. Note: the ECU (European Currency Unit) was the term used prior to the adoption of the Euro as the European currency denomination. Business users are predicted to provide the largest market sector with a predicted two-thirds share of the market. This is a slightly higher figure than that reported in the Analysis/Intercai UMTS Market Forecast study [ANA-97], which predicted 20 million European users providing annual revenues of 27 billion ECU. Irrespective of the differences between the respective reports, both agree that the mobile multimedia market offers huge potential.

The expected revenues generated from personal and broadband communication services via satellite, as presented in the EC's document "EU Action Plan: Satellite Communications in the Information Society" [CEC-97] divide the market sectors into three: satellite, terminal and services. The outcome of this report suggested that a combined total of in excess of $350 billion for services delivered by traditional geostationary satellite, S-PCN and advanced broadband systems could be envisaged over a 10-year period. Similarly, the revenue generated from terminals could be in the region of $200 billion. Clearly, these figures suggest that there is a significant market opportunity for satellite operators, service providers and terminal manufacturers.

A more detailed analysis of the prospective markets for future mobile-satellite communications will be presented in the following sections.

8.3.4 Service Categories

In order to dimension the market, it is essential to have an understanding of the types of services and applications that will be supported by the network. Whereas previously voice would have been the dominant, if not the only service to be considered, the ability to provide multimedia services opens up new opportunities and markets to be addressed.

The types of 3G services that are likely to be available are expected to be aimed at particular niche markets. Typical users of such services will include:

- People in transit and out of range of terrestrial coverage;
- Travellers to regions of the world without service availability or IMT-2000/UMTS roaming agreements in place;
- Individuals or small/medium enterprises (SMEs) located in areas with poor or inadequate terrestrial access. Satellites are expected to be used to complement terrestrial services in both developed and developing countries. Since satellite technology offers practically the sole means to extend broadband network access to wide areas in a short space of time and at modest cost, there are clear economic and social benefits to be gained from communication service introduction via this route.

Such benefits will apply particularly to remote and less developed regions of the world.

Certainly, a very diverse range of services can be met by applications that utilise the combined power of speech, data and images in the context of a ubiquitous service, such as UMTS. Future-generation mobile systems will be capable of providing different types of services to support various applications, such as multimedia mailboxes, the transfer of documents and files containing text, images and voice, messaging services, directory services, database access, advanced traffic telematics applications, transactional applications, video information transfer, and so on.

Two main classes of service are identified in [ITU-93], namely interactive services and distribution services.

Interactive services are further divided into three categories:

- Conversational services: these services operate in real-time, offering bi-directional communication. Example services could include person-to-person telephony, multipoint video conference, video surveillance, remote medical consultation, etc.
- Messaging services: this category of service provides store-and-forward of data and could include, for example, e-mail, SMS, and so on.
- Retrieval services: this category covers the retrieval of stored information on demand from information centres. This category could include ftp access, utility meter reading, in-car road congestion information, etc.

Distribution services are divided into two sub-categories:

- Distribution services without user individual presentation control: these are broadcast or multicast type services distributed from a central source at a predetermined time and in a predetermined order defined by the service provider. Potential services could include TV distribution, message broadcast and digital audio broadcasting.
- Distribution services with user individual presentation control: this category allows the user to control the type of information and the time delivered from a service provider. A typical example would be video-on-demand.

Using the ITU definition, Table 8.1 lists some possible services/applications that could be made available over the satellite component of a 3G network.

Table 8.1 UMTS terminal service profiles

Service	Applications	Laptop		Briefcase		Hand-held Palm-top/ cellular
		Port	Mobile	Port	Mobile	
Messaging services	Electronic mail	✓	✓	✓	✓	✓
	Paging and short messages (voice and/or text)	✓	✓	✓	✓	✓
Conversational services	Telephony/telefax	✓	✓	✓	✓	✓
	Video telephony	×	×	✓	×	×
	Video conference	×	×	✓	×	×
	Video surveillance	×	×	✓	✓	×
Retrieval services	FTP/database retrieval	✓	✓	✓	✓	✓
	Web browsing	✓	✓	✓	✓	✓
Distribution services without user control	Audio broadcast	×	✓	×	✓	×
	Video broadcast	×	✓	×	✓	×
	Vehicular digital information broadcast	×	✓	×	✓	×
User-controlled distribution services	Video-on-demand	✓	×	✓	✓	×

Here, three broad categories of terminal have been considered: lap-top, briefcase and hand-held. Moreover, the non-hand-held terminals have been further sub-divided into portable (port) and mobile. A portable terminal implies that the user will operate via the satellite while stationary. Mobile terminals operate literally on the move.

In the following example of market analysis, it is assumed that lap-top and briefcase terminal types will support services from 16 kbps up to possibly 2 Mbps when stationary. The palm-top terminal will be mainly used for voice, fax and low data rate services and will support data rates up to 64 kbps. Individual vehicular terminals are assumed to support data rates of up to 144 kbit/s in open environments.

8.4 Future Market Forecast

8.4.1 Terminal Classes

3G mobile systems will aim at providing a divergent set of services with a convergent standard. In the context of mobile service provision, it is envisaged that most of the applications/services will be supported by means of cellular terminals. Existing portable terminals already have built-in data capability to avoid the need for modem or data adapters. Current mobile phones on the market have also incorporated functionalities to support e-mail and Internet access. Hence, in as far as predicting the 3G market is concerned, the following analysis will be based on the historical trend in the growth of cellular mobile phones.

Five terminal types are envisaged for S-UMTS/IMT-2000 (from now on simply referred to as S-UMTS): hand-held, vehicular, transportable, fixed and paging receivers. Like other consumer goods, each terminal type is expected to be available in a range of models, their cost being dependent upon several factors including:

1. The market penetration of the terminals, hence the production volume
2. The competition among manufacturers
3. The type of services supported by the terminals
4. The degree of terminal sophistication (e.g. dual-/single-mode, etc.).

In order to promote mass usage of S-UMTS services, hand-held terminals will need to be priced on a par with other non-luxury type domestic items. Marketing of the product will need to take into account three main considerations:

- The terminal price;
- The subscription rate; and
- The call rate.

Each terminal classification will provide a distinct range of S-UMTS services, with capabilities for seamless handover and roaming between networks. Space/terrestrial dual-mode facilities will be required, as will single-mode handsets. The handset will need to make use of an omni-directional type or, at best, hemi-spherical (3 dBi gain) type antenna, since operation will need to be independent of satellite location.

Vehicular terminals will not be so limited by the availability of transmit power and antenna gain. Essentially, vehicular antennas can be classified into those that track the satellite by mechanical means and those that do so electronically. Antennas can be steered in both azimuth and elevation directions, ensuring optimum space to ground links are established

Transportable terminals, which will essentially be aimed at the international business traveller, will be similar in style to those currently being used for the INMARSAT-M system, whilst fixed VSAT type antennas will be used to provide communications to areas without access to the fixed network infrastructure. Paging terminals will be very low gain, receive only devices capable of receiving and displaying alphanumeric messages. The terminals supported by UMTS, whether via the space component (S-UMTS) or the terrestrial component (T-UMTS), can be broadly divided into three classes: portable, mobile and fixed terminals. Each of these three classes can be further subdivided according to the degree of supported mobility and to their usage. Table 8.2 shows a possible segmentation of UMTS terminals.

Table 8.2 Possible satellite-UMTS market segmentation

Terminal		Mobility during operation	Usage
Class	Type		
Portable	Lap-top	No	Individual
	Briefcase	No	Individual
	Palm-top	Personal	Individual
Mobile	Vehicular	Yes	Individual (cars/trucks/ships)
	Aircraft	Yes	Group/individual
	Ships	Yes	Group/individual
Fixed	VSAT	No	Group (e.g. oil platforms)/individual (e.g. residential)

As noted previously, the differentiation in the terminal types (i.e. palm-top, lap-top or briefcase) allows for different service profiles supported by each type of terminal and for different pricing policies. The distinction between individual and group usage also has a profound effect on determining the market segmentation. The degree of mobility supported during operation in each terminal type will distinguish the end-user groups.

8.4.2 Market Segmentation

The fundamental assumption for the identification of S-UMTS markets is that S-UMTS will be complementary to T-UMTS. This means that S-UMTS will provide services to areas where T-UMTS is under-developed or where T-UMTS will not reach due to either economical or geographical reasons. Bearing this in mind, it is possible to identify three major areas in which S-UMTS will play an important role in the provision of mobile telecommunications services in the European context:

- Rural/remote areas not covered by terrestrial-UMTS;
- Maritime, providing services to commercial and private ships;
- Aeronautical, providing business and in-flight entertainment services.

For aeronautical services, the end-users will be the passengers on board aircraft. The services offered by the airline, including in-flight entertainment, will probably be supported by a local area network (LAN) configuration on the aircraft, with a terminal at each passenger seat. The possibility for passengers to plug-in their own terminals into the aircraft's LAN is also envisaged. The net bit rate offered by an aircraft is likely to be of several Mbit/s, and will require operation outside of the existing S-UMTS allocated frequency bands. In this case, the K-/Ka-band would be the next suitable frequency band for operation. Due to their specific market nature, aeronautical services will not be considered in the following analysis.

For maritime services, the market will be mainly targeted at passenger and cargo ships, cruise liners and research vessels. As with aircraft, an on-board LAN configuration can be anticipated, at least for the commercial service industry.

Additionally, offshore platforms, such as oilrigs, will also be a target for S-UMTS terminals, similar to the VSAT configurations widely deployed today. In the following analysis, however, attention will focus on the mobile sectors of the market.

For land mobile satellite services, S-UMTS will mainly cater for rural and remote areas in order to complement T-UMTS. The following markets for S-UMTS land services have been identified:

- Office staff based at rural or remote areas;
- International business travellers travelling from Europe to regions of the world without UMTS type service coverage. International business travellers may require more sophisticated terminals to support high bandwidth capability for such services as video-conferencing facilities, file transfer, and so on;
- Rural population not covered by T-UMTS. For the rural population, the main means of communications will still be telephony. Mobile hand-held (or palm-top) terminals are envisaged to be adequate for this type of user;
- Commercial vehicles, including trucks and lorries, and private cars operating in rural and remote regions. Mobile terminals supporting voice, fax and e-mail facilities, along with in-car navigation, entertainment, and so on will be essential for trucks and cars.

Furthermore, in the early stages of T-UMTS implementation, satellites can play a key role in ensuring wider service availability. As T-UMTS becomes more established in Europe, say by 2010, the satellite's role can be expected to diminish (at least in Europe) to a minor niche market role. Indeed, it could be argued that such a scenario currently exists today with the co-existence with second-generation mobile systems, such as GSM.

Based on the above discussion, a mapping between the gross potential market (GPM) and the terminal segmentation is shown in Table 8.3.

Table 8.3 Gross potential market for different terminal classifications

Terminal		Usage	Gross potential market (GPM)
Class	Type		
Portable	Lap-top	Individual	Rural office staff
	Briefcase	Individual	Business travellers
	Palm-top	Individual	Rural population
Mobile	Vehicular/maritime	Individual	Cars, trucks, ships

In the following analysis, it is assumed that business travellers will constitute 10% of all international travellers and that each business traveller will travel on average four times per year. It is also assumed that office staff make up of 25% of a country's population.

The S-UMTS market will be predicted by using the historical data of cellular mobile phone services. In [HU-97], a penetration curve between the cellular mobile phone and the relative tariff (or the affordability to pay) is developed. New and updated data sets have been gathered to include data for the cellular market up to 1995, and in some cases up to 1997. (Usually, it

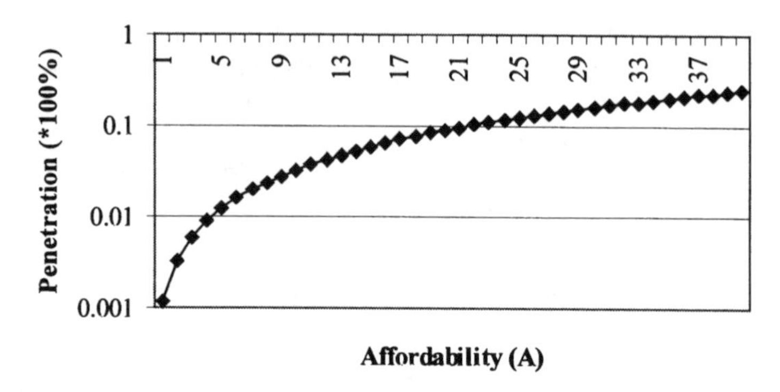

Figure 8.4 Cellular penetration against affordability.

takes organisations like the ITU, 2 or 3 years to assimilate data before it is published.) These new data sets have reflected the rapid growth and demand for mobile services over the past few years. A penetration curve has been derived from the new data and is shown in Figure 8.4.

The penetration curve depicts a global relationship between penetration and affordability. Affordability, *A*, is defined as the ratio of GDP (gross domestic product) per capita to the tariff. However, this curve does not reflect the take-up rate of the market, which is a function of time and differs from country to country due to various factors, such as the GDP per capita and the tariff. The take-up rate determines the growth rate and the penetration of the market. In [HU-97], a logistic model for market penetration prediction has been used to characterise the take-up trend. The characteristics of the logistic model are shown in Figure 8.5.

If such a graph were to be applied to illustrate the demands for mobile technologies in

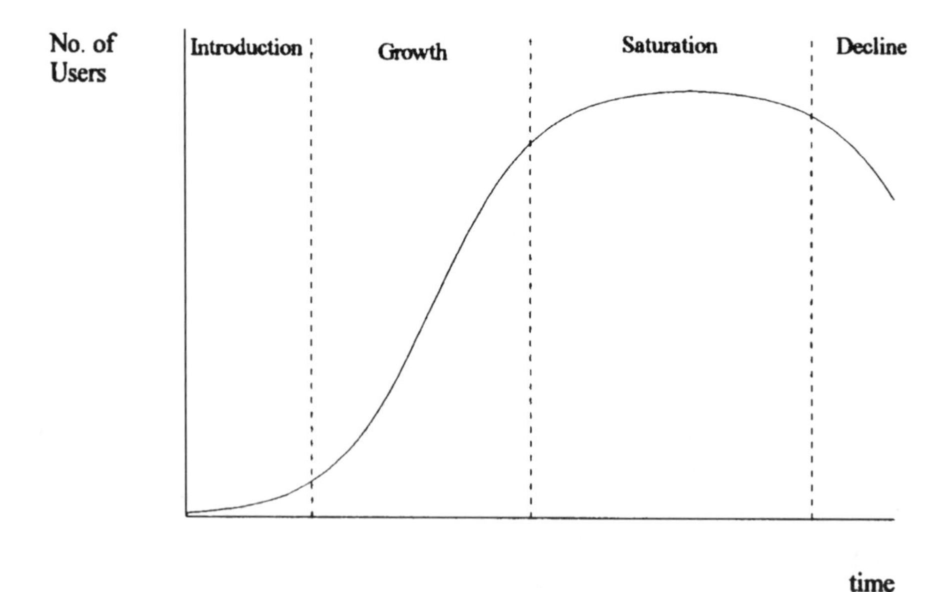

Figure 8.5 Logistic model curve characteristics.

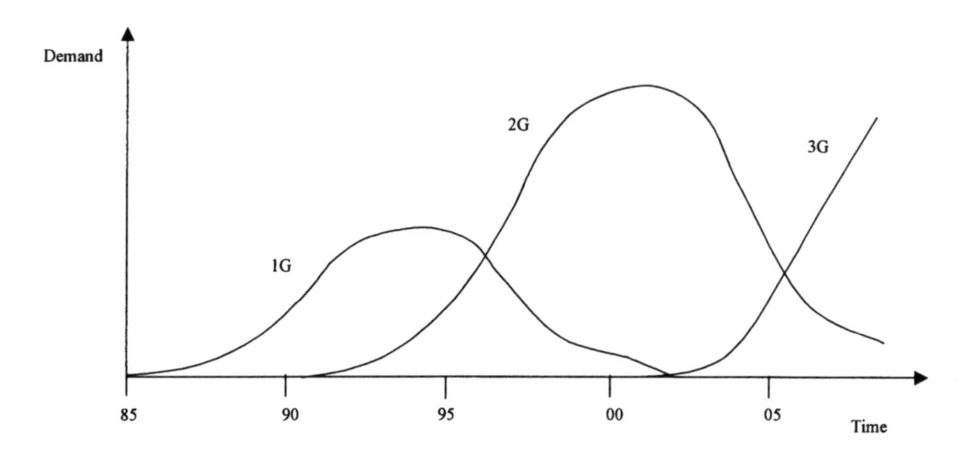

Figure 8.6 Market demand for mobile communications.

Western Europe from 1G to 3G, it would look something like Figure 8.6. Here, a hypothetical market life-span of each particular technology is illustrated.

The equation for the logistic model from [HU-97] is given by:

$$P(t) = \frac{S}{1 + e^{a+bt}} \tag{8.1}$$

where $P(t)$ is penetration at time t and t is the number of years after launch of service; S is the saturation level; a and b are parameters indicating the take-up rate, derived by regression analysis using relevant historical data.

The numerator represents the saturation level, whereas the denominator reflects the yearly take-up rate. The two main economic factors affecting the yearly take-up rate considered here are the yearly tariff and the yearly GDP per capita (GDPCAP). By expressing S in terms of affordability A, equation (8.1) can be rewritten as:

$$P(t) = \frac{cA^d}{1 + e^{a+bt}} \tag{8.2}$$

where c and d are coefficients for the penetration curve shown in Figure 8.4.

A more detailed analysis on the take-up rates for different countries is presented in the next section.

8.4.3 Sizing the Market

From the discussion above, the total market for S-UMTS terminals for individual usage consists of two major components: the portable $M_{port}(t)$ and the mobile $M_{mobile}(t)$markets. In general, the following can represent the total market t years after launch:

$$M_{total}(t) = M_{port}(t) + M_{mobile}(t) \tag{8.3}$$

From Table 8.3, $M_{port}(t)$ and $M_{mobile}(t)$ can be further expanded into the following:

$$M_{port}(t) = M_{briefcase\_port}(t) + M_{lap\text{-}top\_port}(t) + M_{palm\text{-}top\_port}(t) \tag{8.4}$$

$$M_{\text{mobile}}(t) = M_{\text{vehicular\_cars}}(t) + M_{\text{vehicular\_trucks}}(t) + M_{\text{maritime\_ships}}(t) \tag{8.5}$$

where $M_{\text{briefcase\_port}}(t)$ is the total portable briefcase-type terminal market at year t; $M_{\text{lap-top\_port}}(t)$ is the total lap-top-type terminal market at year t; $M_{\text{maritime\_ships}}(t)$ is the total maritime ship terminal market at year t; $M_{\text{mobile}}(t)$ is the total mobile market at year t; $M_{\text{palm-top\_port}}(t)$ is the total palm-top-type terminal market at year t; $M_{\text{port}}(t)$ is the total portable market at year t; $M_{\text{total}}(t)$ is the total S-UMTS terminal market at year t; $M_{\text{vehicular\_cars}}(t)$ is the total car terminal market at year t; $M_{\text{vehicular\_trucks}}(t)$ is the total truck and lorry terminal market at year t.

Each market component in equations (8.4) and (8.5) is calculated by multiplying the gross potential market by the penetration. Referring to the GPM as defined in Table 8.3, the components in equations (8.4) and (8.5) are defined as follows:

$$M_{\text{lap-top\_port}}(t) = N_{\text{office\_staff}} P_{\text{lap-top\_port}}(t) \tag{8.6}$$

$$M_{\text{briefcase\_port}}(t) = (N_{\text{IT}} \times 0.025)\ P_{\text{briefcase\_port}}(t) \tag{8.7}$$

$$M_{\text{palm-top\_port}}(t) = N_{\text{rur\_pop}} P_{\text{palm-top\_port}}(t) \tag{8.8}$$

$$M_{\text{lap-top\_mob}}(t) = (N_{\text{cars}} + N_{\text{trucks}} + N_{\text{ships}}) P_{\text{mob}}(t) \tag{8.9}$$

where N_{cars} is the gross potential market for cars; $N_{\text{office\_staff}}$ is the gross potential market for rural office staff; N_{IT} is the gross potential market for international business travellers; $N_{\text{rur\_pop}}$ is the gross potential market of the rural population; N_{ships} is the gross potential market for ships; N_{trucks} is the gross potential market for trucks and lorries.

A country's market penetration is mainly determined by the applied tariff and the wealth of the country. The tariff includes the terminal charge, the monthly subscription fee and the call charge per minute. The wealth of a country is determined by the GDP per capita (GDPCAP). In the following analysis, GDP, along with other market information, including population, number of cars and trucks are extrapolated on a year-by-year basis over a 14-year period of interest.

Another factor affecting the penetration of S-UMTS services will be the terrestrial roll-out. Since S-UMTS is complementary to T-UMTS, it is expected that it will be more economical to provide satellite services in areas where terrestrial roll-out is not foreseen to be profitable. Table 8.4 shows the criteria used for establishing terrestrial roll-out.

For countries with a GDP of less than 6 kEuro per capita, initial coverage is expected to address major cities with eventual roll-out reaching areas of urban population greater than 1 million after 14 years.

The above criteria have been derived from [KPM-94]. The implication of which is that for high GDP countries, mobile communications services are affordable and hence the penetration of such services is high. It is more economic to provide such services through terrestrial networks in most areas except in very sparsely populated areas. For mid-GDP countries, roll-out of terrestrial mobile services will be cost effective in urban and suburban areas. For low-GDP countries, the roll-out of terrestrial mobile communications will only be limited to major cities where the average income will be higher; and in some areas, the provision of mobile communication services will never be economical through terrestrial networks. In applying the above criteria, it is assumed that most of a country's area is mainly rural. Thus rural population density is evaluated as follows:

Table 8.4 Terrestrial roll-out criteria

GDP per capita (kEuro)	Population density at year of launch (people/km^2)	Population density at and after 14 years of service (people/km^2)
≥22	>10	>3
Between 6 and 22	>100	>30

$$\text{Rural}_{\text{density}} = \frac{\text{Population}_{\text{rural}}}{\text{Area}_{\text{rural}}} \approx \frac{\text{Population}_{\text{rural}}}{\text{Area}} \tag{8.10}$$

where $\text{Area}_{\text{rural}}$ and Area denote the rural area of a country and the area of a country, respectively.

In forecasting the market, the saturation period is assumed to be 14 years. Furthermore, the monthly call minutes shown in Table 8.5 are assumed for each terminal type.

Table 8.5 Monthly call minutes per terminal type

Terminal type	Terminal class	Monthly call minutes
Portable	Lap-top	200
	Briefcase	80
	Palm-top	140
Mobile	Lap-top	80

Flowcharts for the derivation of the mobile and portable markets are shown in Figures 8.7 and 8.8, respectively.

The following illustrates how penetration is calculated for each country.

Consider a country with a GDPCAP of 26 000 Euro and a GDPCAP growth rate of 0.5% at the year of launch. The expected GDPCAP after 14 years is $26\,000 \times (1 + 0.005)^{14} \approx 28\,000$ Euro. If the tariff is 1000 Euro 14 years after service launch, the expected penetration or the saturation level, S, can be deduced from Figure 8.4 to be 15%. However, the penetration has to take into account the take-up rate, which is affected by various factors. As stated before, the take-up rate follows the logistic model. The c and d parameters in the logistic model have to be estimated by making use of historical data. In estimating these parameters, the price difference between the initial launch of S-UMTS services and the historical cellular services has to be taken into account. This is considered as follows:

Let X_{1i} be the GDPCAP at year i obtained from historical data; X_{2j} the estimated GDPCAP at year j; Y_{1i} the tariff of the cellular service at year i obtained from historical data; Y_{2j} the estimated tariff of the new service at year j; p_{1i} the penetration of cellular service at year i obtained from historical data. A factor, k_j, is introduced to account for this difference at year j where k is defined as:

$$k_j = \frac{\left[\frac{x_{2j}}{y_{2j}}\right]^b}{\left[\frac{x_{1i}}{y_{1i}}\right]^b} \tag{8.11}$$

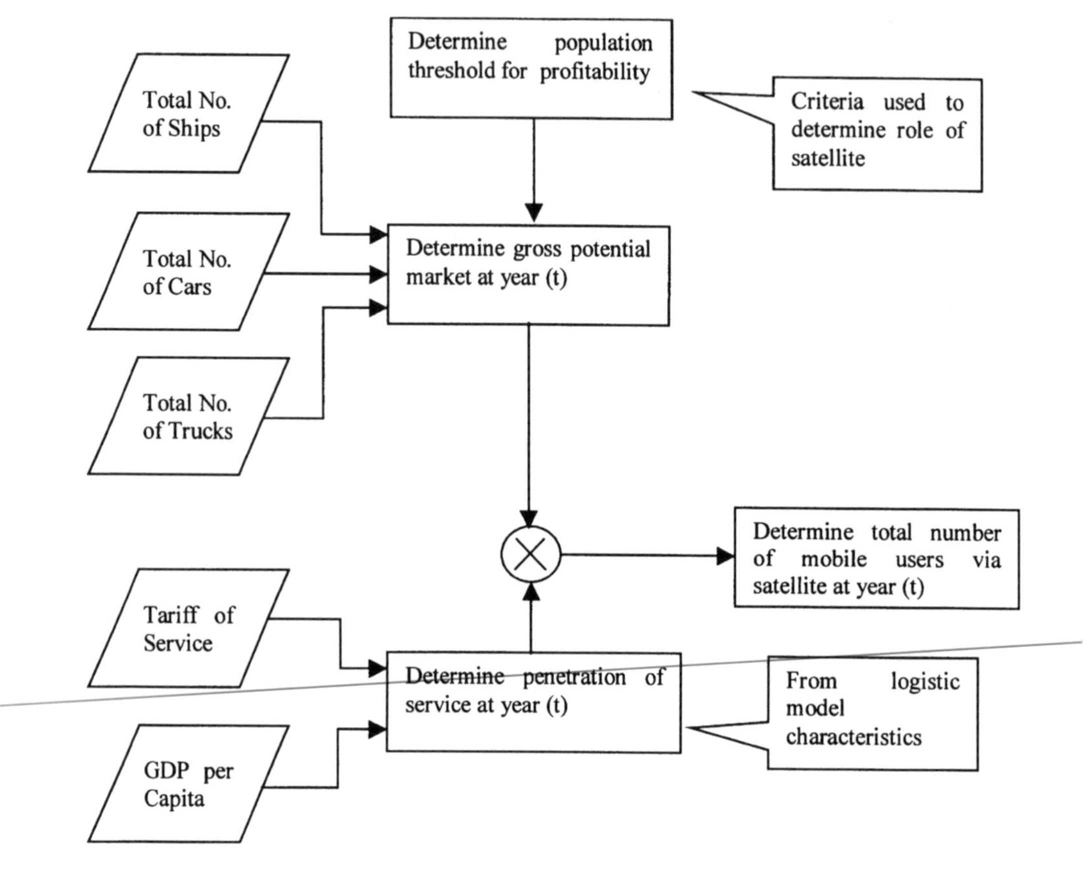

Figure 8.7 Mobile market flowchart.

The take-up for the new service, P_{2j}, at year j is then estimated as $p_{2j} = p_{1i} \times k_j$. This has to be evaluated throughout the whole range of available historical data. Regressional analysis has to be performed in order to obtain the parameters. If the criteria outlined in Table 8.4 are satisfied, the number of subscribers at year t, denoted by $N(t)$, is then obtained as follows:

$$N(t) = (\text{GPM} \times \text{Rur\%}) \times [S/\exp(a + bt)] \tag{8.12}$$

where GPM is the gross potential market as identified in Table 8.3; Rur% is the rural population percentage; and a and b are parameters defined for the logistic model.

8.4.4 Data Sources

An important and necessary step in market forecast methodology is the collection of historical data, ideally spread over a number of years, which are similar in nature to the type of market that is to be predicted. Market prediction is not a concise science, however the more data that are used in the predictive analysis, the more likely it is that the predictions will prove to be accurate in the long term. This is possibly why market predictions for cellular take-up at

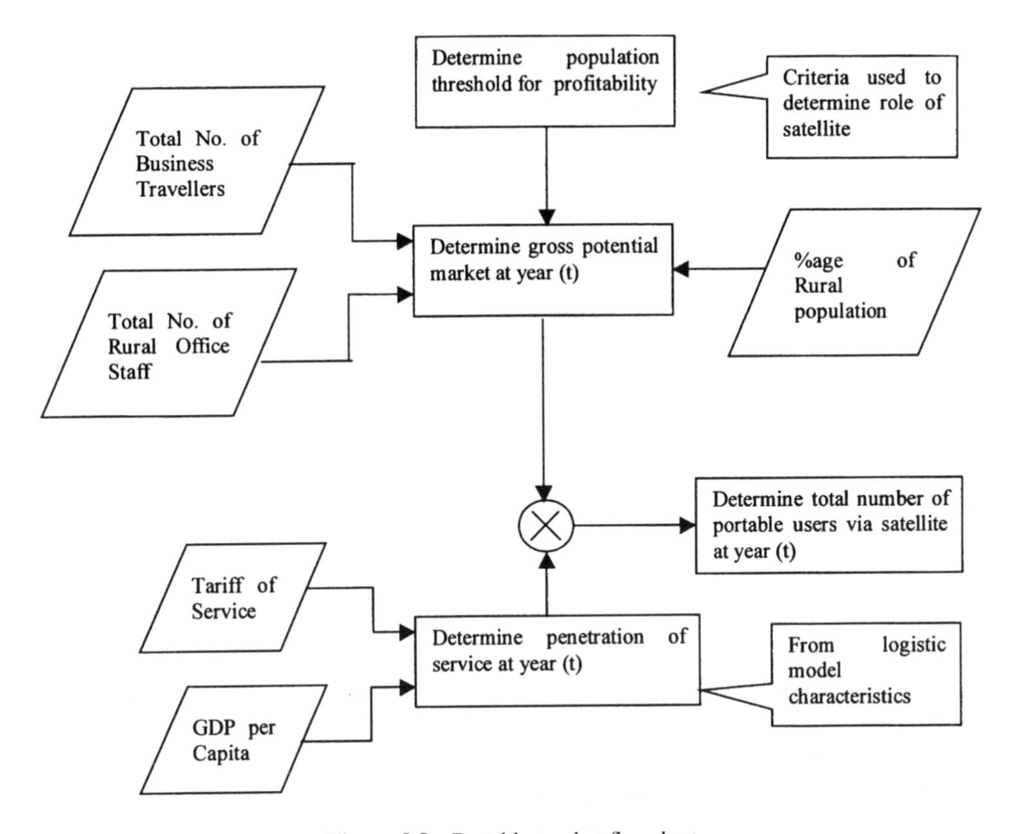

Figure 8.8 Portable market flowchart.

the start of the 1990s were so wide of the mark. At the time, cellular services were still relatively new and the available data upon which predictions were based was likely to be sparse, based on analogue cellular technologies.

The last decade has enabled a large database of cellular subscription growth to be collected and assimilated, from which more accurate predictions can now be derived. This data can be obtained from a wide variety of sources, although the ITU, in terms of telecommunications information, and the United Nations, in terms of socio-economic statistics, are more than adequate. Table 8.6 identifies the data used for the prediction for each user group in this chapter. Although this chapter has considered the European markets as an illustrative example, global issues were also considered, hence the high number of countries reported in Table 8.6.

8.5 Results

8.5.1 Tariff

As will be seen shortly, the size of the potential market is very sensitive to the cost of the service. Based on present costs of mobile-satellite communications, and in order to obtain the market size for different terminals, the tariffs presented in Table 8.7 are assumed for S-

Table 8.6 Data sources

Terminal type	Data for prediction	Number of countries used
Portable	Population	225
	Rural area population	225
	Gross domestic product	225
	Surface area of the region	225
	Number of mobile communications subscribers	220
	Population of cities over a million	58
	Income from telecommunication services	194
	Call charges per 3 min for cellular telephony	180
	Monthly subscription fees for cellular telephony	180
Mobile	Number of cars	134
	Number of trucks	126
	Number of ships	127

UMTS. Comparable T-UMTS tariffs are assumed to be at market rates for current cellular services. These tariffs can be obtained from the ITU Statistical Yearbook, which is published annually.

It is also assumed that there is an annual reduction of 5% in the terminal charges, monthly subscription fees and call charges per minute. The year of launch of service is assumed to be 2002. In selecting these prices the market rates for current or near market mobile satellite receivers have been taken into account.

8.5.2 *Portable Market*

The markets have been predicted for individual EU-15 member states and globally at 14 years after service introduction unless otherwise stated.

Table 8.7 Assumed tariffs at service introduction[a]

Terminal		Terminal charges (Euro)			Monthly subscription fees (Euro)			Call charges per minute (Euro)		
Class	Type	H	M	L	H	M	L	H	M	L
Portable	Lap-top	4500	3000	1500	150	100	50	3.6	2.4	1.2
	Briefcase	4500	3000	1500	37.5	25	12.5	4.5	3.0	1.5
	Palm-top	4500	3000	1500	60	40	20	4.05	2.7	1.35
Mobile	Lap-top	3400	1700	850	150	100	50	3.6	2.4	1.2

[a] H, highest price; M, medium price; L, lowest price.

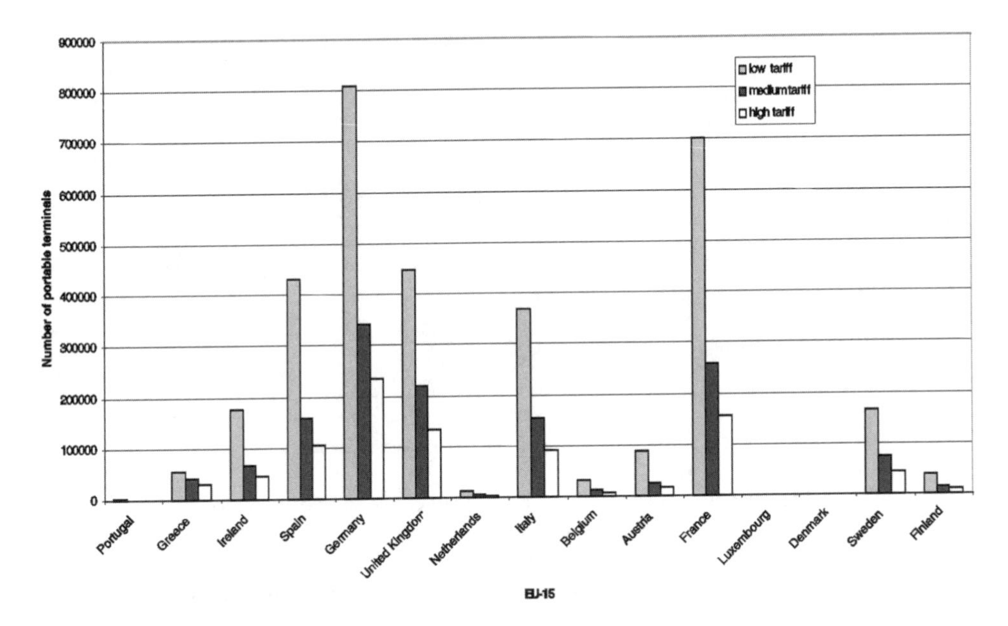

Figure 8.9 Total portable market for EU-15 states.

Figure 8.9 shows the total portable market in the EU-15 states. A further breakdown of the market constituents can be found in [HU-99].

The significance of the tariff can be clearly seen from the results above, particularly in countries that have market potential. In most of the EU-15 countries, the difference between applying the medium tariff, compared to that of the high tariff, can result in 100% difference in the potential market. This is evident in Germany and France, for example. Not surprisingly, in the smaller EU-15 countries, there is no significant market for satellite communications. It can be seen from Figure 8.9 that Germany, France and the UK have the biggest potential market for portable terminals.

Figures 8.10 and 8.11 show the total portable market within the European market, including countries outside of the European Union, and the whole world during a period of 14 years after launch. The distribution over time for both the European and World markets illustrate how the logistic curve parameters affect the market take-up. Both European and World market predictions illustrate the slow build-up in the market take-up rate for portable terminals. The European market for portable terminals is relatively modest, with it taking at least 8 years after service introduction before 1 million terminals are reached. Conversely, if a high tariff policy is adopted, the market may never reach 1 million terminals.

The global distribution of the market share in Figure 8.12 illustrates the importance of the North American and Asian markets and to a lesser extent the EU-15 market.

8.5.3 Mobile Market

The mobile market for S-UMTS will mainly cater for cars, commercial vehicles and ships. Figure 8.13 shows the predicted market for mobile terminals in EU-15 member states. From Figure 8.13, it can be seen that, as with portable terminals, Germany has the biggest potential

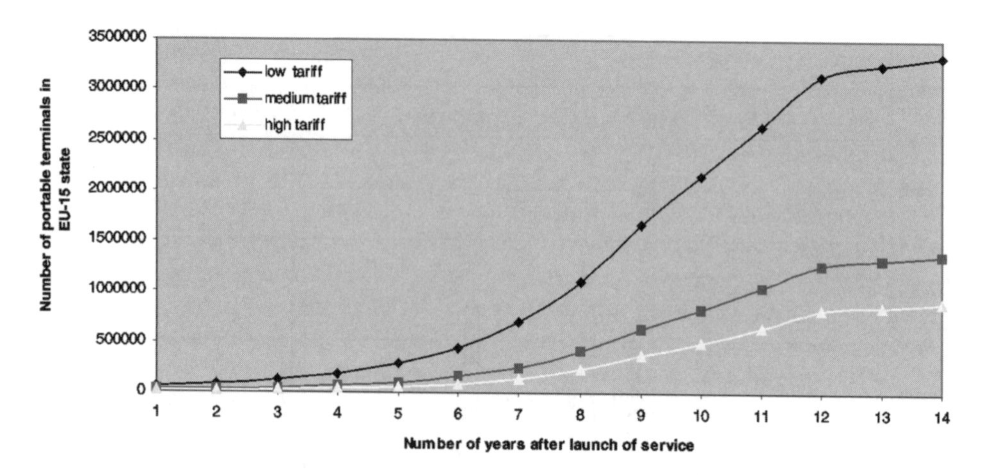

Figure 8.10 Total EU-15 portable market over a 14-year period.

market for mobile terminals within the European Union. France, Spain, Italy, UK and Ireland follow this.

Figure 8.14 illustrates that with a low tariffing policy, half a million terminals can be achieved within about 5 years after service introduction. However, this may take as long as 10 years if a high tariff policy is adopted.

The North American region dominates the global market for mobile terminals (Figure 8.15). Approximately 40 million terminals are predicted for this region 14 years after service introduction, if a low tariff policy is adopted. This market share is heavily influenced by the large number of cars and commercial vehicles in this region in comparison with other parts of the world. On the contrary, areas such as Africa, South America and Russia are not considered suited to this market sector.

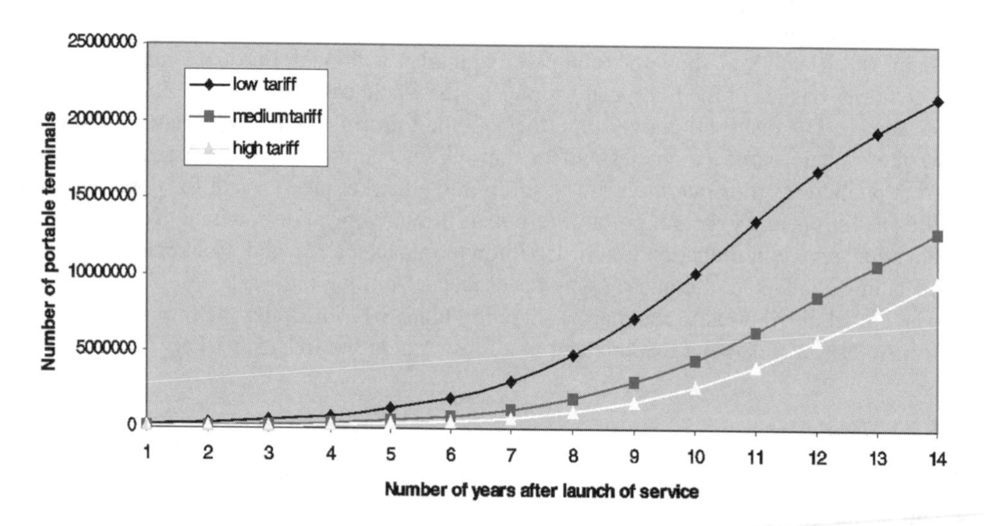

Figure 8.11 Total world market for portable terminals over a 14-year period.

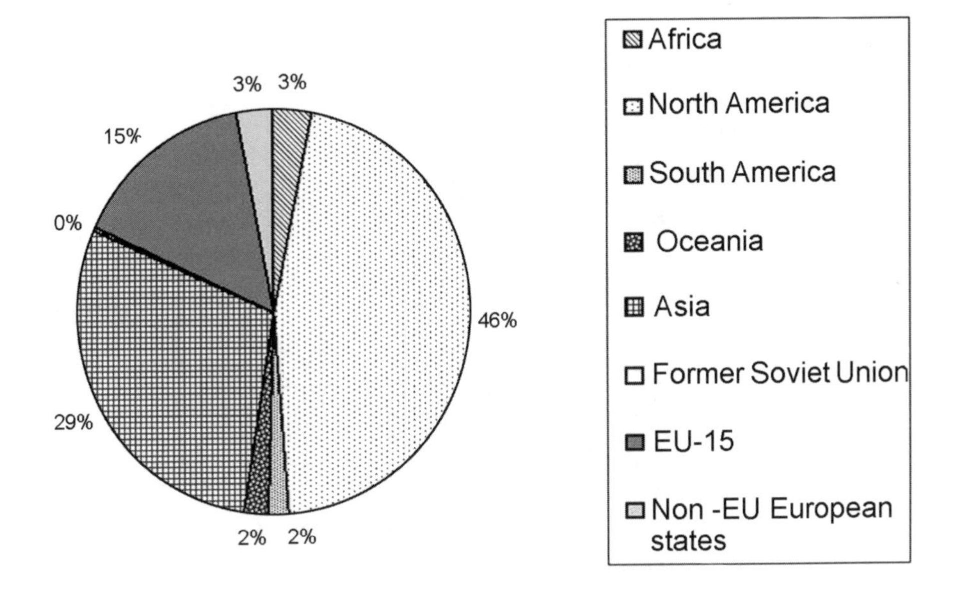

Figure 8.12 Global market share of portable terminals.

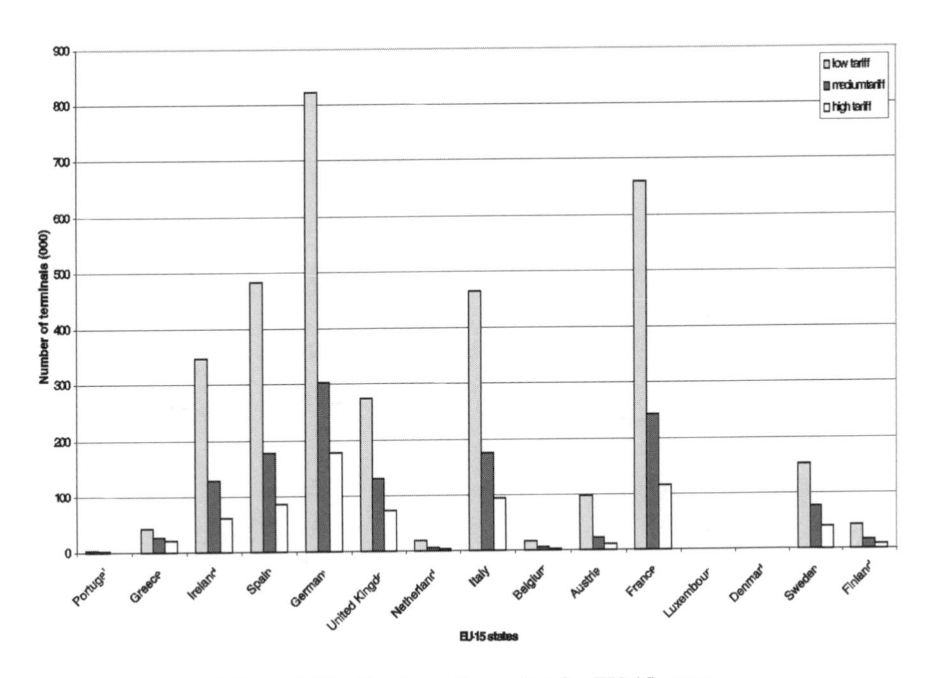

Figure 8.13 Total mobile market for EU-15 states.

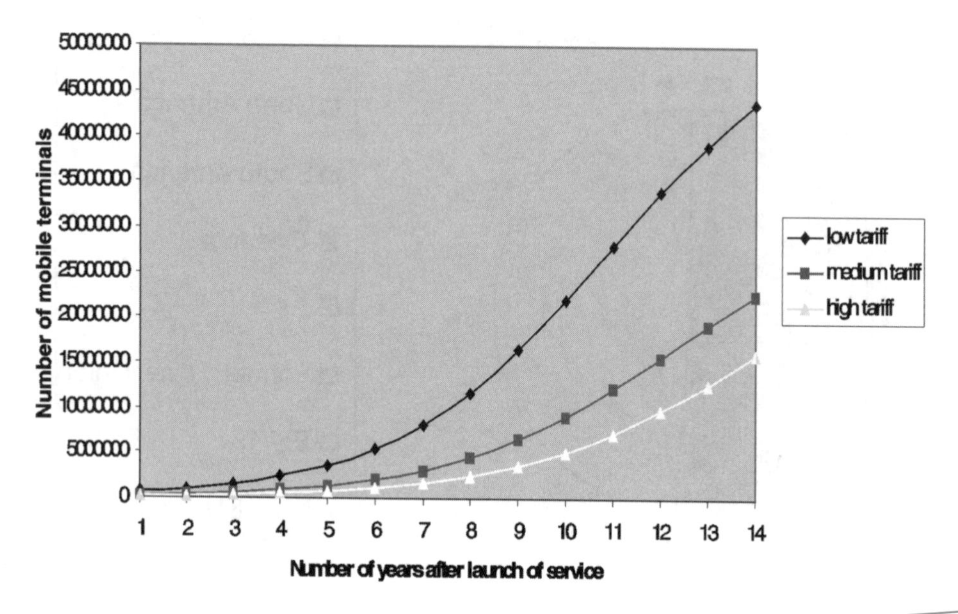

Figure 8.14 Total world market for mobile terminals over a 14-year period.

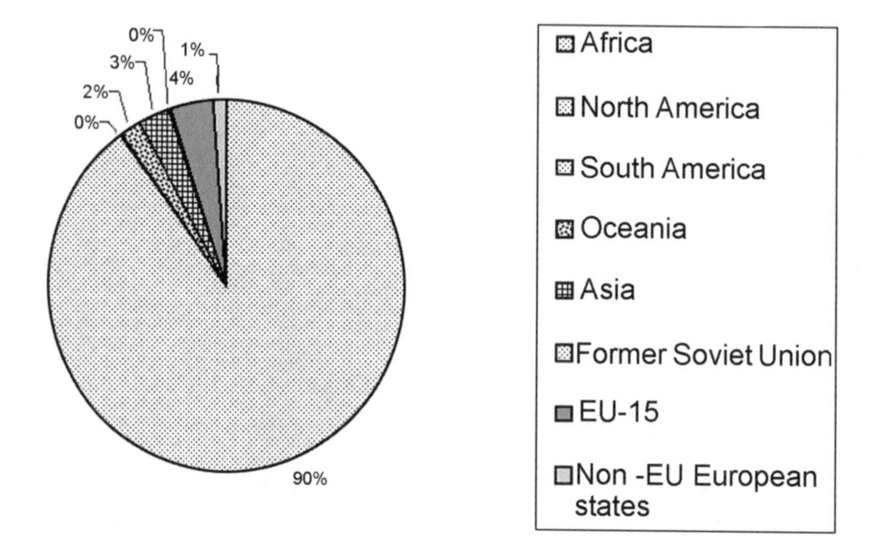

Figure 8.15 Global market share of mobile terminals.

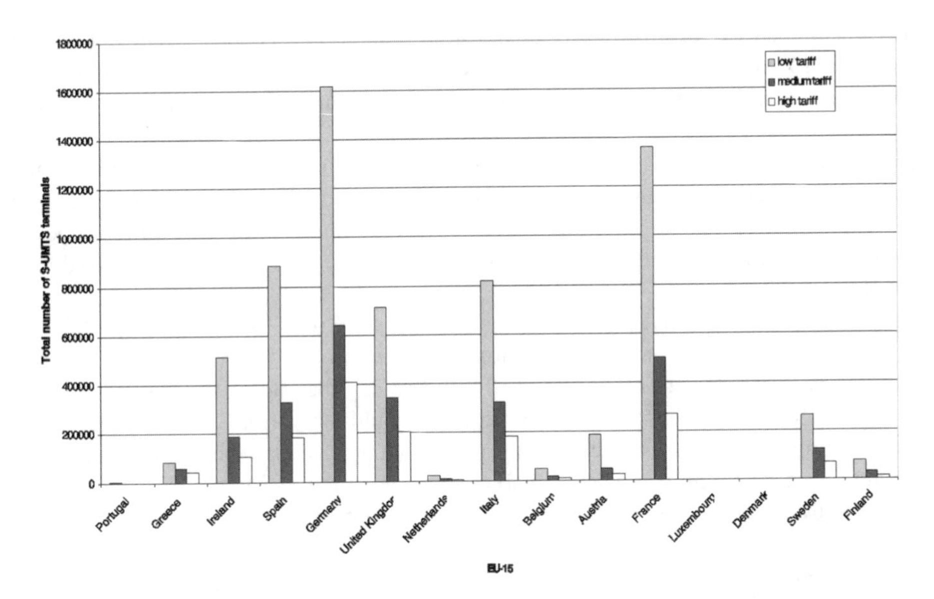

Figure 8.16 Total market for S-UMTS terminals in EU-15 states.

8.5.4 Total Market

Figures 8.16 and 8.17 show the total and take-up of S-UMTS terminals in the EU-15 states, respectively.

The largest market within the EU-15 is Germany, followed by France, Spain and Italy. Germany offers a potential market of just under 1.6 million terminals, a significant share of the total market. This figure, however, could be reduced by as much as 75% if a high tariffing

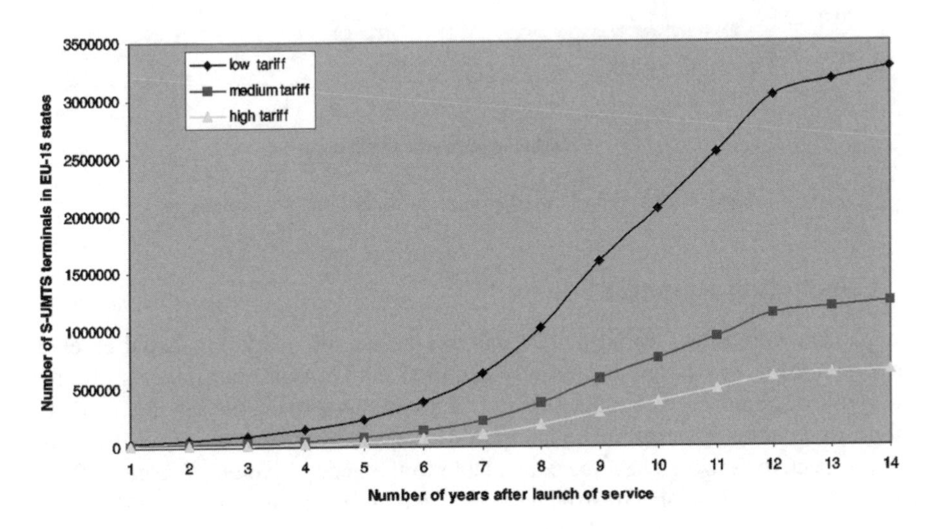

Figure 8.17 Total market take-up for S-UMTS terminals in EU-15 states.

policy is adopted. The results clearly demonstrate that certain countries will not generate significant markets for satellite-UMTS. The take-up of the terminals once again demonstrates the slow pace at which satellite-UMTS terminals will be introduced over the first 5 years or so.

The sensitivity of the EU-15 market is quite apparent. If a high tariffing policy is adopted, less than three-quarters of a million terminals would be sold within the EU-15. By adopting a low tariffing policy, just under 3.5 million S-UMTS terminals are predicted in Europe after 15 years. Even for a low tariff, it would take 8 years before a million terminals would have been sold.

Figure 8.18 illustrates that the total global market for S-UMTS terminals (portable and mobile) will be in the region of 1–3 million by 2005 and this will rise to between 5 and 18 million by 2010, depending on the tariff adopted. By the end of the market prediction period, by 2016, the market for S-UMTS terminals could range from anywhere between 22 and 66 million, depending on the tariffing policy adopted. North America is expected to dominate the market with a predicted market share of in the region of 70% (Figure 8.19). This is a consequence of the number of vehicular terminals.

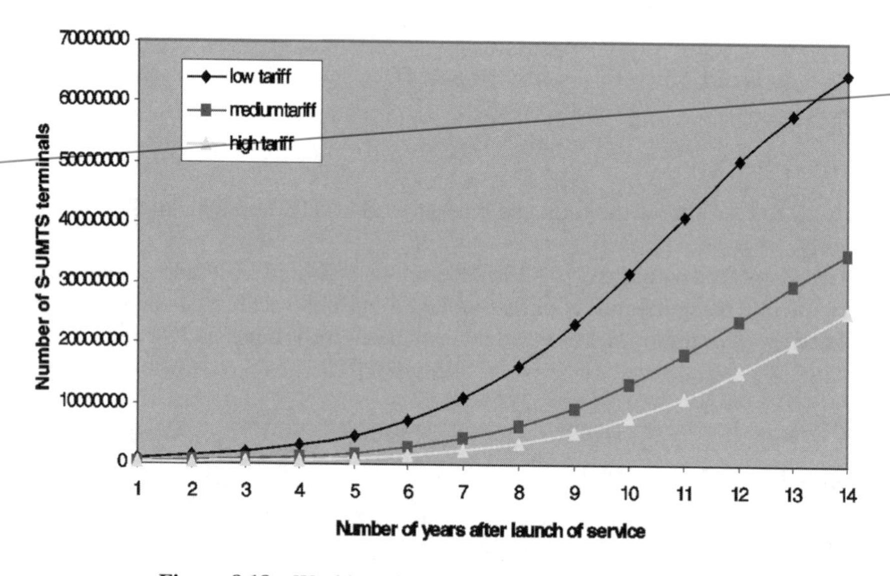

Figure 8.18 World market take-up for S-UMTS terminals.

8.6 Concluding Remarks

In comparison with the terrestrial mobile market, the satellite market potential is relatively modest. Moreover, this is not an instantly acquired market share but rather a progressive build-up of a customer base over a number of years. In this respect, the satellite market is no different from that of its terrestrial counterpart. However, where the two differ is in the time that the respective networks have been available to the general public. If satellite-PCN had had a 10-year market lead on terrestrial networks, perhaps many of us now would be happy to be using a mobile-satellite phone. In reality, however, the situation is very different. Terres-

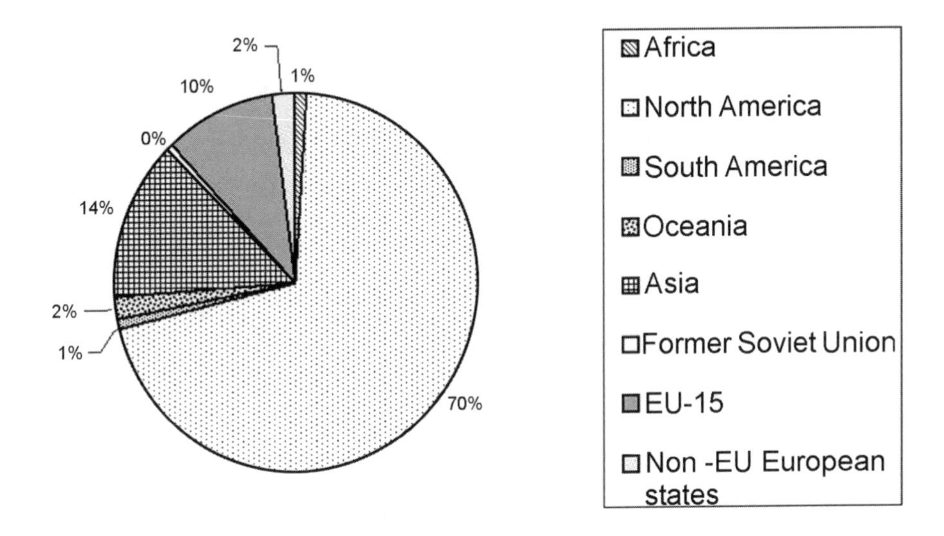

Figure 8.19 Global market share of S-UMTS terminals.

trial mobile networks are well established around the world, and are able to offer a variety of products that suit most peoples needs. Where terrestrial mobile networks are not available, this is due either to demographic or economic considerations, neither of which offer exciting market opportunities for a new, more expensive technology. In a nutshell, there is little evidence of sufficient market demand to support second-generation like services via satellite technologies in such a scenario. Indeed, the economic difficulties experienced by all of the major satellite-PCN operators supports this prognosis.

With this in mind, the introduction of 3G technologies, which provides the opportunity to deliver a vast new range of services and applications creating many new markets, provides the next market opportunity for satellite operators. In this case, the potential for satellite communications appears to be more promising. Systems such as NEW ICO and ELLIPSO and the new generation of INMARSAT satellites are entering the UMTS/IMT-2000 markets roughly at the same time as their terrestrial counterparts. Moreover, the high prices that terrestrial operators have had to pay for their operating licences suggest that in the first few years, at least, T-UMTS/IMT-2000 services may be catered towards the business user, who may be willing to pay a higher price for the additional services on offer. Moreover, T-UMTS/IMT-2000 roll-out is likely to be slow in the first few years, resulting in many gaps in coverage as the network becomes established. Such a scenario is ideally suited to the blanket coverage offered by a satellite network.

Only time will tell whether satellite operators are able to exploit this potential in order to establish mobile-satellite services as a viable alternative to their terrestrial counterparts.

References

[ANA-97] Final Report for EC DGXIII, "UMTS Market Forecast Study", *Analysis/Intercai Report Number 97043*, 12 February 1997.

[CEC-97] "EU Action Plan: Satellite Communications in the Information Society", *Communication from The Commission to The Council, The European Parliament, The Economic and Social Committee and The Committee of the Regions*, Brussels, 5 March 1997, COM(97) 91 Final.

[HU-97] Y.F. Hu, R.E. Sheriff, "The Potential Demand for the Satellite Component of the Universal Mobile Telecommunications System", *IEE Electronic & Communication Engineering Journal*, 9(2), April 1997; 59–67.

[HU-99] Y.F. Hu, R.E. Sheriff, "Evaluation of the European Market for Satellite-UMTS Terminals", *International Journal of Satellite Communications*, 17(5), 1999; 305–323.

[ITU-95] *ITU Statistical Yearbook 1987–1994*, Geneva, 1995.

[ITU-93] *ITU-T Rec. I.211*,"B-ISDN Service Aspects, Series Recommendation – Integrated Services Digital Network (ISDN)", 1993.

[KPM-94] "Satellite Personal Communications and their Consequences for European Telecommunications Trade and Industry", *Report to European Commission*, KPMG, March 1994.

[MCI-00] *Mobile Communications International*, 72, June 2000.

[SAI-94] EC RACE Project SAINT "List of Operational Requirements for the UMTS Satellite Integration", *R2117 TNO DS P102-A1*, May 1994.

[SAT-98] "Cellular Globalization Dampens Prospects for Mobile Satellite", *Satellite International*, 1(17), October 21 1998.

[UMT-97] *A Regulatory Framework for UMTS*, No. 1 Report from the UMTS Forum, 25 June 1997.

[UMT-98] *UMTS/IMT-2000 Spectrum*, No. 6 Report from the UMTS Forum, December 1998.

9

Future Developments

9.1 Introduction

In this concluding chapter, we attempt to look beyond satellite-UMTS/IMT-2000 and in the process highlight some of the key technological drivers that are likely to have an impact on the mobile-satellite industry over the next few years.

Of course, predicting how technology is likely to advance over the next 10 years or so, in such a dynamic and innovative industry, is no easy task. However, there are certain technological drivers and new research initiatives that allow us to identify with some degree of confidence how the industry is likely to evolve with some credibility. One thing is for sure, as the mobile generation matures, the expectancy for high quality, interactive multimedia services delivered at ever increasing data rates is an inevitable consequence of service evolution.

If, rather than writing this chapter at the start of the 21st Century, we were outlining how the mobile industry was likely to evolve a decade ago, we would probably have been some way off with many of our predictions. In this respect, we would not have been alone. On the other hand, certain technological developments known at the start of the last decade have come to fruition with varying degrees of success. For example, at the start of the 1990s, second-generation cellular was on the verge of commercial development, while the concepts of satellite-PCN were starting to be taken seriously for the first time. Moreover, attention had switched to the development of third-generation (3G) network technologies. Ten years on, we now live in an environment where mobile technology is commonplace and satellite-PCN facilities are now starting to gradually become established, although as demonstrated in Chapter 2, the road to commercial reality has been anything but smooth.

As we have seen in the previous chapter, the spectacular success of mobile technology has not been of benefit to satellite-PCN, at least in the developed areas of the world. Confidence in the mobile-satellite industry has gradually been eroded with difficulties with the technology and, significantly, disappointing sales. It is not all gloom and doom, however, within the mobile-satellite community. The latter half of the last decade witnessed a noticeable shift in emphasis away from non-geostationary satellite technology towards larger, multi-spot-beam geostationary satellites. This can be seen with the initiatives described in Chapter 2 including ACeS, THURAYA and INMARSAT-4. Importantly, this new generation of powerful satellites is able to provide hand-held telephony services, compatible with terrestrial cellular systems. Significantly, the dual-mode mobile phones that are used in such networks are now comparable in dimensions to their terrestrial counterparts. Moreover, ETSI are now in the process of finalising the standardisation of the inter-working between geostationary satellites and the GSM network, under its GMR-1 and GMR-2 standards. The importance of standardisation over the last decade has been the key to the success of systems such as GSM, and the move towards the standardisation of the satellite component of GSM and UMTS/IMT-2000 is an important step forward in an industry that is dominated by proprietary solutions.

Perhaps, a decade ago, the influence of mobile Internet access would not have taken up too many paragraphs, however, of all the technological advances over the last 10 years, it is this technological area that many mobile operators are now catering their future market requirements. In such an environment, the mobile-satellite network, like its terrestrial counterpart, will need to operate in a packet-oriented transmission environment, where a high degree of integrity of the transmission, in terms of quality of service (QoS), is required.

One area that is still very much open to discussion is the identification of the "killer" application that will drive the next demand for 3G technologies. While its good to talk, not everyone necessarily feels at home in front of a computer. Clearly, how applications and services evolve over the next few years could have a significant bearing on how the satellite component is utilised in what is intended to be a fully integrated space/terrestrial mobile network.

While the last decade marked a remarkable advancement in the telecommunications infrastructure of the affluent nations of the world, as a consequence, the gap between the "haves" and "have nots" has taken on a greater significance. The fact is that in many parts of the world, the telecommunications infrastructure is not in place simply to establish a telephone call, be it by fixed or mobile means. Figure 9.1 shows the low level of market penetration of cellular mobile communications in Africa at the start of the 21st Century. Clearly, take-up levels are significantly lower than Europe, for example. Perhaps, more significantly, Figure 9.2 shows the corresponding number of fixed telephone lines per 100 inhabitants [ITU-00]. This illustrates the shortage of telecommunications facilities within this part of the world.

The positive influence of telecommunications on the socio-economic development of a region/nation is well known. Of all of the areas in telecommunications that need to be addressed over the next decade, the needs of the developing regions of the world ranks among the highest priorities. The use of satellite communications to establish a telecommunications infrastructure rapidly and cost effectively, has obvious attractions to many regions of the world. Of course, if such a commercial venture is to be viable, the operational costs of such a network should be at such a level that call charge-rates and terminal costs can be offered at a price which would ensure mass market penetration. While the technology may already be available to provide the telecommunications infrastructure to those regions of the world in most need, further advancements in production techniques, combined with innovative business solutions, are required in order to reduce the development and service costs to a level that is affordable to the needy.

9.2 Super GEOs

The introduction of the THURAYA and ACeS geostationary satellite networks marks a significant moment in the mobile-satellite communications industry's development. The

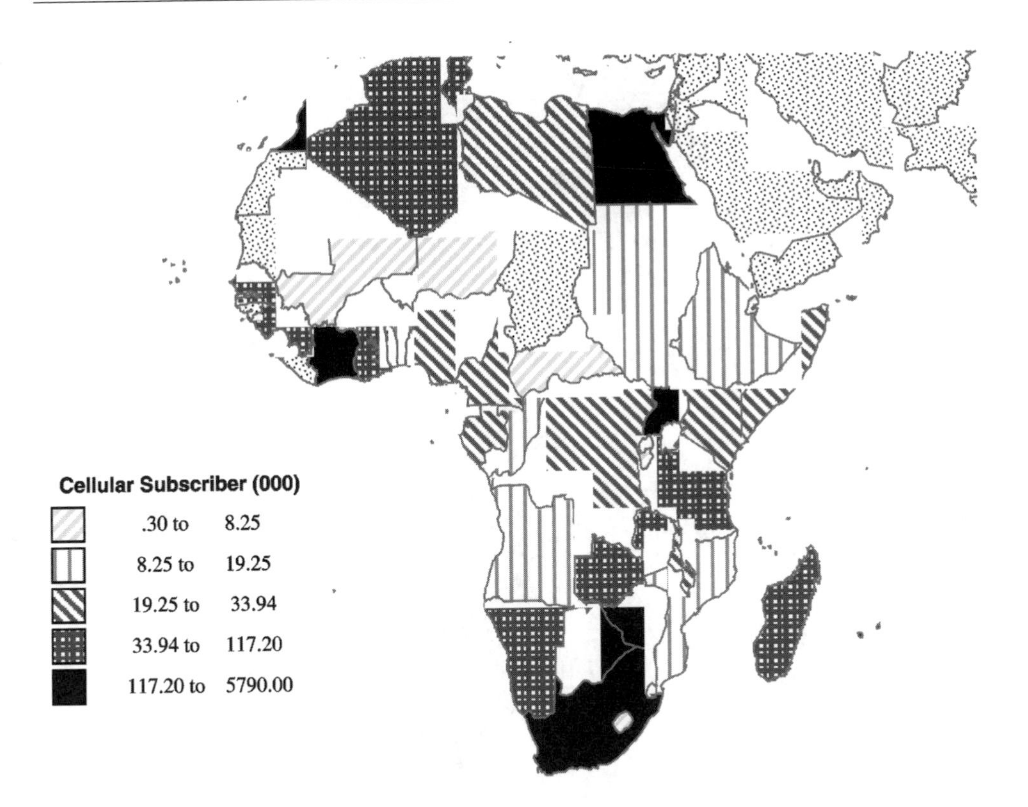

Figure 9.1 Cellular market in Africa mid-2000.

deployment of the INMARSAT-4 satellites in 2004 will further emphasise the importance of geostationary satellite technology to the mobile-satellite industry. These L-band satellites all have predicted life-spans of more than 10 years and are planned to cope with a significant demand for regional mobile services over this period.

THURAYA and ACeS have been developed in particular to service regions of the world with a high traffic demand forecast. As the number of satellite users increase, the requirement for on-board processing to meet the demands of what will effectively be a telephone exchange in the sky will need to expand appropriately. The "digital exchange satellite" highlighted in Chapter 5 is likely to become a reality within the next few years. Moreover, satellites, like their terrestrial counterparts, will move away from circuit-switched delivery towards packet-oriented services. Again, this could have an impact on the available technology that is required on-board the satellite. However, for a packet-oriented transmission scheme to be effective may require the development of a transmission control protocol (TCP) scheme that is able to take into account the special characteristics introduced by the mobile-satellite channel. In such a future scenario, each satellite spot-beam could be thought of as a particular IP sub-network, with each user terminal having its own IP address. The influence of mobile-IP on satellite technologies will be discussed further in a later section.

One of the drawbacks previously cited against geostationary satellite solutions ever achieving mass market penetration was the large, cumbersome mobile terminals needed to make a call. As was noted in Chapter 2, the new generation of high-powered, multi-spot-

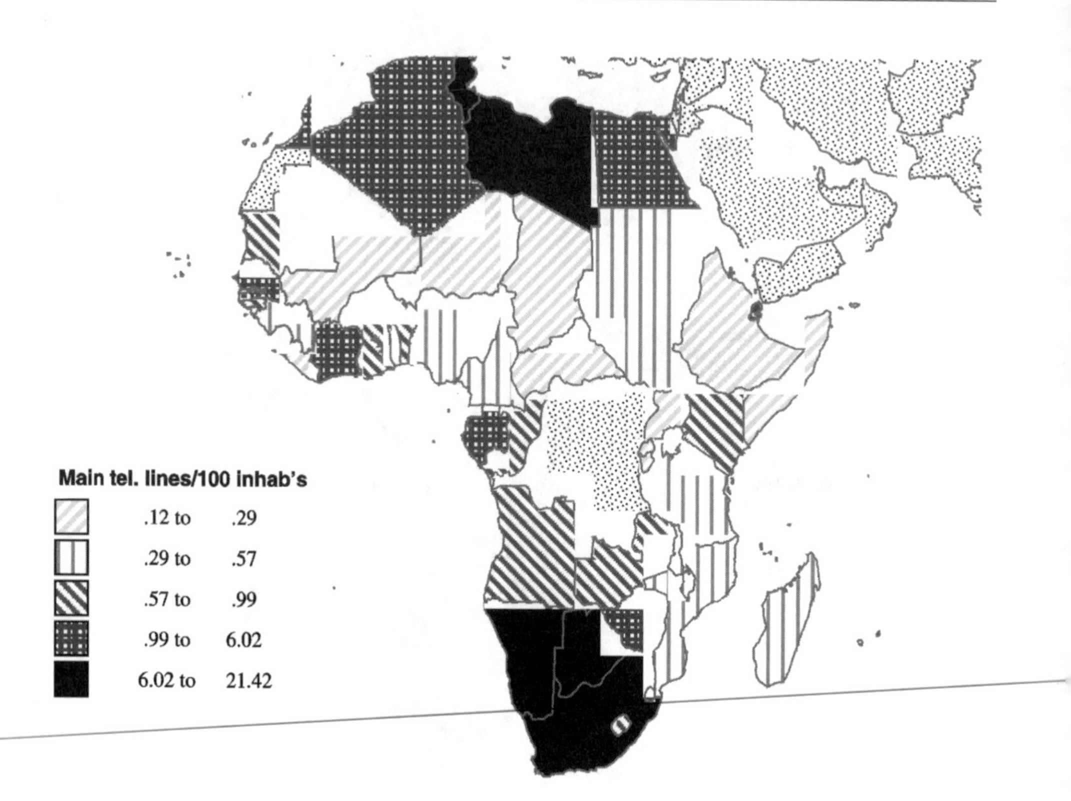

Figure 9.2 Number of telephone lines per 100 inhabitants in Africa.

beam satellites can now enable a mobile terminal to be produced of a dimension similar to that of a GSM phone. In an industry where style as much as anything else dictates the demand for mobile terminals, this reduction in size is clearly likely to be of immense benefit. Moreover, developments in on-board processing power has alleviated the need for a double-hop when making a mobile-to-mobile call. This, of course, reduces significantly the latency on the link.

Geostationary satellites are likely to continue to increase in launch mass, reflecting the need for higher capacity satellites employing a greater number of spot-beams. The increased launch mass of satellites in turn will require larger launch vehicle technology to be developed, as recognised by the continued evolution of the PROTON and ARIANE launchers, for example.

It can be anticipated that several geostationary satellites of similar capabilities will be deployed around the globe in the coming years. Areas of obvious benefit would include China, North and South America, Africa and Central and Eastern Europe. Taking the geostationary satellite network concept a step further, such satellites suitably placed along the Equatorial plane, can form a single global network by incorporating inter-satellite link technology. Such a scenario would enable the satellite network to achieve autonomy from the terrestrial network infrastructure. It has already been seen that IRIDIUM employs ISL technology in its network. However, the static nature of the geostationary satellite configuration clearly offers a far more practical solution in comparison to the dynamic nature of the non-geostationary scenario. The ability to offer high data rates, equivalent to core network type services, over ISLs, while offering access network service data rates over the user to satellite link is an attractive future mobile scenario.

As the capabilities of geostationary satellites increase, operators need to be aware of two important design criteria:

- The extended lifetime of the next generation of satellites, perhaps existing for as long as 15–20 years, implies that any new satellite platform must be designed with significant flexibility in order to be able to adapt to changes in market demand and the evolutionary nature of service delivery. While satellites have traditionally made use of established technologies in order to increase reliability, in future, there will be a need to place a greater emphasis towards more leading edge, state-of-the-art technology, in order to maximise the flexibility and service lifetime of the satellite;
- Bearing the above in mind, the relationship between on-ground and space-borne technology will need to be carefully traded off in order to ensure an optimum design solution.

9.3 Non-Geostationary Satellites

At the start of the 21st Century, the jury is still out on the future of the non-geostationary mobile-satellite concept. However, from the previous chapter, this should not be such a surprise, since these services are still in their infancy and will probably need 5 years or so to mature into anything like a global service. As we have seen, however, non-geostationary satellites have an anticipated life-span of about 7 years, hence there will be a need to replenish constellations at around about the same time that significant inroads into the market should be starting to occur. The only question then is whether the financial backers of these expensive, not only to set-up but also to maintain, high risk networks, have the patience and the belief to wait for a return on their investment.

Presently, there are five non-geostationary "big LEO" or MEO constellations on the table: GLOBALSTAR, IRIDIUM, NEW ICO, CONSTELLATION and ELLIPSO. Should everything go to plan, all of these systems should be providing mobile-satellite services within the next few years. However, there is little evidence to support such an optimistic scenario, and no doubt, there will be further shake-outs in the industry in the coming years. Indeed, it has already been noted how NEW ICO and ELLIPSO have formed a co-operative agreement. In addition to the satellite-PCN solutions, there is also a number of "little LEO" constellations that are addressing market niche areas. As with "big LEOs", the next few years are critical. All of these systems have their roots in proposals developed a decade ago. In recent years, there have been no proposals for new non-geostationary constellations that aim to address specifically the mobile markets *per se* (see Section 9.9 for a discussion on the new satellite navigation system). Given the time from concept to reality, it could be argued that there will be no new players in the non-geostationary mobile market before the end of the decade. The key, here, is to determine how the existing and currently planned systems fare over the next few years.

It is practically impossible to discuss the future of non-geostationary satellites without considering the influence of the cellular market. Low earth orbit satellites, in particular, fare poorly in comparison, since they were developed primarily with the hand-held market in mind. Several technological advances need to be developed in order to sustain the LEO mobile-satellite option. As we have seen, in order to provide global services, a large number of satellites, in excess of 40, is required. This significantly impacts on the cost of the network. While mass satellite production techniques are now in place, as used by GLOBALSTAR, further improvements are required in order to decrease the cost of satellites to something approaching that of "little LEOs". Moreover, from a user terminal's perspective, further advancements are needed in terminal and antenna technologies to facilitate a more aesthetic device.

Service data rates will need to be increased from the present offering of in the region of 9.6 kbit/s up to 64 kbit/s and perhaps as high as 144 kbit/s in some areas, in line with the requirements of S-UMTS/IMT-2000. LEO satellite operators may also need to move away from proprietary radio interface solutions to a common standardised approach, as is being developed for geostationary satellites. The benefits of operating using a standardised solution to facilitate market production techniques, and hence reduce costs, are well known. It is important that satellites follow their terrestrial counterparts in this respect.

IRIDIUM and GLOBALSTAR have adopted different strategies with respect to the networking of the services provided. As we have seen, IRIDIUM has incorporated ISL technology, which has minimised the need for a global terrestrial support infrastructure, whereas GLOBALSAR, by offering its facilities to local service providers, has shifted emphasis towards the local terrestrial network infrastructure. The relative spatial distribution between satellites in polar orbits greatly simplifies the on-board intelligence and antenna design required by satellites in order to perform inter-satellite link connections. Unfortunately, deployment of polar orbits results in a large concentration of satellites over the polar caps, where demand for mobile-satellite services is minimal, at best. In this respect, it can be seen that this is a major drawback of the network and implementation of an inclined orbital configuration could be preferable. However, the dynamic nature of an inclined constellation, seriously complicates the design of a constellation incorporating ISL technology and if such a constellation is to become a reality, advances in on-board processing, routing strategies and antenna design are required.

MEO satellite constellations represent less of a design and implementation challenge, in comparison with LEO satellites, and may provide a more viable long-term alternative to geostationary satellites. The success or failure of NEW ICO and its relationship with TELEDESIC over the next few years will provide an insight into the viability of this technology.

9.4 Hybrid Constellations

ELLIPSO, discussed in Chapter 2, is currently the only satellite network proposing to employ a constellation of different orbital types for real-time services. In this case, the respective contributions of the circular and elliptical orbits are used to optimise coverage over potential traffic hot spots. In more general terms, it has already been noted that satellite networks are now designed to complement their terrestrial counterparts to improve service availability in regions that are not covered by the terrestrial network, hence the hybrid satellite/terrestrial network is already in operation through the likes of GLOBALSTAR and IRIDIUM, ACeS and THURAYA.

When it comes to service delivery, each type of satellite orbit has its own set of drawbacks and advantages. For example, in very simplistic terms, the geostationary orbit could be considered to be more suited to the provision of regionally deployed, non-delay sensitive services, whereas the low Earth orbit in comparison, may be better suited for global, real-time service delivery. Using this simplistic approach, it could be argued that the most optimum solution from a satellite perspective, is a combination of the two, in other words, a hybrid constellation.

Such a scenario may be ideally suited to the needs of next generation services, which for certain applications, such as Web browsing, will be asymmetric in nature. For example, the narrowband, forward link, could be provided over a LEO link (or a terrestrial link such as GPRS for that matter), while the broadband return link, could be provided over the geostationary link. Such a scenario is shown in Figure 9.3. There are many other possibilities. For example, the geostationary satellite, by exploiting its large coverage area, could be catered towards broadcast or multicast services (see Section 9.8 for a possible example), whereas the non-geostationary orbit could be used for unicast services. There is also the opportunity to extend the regional coverage provided by a geostationary satellite globally by using a constellation of non-geostationary satellites. Here, however, a cost-benefit analysis would need to be performed to determine whether the increased level of traffic made available to the network would justify the additional expense of developing a multi-satellite constellation.

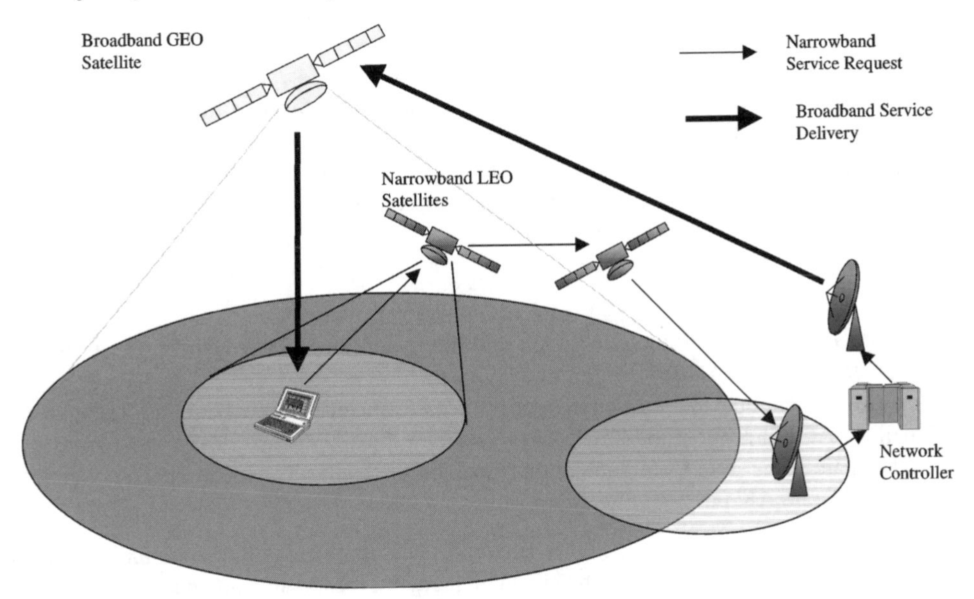

Figure 9.3 Possible future hybrid constellation scenario.

9.5 Mobile-Broadband Satellite Services

The evolution of mobile-satellite services is, in many respects, no different from that of terrestrial mobile networks. The next phase in the development of mobile-satellite networks will be with the provision of broadband multimedia services, along similar lines to those proposed for 3G mobile networks. Although provision for S-UMTS/IMT-2000 is already set aside at L-/S-bands, in order to achieve this on a mass user scale, and fully exploit the broadband capabilities of next generation networks, it will be necessary to move up in

Table 9.1 Allocation of mobile-satellite service frequencies in the Ka-band

Frequency (GHz)	Direction	Status	Region
19.7–20.1	Downlink	Primary	Region 2
19.7–20.1	Downlink	Secondary	Region 1/Region 2
20.1–20.2	Downlink	Primary	World-wide
20.2–21.2	Downlink	Primary	World-wide
29.5–29.9	Uplink	Primary	Region 2
29.5–29.9	Uplink	Secondary	Region 1/Region 3
29.9–30.0	Uplink	Primary	World-wide
30.0–31.0	Uplink	Primary	World-wide
39.5–40.5	Downlink	Primary	World-wide

frequency band to an allocation where sufficient bandwidth is available. The next suitable frequency band is in the Ka-band, the frequency allocation for which is summarised in Table 9.1.

Many, if not all of the technological advances highlighted for super GEOs and non-geostationary satellites apply equally to the case for mobile broadband satellite service delivery. However, the move up in frequency also requires advances in several key technological areas.

Recent trials around the world using geostationary satellites have demonstrated the new possibilities offered in service delivery when moving up in operating frequency to a broader bandwidth. In particular, the following experimental campaigns have shown the viability of providing mobile multimedia services in the Ka-band to aeronautical and vehicular platforms:

- SECOMS/ABATE using the ITALSAT satellite under the EU's Advanced Communications Technologies (ACTS) programme [LOS-98];
- Experiments conducted in America under the Advanced Communications Technology Satellite (ACTS) programme [ACO-99];
- And Japan using the Communications and Broadcasting Engineering Test Satellite (COMETS) [WAK-00].

One of the major barriers that need to be overcome if mobile-satellite communications in these bands are to become viable, is the channel characteristic. As was noted in Chapter 4, the land-mobile channel at higher frequencies is subject to deep fades due to shadowing, rendering the channel an on-off characteristic. Compensation for such fade depths, in excess of 20 dB, is beyond the capabilities of today's power-limited satellites. Moreover, at these frequencies, the impact of hydrometers, in particular rain, can cause significant fade in signal strength for short-periods of time. In the fixed-satellite service, there are several methods that can be used to counteract the effects of rain fading. These include:

- Uplink or downlink power control;
- Adaptive modulation and coding techniques;
- Site and height diversity to change the direction of the transmission path between the Earth station and satellite;

- And satellite-orbit diversity, which again can be used to alter the direction of the transmission path.

Certainly, some of these techniques could be applied to the mobile-satellite scenario. In particular, the use of adaptive modulation and powerful coding techniques, such as turbo codes, could improve the availability of the mobile link. Similarly, open or closed-loop power control techniques could be used to counter-act short-term fading events, however, when used by the satellite, all terminals within a particular spot-beam will be subject to the same power levels, which could result in the wasting of a valuable resource. Moreover, power limits will be restricted by interference capacity considerations. Satellite-orbit diversity is only likely to be effective in a non-geostationary environment, when used in a land mobile environment. Maritime and aeronautical applications, on the other hand, and even land mobile in open environments, could benefit from this approach to counteract the effects of rain fading.

As we have seen with the non-geostationary case, the key to the development of mobile-satellite services is terminal technology. In this respect, the design of an efficient antenna to address the move up in frequency to the 20/30 GHz bands is now a subject of significant research and development. Three types of antenna are presently envisaged: active-phased array; wave-guide slot-array; and reflector. These antennas are electronically or mechanically steered devices, incorporating some form of position location mechanism, in order to maintain a constant elevation angle to the satellite. Here, geostationary satellite technology is assumed. Such devices tend to be suitable for vehicular-mounted applications, although portable applications can also be envisaged.

Of all of the non-geostationary constellations proposed in the early 1990s, only the TELEDESIC constellation aims to provide user links in the Ka-band. As was noted in Chapter 2, TELEDESIC is catered predominantly for fixed-users, although portable and, to a lesser extent, mobile users are also envisaged. In Chapter 3, it was shown that non-geostationary satellite transmissions are subject to Döppler shift and the higher the transmission frequency, the greater the effect, consequently limiting the transmission capacity and increasing terminal complexity. As with geostationary satellites, the major problem with moving up in frequency is the on-off nature of the channel. The original concept of the 840-satellite TELEDESIC system was to use a large number of satellites in order to guarantee a high minimum user to satellite elevation angle, in excess of 40°. It can be imagined that in order to counteract the on-off nature of the channel, a high minimum elevation angle approaching that specified in the original TELEDESIC concept would be required with the subsequent need for a huge number of satellites. With the present fragile confidence in the non-geostationary satellite industry, it is unlikely that such a venture would come to fruition in the near future.

Of all of the niche mobile markets that broadband satellite technology can address, it is perhaps the aeronautical sector that offers the most attractive scenario, and in particular the long-haul inter-continental flight market. Such flights are characterised by long periods over Oceanic or sparsely populated regions, where communication is solely dependent upon satellite communications. Interestingly, although the number of air passengers are increasing on a year-by-year basis by in the region of 5–7%, to meet this demand, aircraft manufacturers are resorting to building fewer, larger aircraft, such as the planned A380 by the Airbus consortium, rather than building more, smaller aircraft. One of the reasons for this approach is simply that in the future, certainly in Europe, the number of new airports that will be developed to meet this demand is likely to be very limited. Moreover, airspace is becoming increasingly congested and is approaching saturation in many areas of the world. For example, at London's Heathrow airport, an aircraft takes-off or lands every 90 s.

Airbus' A380 can seat in the region of 480–650 passengers and will commence service in 2005. In the future, long haul flights of 500–700 seat capacities may well become the norm. On such flights, the need for in-flight entertainment, as well as in-flight business services, through individual terminals connected to each seat will generate a significant traffic demand. Moreover, this new generation of "super-Jumbos" will provide the passenger with a range of leisure facilities on-board, including shopping areas and entertainment facilities, which will also add to the demand for tele-services. Looking further ahead, research is now focusing on the Blended-Wing Body (BWB) aircraft, what essentially amounts to a flying wing. This new passenger concept will provide seating for up to 1000 passengers in an entirely new transport concept, in which passengers will be transported in a spacious but windowless environment. Clearly, there will be a need for new, innovative infotainment facilities in such an environment. Of course, the need for telecommunication services is not restricted to the long-haul flight sector. The inter-continental flights of 2–4 hour duration provide a prime opportunity for the satellite market. Here, again, the size of aircraft is likely to increase to cater for the increased demand. Short-haul flights of 1 h or so are also likely to generate traffic demand, particularly from the business sector, although this may be better served using L-band technology.

Operation in the Ka-band will be needed to meet this demand for new tele-services, including tele-medicine and tele-working. Of course, one of the benefits of aeronautical applications is the fact that aircraft tend to spend most of their time above the cloud layer, after reaching cruise altitude. Hence, the attenuation due to rain no longer comes into the equation. Moreover, line-of-sight to the satellite once in cruise mode can be guaranteed through the optimum location of the aeronautical antenna.

Developments in the maritime leisure industry, in particular with the launch of the new generation of cruise liners, clearly offer similar potential to that of the aeronautical sector for satellite-distributed entertainment services, while scientific and commercial ships would be cater for business as well as entertainment needs. In this context, it can be seen that satellites could play an important role in ensuring that 21st Century citizen can participate in the information society, be it on land, sea or air.

9.6 Mobile IP

The drive towards the establishment of the Information Society will bring together the two most successful of all of the technological advances of the latter quarter of the 20th Century: the Internet and mobile communications. In this respect, fourth-generation (4G) mobile networks will be based upon an all IP-environment [BAS-00]. Significant effort around the world is now underway towards standardising such a mobile environment through such organisations as 3GPP, 3GPP2, Mobile Wireless Internet Forum (MWIF) and Internet Engineering Task Force (IETF). At the time of writing, there are two different approaches to the network architecture, as proposed by 3GPP and 3GPP2. The 3GPP solution for the W-CDMA radio interface is based on an evolution of the GPRS network, with enhancements to the call control functionalities obtained though the introduction of a new network element, called state control function (CSCF), to allow the provision of voice over IP (VoIP) services. The 3GPP2 solution for the cdma2000 radio interface, on the other hand, has adopted a new packet network architecture incorporating mobile IP functionalities in support of packet data mobility. Eventually, it is hoped that a single, harmonised solution will emerge from the two distinct approaches [PAT-00].

In the future mobile network, where satellites will operate alongside terrestrial mobile networks, various categories of service will co-exist, where "best effort", which is presently available over the Internet, will operate alongside guaranteed QoS classes. The move towards an all IP network should facilitate the interworking between satellite and terrestrial mobile networks, since the problem of providing mobile connectivity across the different networks reduces to the level of attaining the appropriate route to direct the packets of information to the appropriate terminal. Such a scenario is illustrated in Figure 9.4, where the satellite forms part of a network comprising a GPRS network [CON-00].

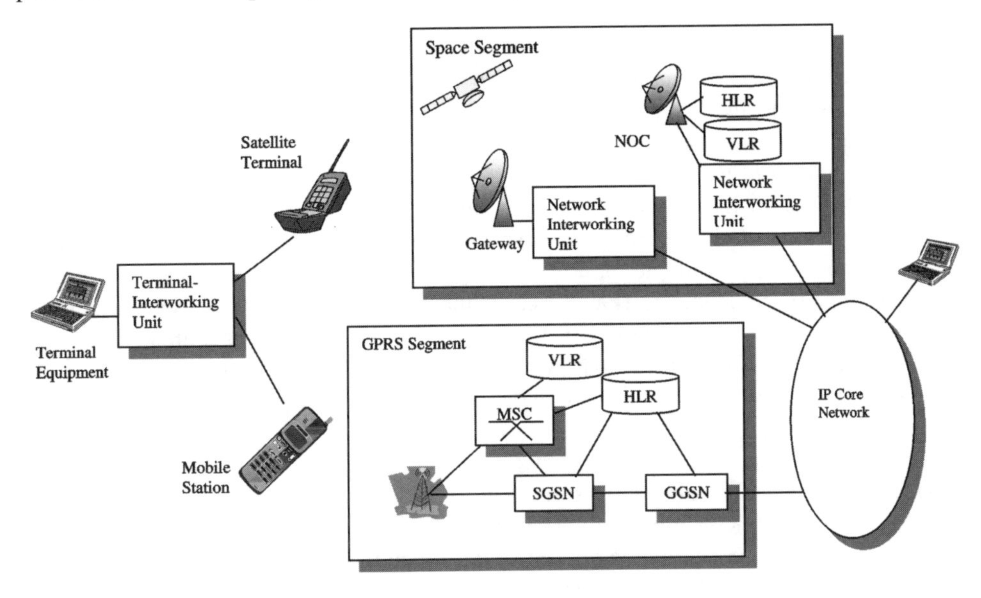

Figure 9.4 Satellite and GPRS integration scenario.

In the mobile IP environment, the mobile terminal has a temporary IP address, known as the care-of-address (CoA), which is associated with a correspondent node of a particular access network. The CoA is in addition to the IP address that is permanently assigned to the mobile terminal, that is its home address, which is stored at its home sub-network. The mobile node's CoA is made known to the home agent. Packets addressed to the mobile node are thus tunnelled by the home agent to the correspondent node identified by the CoA. When a mobile node moves from one access network to another, a new CoA will be assigned by the visiting network operator corresponding to the new correspondent node. The mobile node then makes the home agent aware of the change in the CoA through the transmission of a *binding update* message.

An additional advantage in the adoption of MIPv6 (Mobile Internet Protocol Version 6), which is the final stages of standardisation, should be to provide the opportunity to perform seamless handover between satellite and terrestrial networks. Early trials involving handover between terrestrial networks using MIPv6 have demonstrated the feasibility of such an approach. Indeed, researchers are now addressing the needs of integrated space/terrestrial mobile networks, including the need to perform seamless handover between networks, based upon packet-oriented service delivery [CHA-00].

9.7 Transmission Control Protocol (TCP)

9.7.1 Overview

The transmission of Internet packets from a transmitter to a receiver is primarily achieved using the TCP. TCP, a connection-oriented protocol, provides the functionality to ensure that the transmission rate of data over the network is appropriate for the capabilities of the receiver device, as well as the devices that are used to route the data from the transmitter to the receiver. TCP is also responsible for ensuring that the resources of the network are divided in an equable manner between all users of the network. Applications such as FTP (file transfer protocol) and HTTP (the language of the web) rely on TCP to transport their data over the network as quickly as the network will allow.

The development of the Internet and its technologies is the concern of the Internet Engineering Task Force, which publishes its recommendations on its website under a series of request for comments (RFC) documents. A specific RFC number identifies each RFC. These documents are publicly available for downloading from the IETF's website and can take the form of information documents (FYI RFC) or those that specify Internet standards (STD RFC). This section refers to a number of RFCs related to TCP and the satellite network. The interested reader is recommend to visit the IETF's website for the latest information on the evolving technologies. Further information on the workings of the IETF can be found in Ref. [MAL-94].

As far as a TCP/IP connection is concerned, a mobile-satellite network should be viewed as any other network connection. Given the way that TCP operates, the transmission of IP packets over the mobile-satellite link poses several problems that need to be overcome if services are to be delivered efficiently. The major difficulty is due to the latency of the link, which when combined with a bursty-error channel, and the characteristics of the TCP protocol itself, can result in an inefficient means of transmission. This is because TCP operates using a conservative congestion control mechanism, whereby new data can be transmitted only when an ACK from a previous transmission has been received. With this in mind, the need for a high quality link between the satellite and the mobile terminal is re-emphasised, since packets in error are presently deemed to be due to congestion on the network. Hence, in this case, TCP responds by reducing its transmission rate accordingly. It can be seen that TCP operates on the basis of "best effort", based on the available resources of the network.

When starting transmission, TCP enters the network in a restrained manner, whereby the initial rate of transmission is carefully controlled to avoid overloading the network with traffic. TCP achieves this by employing congestion control mechanisms, namely: slow start, congestion avoidance, fast retransmit and fast recovery. Slow start is used, as its name suggests, at the start of transmission or after congestion of the network has been detected and the data transmission is reduced. Congestion control is used to gradually increase the transmission rate once the initial data rate has been ramped-up using the slow start algorithm. The fast transmit and fast recovery algorithms are used to speed up the recovery of the transmission rate after congestion in the network has been detected.

9.7.2 *Congestion Window and Slow Start Threshold*

The operation of these mechanisms is controlled by two state variables: congestion window (*cwnd*) and slow start threshold (*ssthresh*). *cwnd* defines the upper bound on the amount of traffic, in terms of the amount of data that a sender can transmit without receiving an ACK message from the receiver. Data that has been sent but not yet acknowledged is termed the *flight size* [ALL-99a].

Another important parameter is the receiver advertised window, *rwnd*, which is the maximum amount of data, in bytes, that can be buffered at the receiver. The maximum transmission rate is determined by the minimum of *cwnd* and *rwnd*, thus ensuring that the receiver is not overloaded with data that it is not able to acknowledge.

ssthresh is used to determine which algorithm is used to vary *cwnd* when either the slow start or congestion avoidance algorithm is used. As noted earlier, data senders, upon starting transmission, employ the slow start mechanism. Slow start operates by setting the value of *cwnd* initially to no more than two segments, which is less than or equal to twice the sender maximum segment size (SMSS). A segment refers to any TCP/IP data or acknowledgement packet. The default TCP maximum segment size is 536 bytes [POS-83]. The SMSS can be based on the receiver maximum segment size (RMSS), the size of the largest segment the receiver is willing to accept, or the maximum transmission unit, this being the maximum packet size that can be transported over a network before IP packets are fragmented. Fragmentation implies that IP packets are broken up during transmission and are then put back together at the receiver. Such a process could decrease the transmission rate, as extra processing is required at the receiver to put back together fragments before an ACK can be generated. The maximum transmission unit that can be transported over the network can be determined using the maximum transmission unit (MTU) algorithm [MOG-90].

ssthresh may be set to an arbitrarily high value, which in some implementations is the receiver's advertised window, *rwnd*.

From the above description of *cwnd*, it can be seen that after transmitting the first segment, the data sender remains idle until an ACK message has been received. For a geostationary satellite, this would take in the region of roughly 500–570 ms. Once an ACK message has been received, *cwnd* is increased by one segment and in this case, two data segments are transmitted. Assuming these data are successfully acknowledged with two ACKs, *cwnd* is increased by a further two segments resulting in four data segments being transmitted. This process continues, increasing the number of segments in *cwnd* each time by the number of received ACKs, until either a packet is lost due to perceived or actual network congestion, or *cwnd* exceeds *ssthresh*, at which point the congestion avoidance algorithm takes over. When *cwnd* equals *ssthresh*, either slow start or congestion control can be used.

The congestion avoidance algorithm increases *cwnd* but at a much slower rate than that associated with the initial ramp-up employed by the slow start algorithm. In this case, *cwnd* is increased by one segment per round trip time (RTT).

9.7.3 *Loss Recovery Mechanisms*

A lost packet is detected when an ACK signal has not been received within the expected time, based on knowledge of the RTT. This is termed the retransmission timeout (RTO). When a packet is lost, the value of *ssthresh* is set to either half the flight size or twice SMSS, which ever is the greater. The value of *cwnd* is re-set to one segment. The slow start algorithm then starts once more until *cwnd* reaches *ssthresh*, at which stage the congestion avoidance algorithm takes over once more.

TCP allocates a unique sequence number to each transmitted byte. The receiver, when acknowledging correct reception, identifies the sequence number of the byte, up to which all other data is assumed to be correctly received. Sequence numbers are derived from a 32-bit wraparound sequence space. To ensure that each byte on the network has a unique identity, no two bytes are allowed to have the same sequence number. The lifetime of a byte on the network is assumed to be 2 minutes, resulting in a maximum bit rate of 286 Mb/s [PAR-97]. In order to increase this rate, the IETF devised a means of adding an optional time-stamp to the 32-bit sequence number. Comparing time stamps using the PAWS (protection against wrapped sequence numbers) algorithm allows the differentiation between two bytes with the same sequence number transmitted within 2 minutes of each other. With this option data rates of between 8 Gbit/s and 8 Tbit/s can be achieved, depending on the granularity of the time stamp.

The fast retransmit and fast recovery algorithms make use of the TCP acknowledgement characteristics, which always acknowledge the highest in-order segment correctly received, to determine which segments have been lost and then to subsequently arrange the re-transmissions [ALL-99a]. This is known as *TCP Reno*. When a segment is lost or received out of order, TCP acknowledges the highest order received segment, hence at the transmitter, duplicate ACKs can be used to detect lost segments. The fast retransmit algorithm uses three duplicate ACKs to represent a lost segment, at which point the identified segment is re-transmitted without waiting for the RTO and *ssthresh* is then set to either half the flight size or twice SMSS, which ever is the greater. *cwnd* is then increased by one segment for each duplicate ACK received. Once the re-transmitted segment has been acknowledged, *cwnd* is re-set back to the value of *ssthresh* and congestion avoidance is then used to send new data.

When more than one segment is lost in a multi-segment transmission, the fast retransmit/fast recovery algorithms can be used to re-transmit one of the segments, however, the loss of the other segments is usually detected by the expiry of the RTO, upon which the slow start algorithm takes over. This then requires all of the segments sent after the detected lost segment to be re-transmitted, irrespective of whether these segments were received correctly or not. To improve the efficiency of the transmission, a TCP option, which allows the means of providing selected acknowledgement (SACK) has been devised [MAT-96]. In this instance, the receiver is able to identify the segments that were incorrectly received in its SACK message. The transmitter is therefore able to re-transmit only the segments that were lost. This approach could be particularly useful in a mobile-satellite environment, where packet loss due to the transmission environment rather than network congestion, is likely to lead to randomly distributed error patterns at the receiver.

The NewReno modification to the fast retransmit and fast recovery algorithms (TCP Reno) has been proposed to counteract multiple packet drops where the SACK option is not available [FLO-99].

9.7.4 Future Work

Most research on TCP/IP over satellite has been performed for fixed networks, where usually a good quality link can be guaranteed, based on line-of-sight design criteria [PAR-97]. The performance of TCP/IP over mobile-satellite links is an area of on-going research, requiring several areas that need to be addressed in order to take into account the particular transmission characteristics introduced by the network.

It is noted that a means of recognising that corruption of transmitted packets is not necessarily due to network congestion needs to be defined [ALL-99b]. Moreover, the performance of TCP over non-geostationary satellite links, where inter-satellite handover may result in packet loss also needs to be considered. The use of asymmetric channel assignments over the satellite link, which is considered to be an important requirement for future-generation mobile-satellite links, may also effect the performance of TCP, when the narrowband link may be subject to network congestion.

9.8 Fixed-Mobile Convergence

The convergence of mobile communications and Internet technology has opened many opportunities to deliver new multimedia services to mobile users. The interactive nature of the Internet has paved the way for next generation mobile communication systems to support interactivity. Apart from the convergence between mobile and Internet technologies, the other major technological driver is the convergence between mobile and fixed technologies. By supplementing broadcasting systems with a narrowband uplink, new interactive services can be facilitated in digital video broadcast (DVB) and digital audio broadcast (DAB) systems. This is foreseen for fixed network operation and could equally be adapted onto a mobile network such as the UMTS, thus demonstrating the concept of convergence of personal mobile communications, Internet and broadcasting technology.

Certainly, in the future environment, the bringing together of all three technologies will provide a powerful alliance, allowing a new range of services to be delivered in an efficient manner and at the desired QoS. Clearly, the ability of fixed networks to deliver services on a broadcast (everyone) or multicast (closed user group) basis within a dedicated coverage area could open up several new types of location specific services that could be made available to the mobile user.

Currently, there is significant interest in the development of the multimedia car concept, whereby the mobile user can gain access to information and entertainment services (infotainment) which would not otherwise be available through the cellular network alone [STA-00]. In addition to the usual multimedia services (video telephony, web browsing, games, and so on), location specific services can be catered for, such as in-car navigation, localised media coverage, and so on. Presently, service delivery is envisaged entirely by terrestrial means. In such a scenario, the user requests access to a particular service through the local wireless network, for example GPRS. Here, in addition to the cellular network, a DVB/DAB-T (terrestrial) cellular coverage also exists and is overlaid onto the cellular network. Upon receiving a request for a DVB/DAB-T type service, the cellular operator will then negotiate with its network controller, which then organises for the required data to be encapsulated and broadcast over the corresponding DVB/DAB-T cell.

Developments in digital video/audio broadcasting are, of course, not solely limited to the terrestrial domain. Digital services on offer by satellite are now making significant in-roads into the home entertainment market. As with cellular networks, satellites again can play an important role in providing coverage in the scenario outlined above to regions not covered by DVB/DAB-T services. Aeronautical and maritime markets are two obvious business cases.

Significantly, there are several technological drivers in the satellite communications industry, which when combined could allow an efficient and effective service delivery to be made via satellite communications in their own right.

The discussion in the previous sections illustrated how hybrid satellite communications could be used for innovative service delivery, in particular in order to cater for the asymmetric nature of many of the new services on offer. The possibility of providing the initial service request through a non-geostationary satellite, as opposed to the cellular network, then providing the broadcast service through the geostationary satellite is an interesting concept. Moreover, the reduction in the spot-beam size of the next generation of geostationary satellites, possibly to less than 50 km in diameter, opens up the possibility of providing localised services approaching parity with the locality available through an equivalent terrestrial network. The other major development in satellite broadcast technology is the standardisation of the DVB-RCS (DVB-return channel via satellite), which allows the user to communicate directly with the broadcast satellite network through an assigned return channel (Figure 9.5). This greatly simplifies the overall network architecture and the associated network management procedures, in that now all communications take place over the same access network. Presently, the DVB-RCS has been specified for indoor use only, however, it can be envisaged that over the coming years, research and standardisation efforts will be directed towards establishing a corresponding mobile standard. This could open up significant new opportunities in the mobile-satellite sector.

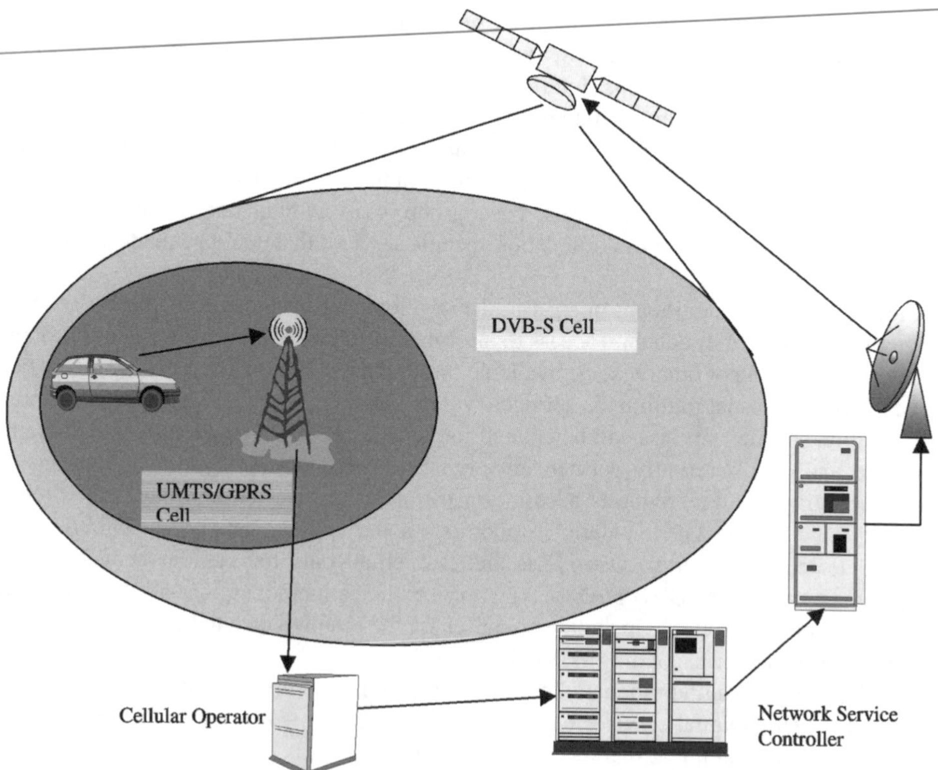

Figure 9.5 A possible hybrid direct video broadcast service scenario.

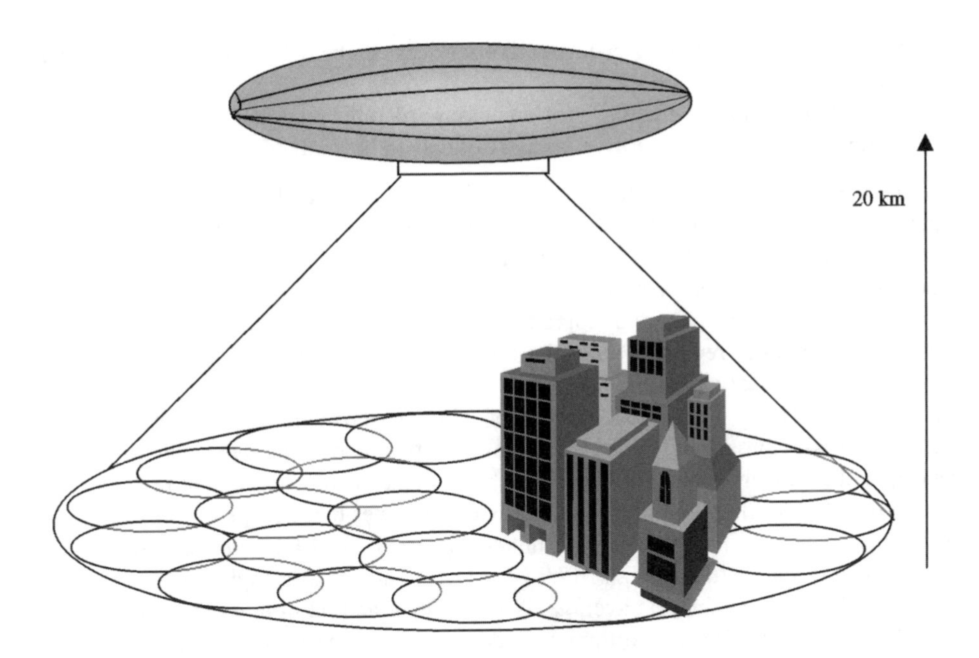

Figure 9.6 Stratospheric platform scenario.

9.9 High Altitude Platforms

In Chapter 1, it was noted that for the first time at WRC 2000, a spectrum was allocated to high altitude platforms (HAPs) for the provision of terrestrial-UMTS services. Previously, at WRC 97, the first spectrum to be allocated to HAPs was made in the EHF band, specifically in the 47 GHz range. HAPs, also commonly referred to has stratospheric platforms, can take the form of either manned/unmanned aircraft or airships. These platforms are positioned at a fixed location in the stratosphere, some 20 km above the Earth and can be thought of as a very high transmission tower (Figure 9.6).

Such a mode of service delivery offers several advantages:

- Coverage can be rapidly set-up over any location and can be just as easily removed or re-located. In this respect, networks can be considered to be local to a particular coverage area. This considerably reduces the components of the network, which in turn reduces the development costs. Moreover, there is no need for any launch vehicle, platforms make their way to the allotted position in the sky using their own resources.
- A high link margin can be established, allowing penetration into buildings. Available data rates are comparable with those of UMTS/IMT-2000 and higher.
- A high elevation angle can be achieved to the mobile user, allowing line-of-sight operation in most environments. Consequently, higher gain, narrow beamwidth antennas can be employed to further improve the link margin.
- The coverage pattern projected onto the Earth is cellular-like, in that the cell pattern remains fixed with respect to the location on the Earth. Efficient frequency re-use schemes can be employed to maximise network capacity.
- The relatively short round-trip delay of less than 1 ms ensures that latency is not a

problem. Hence, broadband user-interactive services can be delivered with a high degree of QoS.

- The cost of providing coverage over an area is considerably less than terrestrial or satellite counterparts, enabling competitive pricing policies to be applied.

On the negative side, the position of the platform must be carefully maintained within a particular location to ensure that coverage on the Earth is fixed. The stratosphere is characterised by strong winds; hence some means of controlling the location of the platform is needed. Unmanned platforms need maintaining, which requires them to return to Earth periodically for fuelling, payload upgrades, and so on. This could be achieved during times of low demand; replacing one platform with another can alleviate interruption in service. Manned platforms, on the other hand, operate in rotational shifts.

SKY STATION™ is a planned network of 250 solar-powered, airships, which will be targeted to provide broadband communications over the major populous areas of the world. At a height of 21 km above the Earth, each airship provides a coverage area of in the region of 19 000 km^2. User terminals will be able to provide data rates in the range of 2–10 Mbit/s, at a cost of only a few cents per minute. Transmission will be in the EHF band, specifically 47.2–47.5 GHz (Stratosphere to Earth) and 47.9–48.2 GHz (Earth to Stratosphere).

A SKY STATION airship, on average, is 157 m long and 62 m in diameter at its widest part. The SKY STATION consortium includes Alenia AeroSpazio SpA of Italy, who is responsible for the development and production of the payload, employing the mass production techniques developed for the GLOBALSTAR constellation. SKY STATION plans to start deployment of its network in 2002.

The basis for the aircraft technology has its roots in the military. In particular, piloted aircraft used for surveillance operate within the stratosphere, although their endurance tends to be limited to a few hours, while military experiments have also been performed using unmanned aircraft. An alternative solution to that of SKY STATION is the HALO™ Network, proposed by Angel Technologies and its partners. The HALO (high altitude long operation) approach uses manned aircraft at 15 km above the Earth, with continuous coverage being achieved through the use of three HALO aircraft per coverage area, operating in shifts. Available data rates are similar to those offered by SKY STATION.

The European Space Agency (ESA) has also considered the feasibility of providing stratospheric service delivery under its HALE (high altitude long endurance) study conducted in 1996. This study considered the use of unmanned aircraft operating in the region of 15–30 km above the Earth. Furthermore, the concept of using an unmanned spacecraft as a GSM base station in the sky is under investigation under the EC's IST project HeliNet, which incorporates the HeliPlat solar-powered platform [MON-01]. This platform supports a payload of 100 kg, with total power of 800 W. In the US, NASA's Dryden Flight Research Centre's Environmental Research Aircraft and Sensor Technology (ERAST) project aims to develop remotely-piloted, ultra-high altitude, long endurance flight capabilities for science and commercial applications. The $15 million dollar Helios prototype is the latest light-weight, all-wing aircraft to evolve under this project, its predecessors being Centurion and Pathfinder. The Helios prototype aims to achieve a 100 000 ft altitude using solar panels as a primary power source, and 4 days flight endurance above 50 000 ft, the plan being to achieve the 4-day duration by 2003. Half of this goal was virtually achieved in July 2001, when Helios reached a height of 96,500 ft, breaking the previous height record for a non-rocket aircraft of

80,201 ft established by Pathfinder Plus in 1998. The eventual goal of the Helios production aircraft is to develop the capability of unmanned continuous flight for up to 6 months at a time.

From the above, it can be seen that HAPs provide most, if not all of the advantages of a satellite network, without the same degree of technological risk, and at a much reduced cost. Hence, the HAP could be considered to be a new threat to the fragile, satellite market. Since HAP technology is still new, it will take a few years to see just how much of a threat to satellite technology this mode of delivery actually is. If services can be delivered with the required QoS and at a cost to the user comparable with existing services, or possibly even cheaper, then certainly this type of service delivery cannot be ignored in any future market projections.

9.10 Location Based Service Delivery

The ability to deliver services to the mobile user based on their current location opens up many new opportunities for innovative applications. The use of satellites to provide positional information is now widespread, particularly in its use for aircraft navigation, while its applications for business and leisure use are now becoming commonplace. At present, the world's satellite navigation capabilities are provided by two constellations of MEO satellites, namely GPS and GLONASS. GPS is a US Government owned constellation, developed primarily for military use, but also widely used for civilian applications, albeit at a reduced degree of accuracy. GLONASS is the Russian equivalent. As civil applications become more dependent upon positional information provided by satellites, the reliance on the use of military systems to provide such information is not ideal. Moreover, in Europe there is a desire to obtain its own independent satellite navigation network of satellites.

With this in mind, the European Union (EU) in association with the European Space Agency initiated a development programme in 1998, which will result in a publicly owned stand-alone navigation system becoming operational by 2008. This will be known as GALILEO, which will have the additional capability to interwork with GPS and GLONASS. This will mark the second phase of Europe's Global Navigation Satellite System (GNSS-2). The first phase, GNSS-1, will be introduced in 2004 under the name European Geostationary Navigation Overlay Service (EGNOS) [GAU-01].

EGNOS will enable the positional accuracy provided by GPS and GLONASS to be improved from approximately 20 m to 5 m. This is achieved using a network comprising of three regionally deployed geostationary satellites, in the region of 30 ranging and integrity monitoring stations (RIMS), four master control centres and six uplink stations. The space segment will be provided by two INMARSAT-3 satellites, located over the IOR and AOR-E coverage regions, respectively (see Chapter 2), and ESA's ARTEMIS satellite. The RIMS are used to compare accurate measurements of the GPS and GLONASS satellites with positional measurements derived from their transmitted signals. This data is sent to the master control centres, which calculate the accuracy of the positional information provided by the navigational satellites. This information, termed integrity data, which includes the accuracy of the atomic clocks on-board the satellites, along with information on ionospheric disturbances, is then distributed to an uplink station for transmission to one of the geostationary satellites. This is achieved by modulating the integrity data onto the satellite's ranging signal. The subsequent relaying of the signal to an EGNOS receiver can then be decoded and used to improve the accuracy of the positional information. Similar systems are planned for North America (WAAS – Wide Access Augmentation System) and Japan (MSAS – Multi-transport Satellite Based Augmentation System) for aeronautical navigation, with services being anticipated by 2006.

The second phase of GNSS will commence with a fully operational GALILEO system, which will come on-line in 2008. The phased launch programme will allow an initial service being offered in 2006. The system will cost in the region of 3.2 billion Euro to develop. The constellation comprises of 30 satellites (including three in-orbit spares), positioned in MEO orbits at 23222 km above the Earth. The satellites are equally divided into three orbital planes, which are inclined at 56°. Satellites transmit in dual-frequency bands, which enable a positional accuracy of down to 4 m to be achieved.

GALILEO is intended to allow inter-operation with GPS and GLONASS, the former of which is also planned for a service upgrade at about the same time as GALILEO implementation. Thus, the same item of equipment could be used to receive transmissions from all three satellite constellations, increasing reliability and accuracy.

The GALILEO network will comprise a navigation system control centre (NSCC), a number of orbitography and synchronisation stations (OSS) distributed around the world, and tracking, telemetry and command stations (TT&C) [BEN-00]. The OSS will be used to receive navigation messages from each GALILEO satellite, as well as collecting their one-way pseudo-range measurements. Data will be sent to the NSCC for processing. The TT&C stations will be used to relay information generated by the NSCC, derived from the data received from the OSS (e.g. satellite ephemeris data), to the relevant satellite to be incorporated into its navigation data message. The data message incorporates the relative satellite epoch time, satellite ID and status, ephemeris information, as well as a satellite constellation almanac.

GALILEO will incorporate the possibility to alert users of when satellite transmissions are operating outside of specification. This is particularly aimed at safety-critical operation. This will be achieved by the use of an independent network of integrity monitoring stations (IMS), which will be used to perform measurements on satellites and to relay data to the integrity centre, where the measurements are processed. Subsequently, satellite integrity information is then forwarded to an uplink station for incorporation into the satellite navigation data message. The user's receiver can then use this information to either ignore messages from a particular satellite or to reduce the influence of a particular satellites transmission.

GALILEO will operate using three levels of service category: The basic signal will be made available to all users. This will provide a horizontal positional accuracy of 10 m for better than 70% of the time. A commercial service offering higher accuracy and availability will also be available. The final category of use is for safety critical applications, where a 4 m vertical accuracy is specified.

9.11 Concluding Remarks

The last decade proved to be a turbulent time for the mobile-satellite industry. However, at the start of the 21st Century, the evolution of mobile communications towards the establishment of the information society opens up many new opportunities for innovative service delivery. As far as the last decade was concerned, in the end, the satellite communications industry was left to play a bit-part role in the mobile communications success story. There is now a fresh, new enthusiasm within the industry with the introduction of regionally-based geostationary satellite networks, while the satellite-PCN markets continue to be redefined in an attempt to address particular niche markets.

Introducing new technologies will always be associated with an element of risk. In fact, the difficulties experienced by the mobile-satellite industry at the start of the new millennium have been mirrored by other technological sectors. In particular, the spectacular rise and fall of numerous Internet-based companies, specialising in providing niche market, entrepreneurial solutions to perceived user needs has continued to take-up numerous columns within the business supplements. This could suggest that perhaps we are not yet ready to migrate on mass to cyberspace.

The mobile communications sector is now also starting to raise some serious questions from their financial backers. In 2000, the production of GSM cellular phones decreased for the first time, indicating a slowdown in demand. Indeed, referring back to Chapter 8, this should not be surprising, as every market reaches a point of saturation. Perhaps of more concern, however, is the effect the 3G auctions have had on the mobile communications industry. The huge amount of investment required to deploy 3G technologies has placed severe financial strain on most, if not all of the mobile players. Such investments make the cost of establishing a mobile-satellite network pale to insignificance. The fact that the demand for 3G services is still unclear, and will probably take 5–7 years to build up a strong foothold in the global market, could mean that many of the 3G investors will take a number of years before seeing a return on their investment. In this respect, the analogy with satellite-PCN is clear.

In terms of entering into the 3G market, the mobile-satellite communications industry is in much better shape than when it entered the race for 2G technologies, some 6 years after the firing of the start gun. Indeed, the introduction of systems such as NEW ICO and the launch of Inmarsat's Broadband Global Access Network (BGAN) services should be able to cater towards the UMTS/IMT-2000 user shortly after service introduction. Moreover, the gradual shift in emphasis away from voice-dominated services to a data-centric environment could favour the satellite industry. This is especially the case where the need for high volume, delay-insensitive material, such as large documents or images is required. In the longer term, the move towards an all IP environment should provide the opportunity to provide seamless interworking between satellite and terrestrial access networks.

References

[ACO-99] R.J. Acosta, R. Bauer, R.J. Krawczyk, R.C. Reinhart, M.J. Zernic, F. Gargione, "Advanced Communications Technology Satellite (ACTS): Four-Year System Performance", *IEEE Journal on Selected Areas in Communications*, 17(2), February 1999; 193–203.

[ALL-99a] M. Allman, V. Paxson, "TCP Congestion Control", *Internet Engineering Task Force RFC 2581*, April 1999.

[ALL-99b] M. Allman, D. Glover, L. Sanchez, "Enhancing TCP over Satellite Channels using Standard Mechanisms", *Internet Engineering Task Force RFC 2488*, January 1999.

[BAS-00] K. Basu, A.T. Campbell, A. Joseph, "IP-Based Mobile Telecommunications", Guest Editorial, *IEEE Personal Communications*, 37(8), August 2000; 8–9.

[BEN-00] J. Benedicto, S.E. Dinwiddy, G. Gatti, R. Lucas, M. Lugert, *GALILEO: Satellite System Design and Technology Developments*, European Space Agency, November 2000.

[CHA-00] P.M.L. Chan, F. Di Cola, R.E. Sheriff, Y.F. Hu, "Handover with QoS Support in Multi-Segment Mobile

Broadband Networks", *Proceedings of European Personal & Mobile Satellite Workshop*, London, 18 September 2000; 92–101.

[CON-00] P. Conforto, G. Losquadro, C. Tocci, M. Luglio, R.E. Sheriff, "SUITED/GMBS System Architecture", *Proceedings of IST Mobile Communications Summit 2000*, Galway, 1–4 October 2000; 115–122.

[FLO-99] S. Floyd, T. Henderson, "The NewReno Modification to TCP's Fast Recovery Algorithm", *Internet Engineering Task Force RFC 2582*, April 1999.

[GAU-01] L. Gauthier, P. Michel, J. Ventura-Traveset, J. Benedicto, "EGNOS: The First Step in Europe's Contribution to the Global Navigation Satellite System", *ESA Bulletin*, 105, February 2001; 35–42.

[ITU-00] "Year Book of Statistics: Telecommunication Services Chronological Time Series 1989–1998", *International Telecommunication Union*, February 2000.

[LOS-98] G. Losquadro, "End-to-End Broadband Mobile Satellite Services Demonstration at Ka Band", *Proceedings of 3rd ACTS Mobile Communications Summit*, 2, Rhodes, Greece, 8–11 June 1998; 604–609.

[MAL-94] G. Malkin, "The Tao of IETF. A Guide for new Attendees of the Internet Engineering Task Force", *Internet Engineering Task Force RFC 1718*, November 1994.

[MAT-96] M. Mathis, J. Mahdavi, S. Floyd, A. Romanow, "TCP Selective Acknowledgement Options", *Internet Engineering Task Force RFC 2018*, October 1996.

[MOG-90] J. Mogul, S. Deering, "Path MTU Discovery", *Internet Engineering Task Force RFC 1191*, November 1990.

[MON-01] M. Mondin, F. Dovis, P. Ulassano, "On the Use of HALE Platforms as GSM Base Stations", *IEEE Personal Communications*, 8(2), April 2001; 37–44.

[PAR-97] C. Partridge, T. Shepard, "TCP Performance Over Satellite Links", *IEEE Network*, 11(5), September/October 1997; 44–49.

[PAT-00] G. Patel, S. Dennett, "The 3GPP and 3GPP2 Movements Toward an All-IP Mobile Network", *IEEE Personal Communications*, 7(4), August 2000; 62–64.

[PER-96] C. Perkins (Ed.), "IP Mobility Support", *Internet Engineering Task Force RFC 2002*, October 1996.

[POS-83] J. Postel, "The TCP Maximum Segment Size and Related Topics", *Internet Engineering Task Force RFC 879*, November 1983.

[STA-00] E. Stare, P. Robertson, P. Krummenacher, P. Christ, "The Multimedia Car Platform", *Proceedings of IST Mobile Summit*, Galway, Ireland, 1–4 October 2000; 421–427.

[WAK-00] H. Wakana, H. Saito, S. Yamamoto, M. Ohkawa, N. Obara, H.-B. Li, M. Tanaka, "COMETS Experiments for Advanced Mobile Satellite Communications and Advanced Satellite Broadcasting", *International Journal. of Satellite Communications*, 18(2), March–April 2000; 63–85.

Appendix A

Acronyms

1G	First-Generation
2G	Second-Generation
3G	Third-Generation
4G	Fourth-Generation
3GPP	3G Partnership Project
ABATE	ACTS Broadband Aeronautical Terminal Experiment
ACCORD	ACTS Broadband-Communcation Joint Trials and Demonstrations
ACK	Acknowledgement
ACTS	Advanced Communications Technologies and Services
ACTS	Advanced Communications Technologies Satellite
ADPCM	Adaptive Differential Pulse Code Modulation
AFD	Average Fade Duration
AGCH	Access Grant Channel
AHM	Authentication Handler Mobile Terminal
AHN	Authentication Handler Visited Network
AHU	Authentication Handler User
AMPS	Advanced Mobile Phone Services
AOR-E	Atlantic Ocean Region – East
AOR-W	Atlantic Ocean Region – West
APC	Adaptive Predictive Coding
APN	Access Point Name
ARFCN	RF Channel Number
ARQ	Automatic Repeat Request
ASE	Application Service Element
ASK	Amplitude Shift Keying
AUC	Authentication Centre
AWGN	Additive White Gaussian Noise
B-O-QAM	Binary-O-QAM
BACH	Basic Alerting Channel
BCC	Base Station Colour Code
BCCH	Broadcast Control Channel
BCF	Bearer Control Function
BCH	Bose–Chadhuri–Hocquenghem
BER	Bit Error Rate

BFN	Beam Forming Network
BGAN	Broadband GAN
BPSK	Binary PSK
BS	Base Station
BSC	Base Station Controller
BSIC	Base Station Identity Code
BSS	Base Station System
BSSMAP	BSS Management Application Part
BSSMP	BSS Mobile Part
BTS	Base Transceiver Station
BTSM	BTS Management
BWB	Blended-Wing Body
CAI	Common Air Interface
CBR	Carrier and Bit timing Recovery
CCCH	Common Control Channel
CCE	Capabilities Check Environment
CCF	Call Control Function
CCT	Capabilities Check Terminal
CDG	CDMA Development Group
CDMA	Code Division Multiple Access
CDPD	Cellular Digital Packet Data
CEFM	Combined EFM
CEPT	Conference of European Posts and Telegraphs
CGI	Cell Global Identification
CI	Carrier Information
CI	Cell Identity
CI	Control Information
CIMS	Customer Information Management System
CM	Connection Management
CN	Core Network
CoA	Care-of-Address
COMETS	Communications and Broadcasting Engineering Test Satellite
CONUS	Continental United States
CPFSK	Continuous Phase Frequency Shift Keying
CSCF	Call State Control Function
CT-2	Cordless Telephone-2
D-AMPS	Digital AMPS
DAB	Digital Audio Broadcasting
DAB-T	DAB-Terrestrial
DCA	Dynamic Channel Assignment
DCCH	Dedicated Control Channel
DCS1800	Digital Cellular System 1800
DDB	DataBase
DECT	Digital Enhanced Cordless Telecommunications
DMT	Dual-Mode Terminal
DPC	Destination Point Code

DPSK	Differential PSK
DQPSK	Differential QPSK
DR	Directory Register
DS-CDMA	Direct Sequence-CDMA
DS-SSMA	Direct Sequence-SSMA
DSS	Direct Spreading Sequence
DTAP	Direct Transfer Application Part
DVB	Digital Video Broadcast
DVB-RCS	DVB-Return Channel via Satellite
DVB-T	DVB-Terrestrial
E-TACS	Extended-TACS
ECSD	Enhanced HSCSD
EDGE	Enhanced Data Rates for GSM Evolution
EFM	Empirical Fading Model
EGC	Enhanced Group Call
EGNOS	European Geostationary Navigation Overlay Service
EGPRS	Enhanced GPRS
EIR	Equipment Interface Register
EIRP	Effective Isotropic Radiated Power
EMS	European Mobile Satellite
ERAST	Environmental Research Aircraft and Sensor Technology
ERS	Empirical Roadside Shadowing
ESA	European Space Agency
ETACS	Extended TACS
ETSI	European Telecommunications Standards Institute
EU	European Union
EUTELSAT	European Telecommunication Satellite Organisation
F-FH	Fast FH
FACCH	Fast Associated Control Channel
FCA	Fixed Channel Assignment
FCC	Federal Communications Commission
FCC	Fundamental Code Channel
FCH	Frequency Correction Channel
FDD	Frequency Division Duplex
FDM	Frequency Division Multiplex
FDMA	Frequency Division Multiple Access
FEC	Forward Error Correction
FES	Fixed Earth Station
FH	Frequency Hopping
FM	Frequency Modulation
FOCC	Forward Control Channel
FPLMTS	Future Public Land Mobile Telecommunication Systems
FS	Fixed Station
FSK	Frequency Shift Keying
FSL	Free Space Loss
FTP	File Transfer Protocol

FVC	Forward Voice Channel
GAN	Global Access Network
GBCCH	GPS Broadcast Control Channel
GCA	Guaranteed Coverage Area
GCC	Gateway Control Centre
GDP	Gross Domestic Product
GDPCAP	GDP per Capita
GEO	Geostationary Orbit
GERAN	GSM/EDGE Radio Access Network
GES	Gateway Earth Station
GFSK	Gaussian FSK
GGSN	Gateway GSN
GMR	Geo Mobile Radio
GMSC	Gateway MSC
GMSK	Gaussian-filtered MSK
GMSS	Gateway Message Switching System
GNSS	Global Navigation Satellite System
GOCC	Ground OCC
GPM	Gross Potential Market
GPRS	General Packet Radio Service
GPS	Global Positioning System
GRAN	Generic Radio Access Network
GSC	Gateway Station Control
GSM	Global System for Mobile Communications
GSN	GPRS Support Node
GSO	Geostationary Orbit
GTP	GPRS Tunnelling Protocol
GTS	Gateway Transceiver Subsystem
GWS	Gateway Subsystem
HALE	High Altitude Long Endurance
HALO	High Altitude Long Operation
HAP	High Altitude Platform
HEO	Highly Elliptical Orbit
HLR	Home Location Register
HMI	Human Machine Interface
HO	HandOver
HPA	High Power Amplifier
HSCSD	High Speed Circuit Switched Data
HSD	High Speed Data
HTTP	Hypertext Transfer Protocol
IAM	Initial Address Message
ICA	Intermediate Coverage Area
ICO	Intermediate Circular Orbit
IETF	Internet Engineering Task Force
IF	Intermediate Frequency
IMEI	International Mobile Equipment Identity

IMSI	International Mobile Subscriber Identity
IMUI	International Mobile User Identity
IMUN	International Mobile User Number
IMT-2000	International Mobile Telecommunications – 2000
IN	Intelligent Network
INMARSAT	International Maritime Satellite Organisation
INSURED	Integrated Satellite-UMTS Real Environment Demonstrator
IOL	Inter-Orbit Link
IOR	Indian Ocean Region
IP	Internet Protocol
IS	Interim Standard
ISC	International Switching Centre
ISDN	Integrated Services Digital Network
ISL	Inter-Satellite Link
IST	Information Society Technologies
ISUP	ISDN User Part
ITU	International Telecommunications Union
IWU	Interworking Unit
JD	Julian Date
JDC	Japanese Digital Cellular system
LA	Location Area
LAC	LA Code
LAI	LA Identifier
LAN	Local Area Network
$LAPD_m$	Link Access Protocol on D_m Channel
LCR	Level Crossing Rate
LE	Local Exchange
LEO	Low Earth Orbit
LES	Land Earth Station
LHCP	Left Hand Circular Polarisation
LNA	Low Noise Amplifier
LO	Local Oscillator
LOS	Line-of-Sight
LS	Location Server
MAHO	Mobile Assisted HO
MAP	Mobile Application Part
MAP	Maximum a Posteriori
MBCF	Mobile BCF
MCC	Mobile Country Code
MCF	Mobile Control Function
MCHO	Mobile Controlled HO
MCPC	Multiple Channels Per Carrier
MCT	Mobile Communications Terminal
ME	Mobile Equipment
MERS	Modified ERS
MEO	Medium Earth Orbit

MIPv6	Mobile IP Version 6
MM	Mobility Management
MNC	Mobile Network Code
MoU	Memorandum of Understanding
MS	Mobile Station
MSAS	Multi-Transport Satellite Based Augmentation System
MSC	Mobile Switching Centre
MSK	Minimum Shift Keying
MSRN	MS Roaming Number
MT	Mobile Terminal
MTP	Message Transfer Part
MTRN	Mobile Terminal Roaming Number
MTU	Maximum Transmission Unit
MWIF	Mobile Wireless Internet Forum
N-AMPS	Narrowband-AMPS
NACK	Negative ACK
NAHO	Network Assisted HO
NCC	Network Control Centre
NCC	PLMN Colour Code
NCHO	Network Controlled HO
NGSO	Non-Geostationary Satellite Orbit
NMS	Network Management Station
NMSS	Network Management and Switching Subsystem
NMT	Nordic Mobile Telephone
NRSC	National Remote Sensing Centre
NSCC	Navigation System Control Centre
NSP	Network Service Part
NTT	Nippon Telegraph and Telephone
O-QAM	Offset-QAM
OAM	Operation and Maintenance
OBP	On-Board Processing
OCC	Operations Control Centre
OMAP	Operation, Maintenance and Administration Part
OOK	On-Off Keying
OPC	Originating Point Code
OQPSK	Offset QPSK
OSI	Open Systems Interconnection
OSS	Operation Subsystem
OSS	Orbitography and Synchronisation Station
P-TMSI	Packet-TMSI
PACCH	Packet Associated Control Channel
PAGCH	Packet AGCH
PAWS	Protection Against Wrapped Sequence Numbers
PBCCH	Packet Broadcast Control Channel
PBX	Private Business Exchange
PC	Personal Computer

PCCCH	Packet Call Control Channel
PCH	Paging Channel
PCN	Personal Communications Network
PCS	Personal Communication Services
PCU	Packet Control Unit
PDC	Personal Digital Cellular System
PDCCH	Packet Dedicated Control Channel
PDCH	Packet Data Channel
PDF	Probability Density Function
PDP	Packet Data Protocol
PDTCH	Packet Data TCH
PFD	Power Flux Density
PGSC	PSTN Gateway Switching Centre
PHS	Personal Handyphone System
PN	Pseudo-Noise
PNC	Packet Notification Channel
POR	Pacific Ocean Region
POT	Plain Old Telephone
PPCH	Packet Paging Channel
PRACH	Packet RACH
PRMA	Packet Reserved Multiple Access
PS	Portable Station
PSK	Phase Shift Keying
PSDN	Public Switched Data Network
PSTN	Public Switched Telephone Network
PTCCH/D	Packet Timing Advance Control/Downlink
PTCCH/U	PTCCH/Uplink
PTM-M	Point-to-Multipoint-Multicast
PTT	Public Telegraph & Telephony
Q-O-QAM	Quaternary-O-QAM
QAM	Quadrature Amplitude Modulation
QoS	Quality of Service
QPSK	Quadrature PSK
RACE	Research and Development in Advanced Communications in Europe
RACH	Random Access Channel
RAINBOW	Radio Access Independent Broadband On Wireless
RC	Radio Configuration
RCS	Recursive Systematic Convolutional
RECC	Reverse Control Channel
RF	Radio Frequency
RFC	Request for Comment
RFCN	RF Channel Number
RHCP	Right Hand Circular Polarisation
RHT	Registration Handler Terminal
RHVN	Registration Handler Visited Network
RIMS	Ranging and Integrity Monitoring Station

RLM	Rice-Lognormal Model
RMSS	Receiver Maximum Segment Size
RNC	Radio Network Controller
RNS	Radio Network Subsystem
RPE-LPC	Regular Pulse Excitation – Linear Predictive Coder
RR	Radio Resource
RS	Reed-Solomon
RSC	Recursive Systematic Convolutional
RTO	Retransmission Time Out
RTT	Round Trip Time
Ru	Uncertainty Radius
RVC	Reverse Voice Channel
S-AGCH	Satellite-AGCH
S-BBCH	Satellite-Beam Broadcast Channel
S-BCCH	Satellite-BCCH
S-BSS	Satellite-BSS
S-CDMA	Synchronous-CDMA
S-FACCH	Satellite-FACCH
S-FH	Slow FH
S-HBCCH	Satellite-High Margin BCCH
S-HMSCH	Satellite-High Margin SCH
S-HPACH	Satellite-High Penetration Alerting Channel
S-IMT-2000	Satellite-IMT-2000
S-PCH/R	Satellite-PCH/Robust
S-PCN	Satellite-Personal Communication Network
S-PCS	Satellite-PCS
S-RACH	Satellite-RACH
S-SACCH	Satellite-SACCH
S-SCH	Satellite-SCH
S-SDCCH	Satellite-SDCCH
S-TCH/E	Satellite-TCH Eighth-rate
S-TCH/F	Satellite-TCH Full-rate
S-TCH/H	Satellite-TCH Half-rate
S-TCH/Q	Satellite-TCH Quarter-rate
S-UMTS	Satellite-UMTS
SACCH	Slow Associated Control Channel
SACCH	Slow Access Control Channel
SACK	Selected ACK
SAINT	Satellite Integration in the Future Mobile Network
SAN	Satellite Access Node
SAT	Supervisory Audio Tone
SCADA	Supervisory Control and Data Acquisition
SC	Subscriber Communicator
SCC	Satellite Control Centre
SCC	Supplemental Code Channel
SCCP	Signalling Connection Control Part

SCF	Satellite Control Facility
SCH	Synchronisation Channel
SCPC	Single Channel Per Carrier
SCSS	S-PCN Cell Site Switch
SDB	S-PCN Data Base
SDB	Security Database
SDCCH	Stand-alone Dedicated Control Channel
SDF	Service Data Function
SDMA	Space Diversity Multiple Access
SDPSK	Symmetric DPSK
SECOMS	Satellite EHF Communications for Mobile Multimedia Services
SGCSS	S-PCN Gateway Cell Site Switch
SGSN	Servicing GSN
SID	Subscriber Identity Device
SIM	Subscriber Identity Module
SINUS	Satellite Integration into Networks for UMTS Services
SLR	Satellite Location Area
SLS	Signalling Link Selection
SME	Small Medium Enterprise
SMS	Short Message Service
SMSC	SMS Centre
SMSS	Sender Maximum Segment Size
SNMC	Service Network Management Centre
SOCC	Satellite OCC
SP	Signalling Point
SPI	Service Provider Identity
SS-TDMA	Satellite Switched - TDMA
SSF	Service Switching Function
SSMA	Spread Spectrum Multiple Access
SSN	Subsystem Number
SSN7	Signalling System Number 7
SSPA	Solid State Power Amplifier
STP	Service Transfer Part
STP	Service Transfer Point
SUITED	Multi-Segment System for Ubiquitous Access to Internet Services and Demonstrator
T-UMTS	Terrestrial-UMTS
TA	Terminal Adapter
TA	Timing Advance
TACCH	Terminal-to-Terminal Associated Control Channel
TACS	Total Access Communications System
TCE	Traffic Channel Equipment
TCH	Traffic Channel
TCH/F	Full-Rate Traffic Channel
TCH/H	Half-Rate Traffic Channel
TCAP	Transaction Capabilities Application Part

TCP	Transmission Control Protocol
TDD	Time Division Duplex
TDM	Time Division Multiplex
TDMA	Time Division Multiple Access
TE	Terminal Equipment
TIA	Telecommunications Industry Association
TIMSI	Temporary IMSI
TMSI	Temporary Mobile Subscriber Identity
TMTI	Temporary Mobile Terminal Identity
TN	Time-slot Number
TP	Terminal Position
TTA	Telecommunication Technology Association
TT&C	Telemetry, Tracking and Command
TUP	Telephone User Part
TWTA	Travelling Wave Tube Amplifier
UE	User Equipment
UHF	Ultra High Frequency
UMTS	Universal Mobile Telecommunications Service
UPC	User Profile Check
UPT	Universal Personal Telecommunication
URAN	UMTS Radio Access Network
USHM	User Session Handler Mobile Terminal
USHV	User Session Handler Visited Network
USIM	User Services Identity Module
UTRA	UMTS Terrestrial Radio Access
UTRAN	UTRA Network
UW	Unique Word
UWC	Universal Wireless Consortium
VHE	Virtual Home Environment
VHF	Very High Frequency
VIRTUOUS	Virtual Home UMTS on Satellite
VITA	Volunteers in Technical Assistance
VLR	Visitor Location Register
VoIP	Voice over IP
VSAT	Very Small Aperture Terminal
VSELP	Vector Sum Excited Linear Prediction
W-CDMA	Wideband-CDMA
WAAS	Wide Access Augmentation System
WAP	Wireless Application Protocol
WARC	World Administrative Radio Conference
WLL	Wireless Local Loop
WRC	World Radio Conference
XPD	Cross Polar Discrimination
XPI	Cross Polar Isolation

Appendix B

Symbols

Chapter 1

α	Angle of arrival of incident wave
λ	Wavelength
C	Carrier power
D	Frequency re-use distance
f_d	Döppler shift
I	Interference power
N	Cell cluster size
q	Co-channel interference reduction factor
R	Cell radius
v	Mobile velocity

Chapter 3

ξ	Azimuth angle
θ	Elevation angle
θ_{min}	Minimum elevation angle
α	Right ascension angle
α_j	Right ascension for the *j*th orbital plane in an inclined orbit constellation
δ	Declination angle
μ	Kepler's constant
μ_m	Gravitational constant of the Moon
μ_s	Gravitational constant of the Sun
Ω	Right ascension of ascending node
Ω	Solid angle
Φ	Angular half-width of ground swath with single satellite coverage
ω	Argument of perigee
ϖ	Angular velocity of a satellite
γ	Tilt angle or nadir angle
γ_j	Initial phase angle of the *j*th satellite in its orbital plane at $t=0$
χ_j	Time-varying phase angle for all satellites of an inclined orbit constellation
$\Re_{jk}$	Angular angle between any two satellites in the *j*th and *k*th planes of an inclined orbit constellation
φ_{ijk}	Quidistance arc range from the midpoint of the spherical triangle
ρ	Atmospheric density

φ	Central angle or coverage angle at the centre of the Earth
a	Semi-major axis of an ellipse
A	Cross-sectional area
b	Semi-minor axis of an ellipse
B	Ballistic coefficient
C_D	Drag coefficient
$\mathbf{D}$	Drag force on a satellite
e	Eccentricity of orbit
$\mathbf{F}$	Vector sum of all forces acting on mass m
$\mathbf{F}_m$	Vector force acting on a satellite of mass m
G	Universal gravitational constant
h	Altitude of satellite above Earth's surface
$\mathbf{h}$	Orbital angular momentum of the satellite
i	Inclination angle of the orbital plane
J_n	Harmonic coefficient of the Earth of degree n
L	Specified latitude
L_g	Ground station's latitude
$L_{g,max}$	Maximum latitude that is visible to a satellite
l_g	Ground station's relative longitude
L_s	Latitude of satellite
M	Mass of body
M	Mean anomaly
N	Number of satellites in a constellation
N_{A-R}	Total number of satellites required for providing multiple satellite coverage as defined by Adams and Riders
N_{B-1}	Total number of satellites to provide global single-satellite coverage as defined by Beste
p	Semilatus rectum
p	Number of orbital planes
P_n	Legendre polynomial of degree n
$\Re_{jk}$	Angular range between any two satellites in the ith and jth planes
r	Radius of a geostationary orbit
r_a	Apogee radius
r_m	Moon's orbital radius
r_p	Perigee radius
r_s	Distance of a geostationary satellite from the sun
R	Slant range
R_E	Radius of the Earth
r	Distance between two bodies
$\mathbf{r}$	Vector acceleration of mass
s	Number of satellites per plane
T	Orbital period
T_e	Elapsed time since the x-axis using a geocentric-equatorial co-ordinate system coincided with the x-axis using a rotating Earth system
ϑ	True anomaly of satellite

v	Velocity of satellite
v_a	Velocity at the apogee
v_p	Velocity at the perigee

Chapter 4

θ	Elevation angle
Δ	Delay spread
φ	Latitude
ϕ	Off-boresight angle
γ_o	Specific attenuation due to dry air
γ_R	Specific attenuation due to rain
γ_w	Specific attenuation due to water vapour
μ_s	Mean of the shadowed component
ρ	Correlation coefficient between multipath and shadowing
σ_m^2	Mean received scattered power of the diffuse component due multipath propagation
σ_s	Standard deviation of the shadowed component
τ	Polarisation tilt angle relative to the horizontal
$A_{fq}(P, \theta, f)$	Fade exceeded for an outage probability P at an elevation angle θ and at frequency f of the frequency band f_q
A_{XR}	Axial ratio
A	Antenna aperture area
A_e	Effective collecting area of antenna
$A_{0.01}$	Attenuation exceeded for 0.01% of an average year
A_{bb}	Attenuation due to the bright band
$A_C(p)$	Attenuation due to clouds for a given probability
A_o	Attentuation due to oxygen
A_w	Attentuation due to water vapour
A_G	Slant path attenuation due to oxygen and water vapour
A_p	Attenuation exceeded for p% of an average year
$A_{rain}()$	Attenuation due to rain
$A_S(p)$	Attenuation due to scintillation for a given probability p
$A_T(p)$	Total attenuation for a given probability p
$A_{water\ vapour}()$	Attenuation due to a saturated water vapour column
B_c	Channel bandwidth
c	Speed of light
C_θ	Elevation angle-dependent term used in calculation of XPD due to rain
C_σ	Canting angle term used in calculation of XPD due to rain
C_τ	Polarisation improvement factor used in calculation of XPD due to rain
C_a	Rain dependent term used in calculation of XPD due to rain
C_f	Frequency dependent term used in calculation of XPD due to rain
C_{ice}	Ice dependent term used in calculation of XPD due to rain

E_{CP}	Received co-polarised electric field strength
E_{XP}	Received cross-polarised electric field strength
E_{Min}	Minimum electric field strength of a polarised wave
E_{Max}	Maximum electric field strength of a polarised wave
f	Frequency
f_c	Unmodulated carrier frequency
f_d	Döppler shift of diffuse component
f_L	Frequency in the L-band
f_m	Fade rate
f_S	Frequency in the S-band
f_{UHF}	Frequency in the UHF-band
Ho	Rain height as defined by Leitao–Watson
Hm	Thickness of the melting layer
h_0	Mean 0° isotherm height
h_o	Equivalent height for dry air
h_R	Mean rain height
h_s	Height above sea level of Earth station
h_w	Equivalent height of water vapour
L_E	Effective path length
L_G	Horizontal projection of the slant path
L_S	Slant path length
N_T	Total electron content
N_{wet}	Wet term of the radio refractivity
P	Percentage of distance travelled over which fade occurred/outage probability
p	Percentage of an average year for which an attenuation is exceeded
p_w	Equivalent worst month percentage of an average year for which an attenuation is exceeded
r	Signal envelope
R_E	Effective radius of the Earth
$r_{0.01}$	Horizontal reduction factor for 0.01% of an average year
R	Rain rate
$R_{0.01}$	Rain fall rate exceeded for 0.01% of the average year at a location
T_N	Average fade duration
$v_{0.01}$	Vertical reduction factor for 0.01% of an average year
XPD	Cross polar discrimination
XPD_P	XPD not exceeded for p% of the time
XPD_{rain}	Cross polar discrimination due to rain

Chapter 5

θ_{3dB}	3-dB beamwidth of an antenna
η	Antenna efficiency
λ	Wavelength
A_{rain}	Attenuation due to rain
$A(t)$	Signal pulse shape

b_g	Equivalent number of bits in TDMA guard band
b_p	Number of bits in TDMA frame pre-amble sequence
b_r	Number of bits in TDMA reference burst
B	Equivalent noise bandwidth
BO_{ip}	Input backoff
BO_{op}	Output backoff
$\mathbf{c}$	Code sequence matrix
$c(p)$	Codeword polynomial
C	Carrier power
C	Capacity of channel
C/I	Carrier-to-interference ratio
C/Im	Carrier-to-intermodulation ratio
C/N_{down}	Downlink carrier-to-noise ratio
C/N_{tot}	Total carrier-to-noise ratio
C/N_{up}	Uplink carrier-to-noise ratio
d_{min}	Smallest value of Hamming distance
D	Diameter of a parabolic antenna
e_j	Error vector
E	Number of errors in received code word that can be corrected
E_b/N_o	Energy per bit to noise power spectral density ratio
E_b/N_{0_Down}	Downlink energy per bit to noise power spectral density ratio
E_b/N_{0_Total}	Total energy per bit to noise power spectral density ratio
E_b/N_{0_Up}	Uplink energy per bit to noise power spectral density ratio
E_c/N_o	Code symbol energy per bit to noise power spectral density ratio
$EIRP_d$	Downlink EIRP
$EIRP_{sat}$	Satellite EIRP when at saturation
e_j	Error vector
F	Noise figure
F_n	Noise figure of device n
$G(\theta, \phi)$	Antenna gain in direction (θ, ϕ)
G	Antenna gain
G	Generator matrix
G_n	Gain of device n
G_r	Antenna receive gain
G_t	Antenna transmit gain
$g(p)$	Generator polynomial
I_m	Identity matrix
k	Boltzmann Constant (1.38×10^{-23} J/K)
k	Number of data symbols of s-bits used in RS code
L	Gain of lossy device
$\mathbf{m}$	Message sequence matrix
M	Number of states in PSK modulation
$m(p)$	Message polynomial
n	Symbol code word length used in RS code
n	Dimension of a cyclic shift register

N_E	Number of TDMA transmitting Earth stations
N_R	Number of reference Earth stations
P	Parity check matrix
P	Maximum period of repetition of a cyclic code generator
$P(\theta, \phi)$	The power radiated per unit solid angle in direction (θ, ϕ)
P_b	Probability of bit error
PFD_{sat}	Uplink PFD to saturate satellite transponder
PFD_{up}	Uplink PFD
PG	Processing gain of
P_s	Probability of symbol error
P_t	Transmit power
P_T	The total power radiated by an isotropic power source
R	Distance between transmitter and receiver
R_b	Information bit rate
R_C	Chip rate
R_D	Data rate
R_F	TDMA frame bit-rate
r	Received code sequence matrix
S	Symbol size in bits used in RS codes
S_i/N_i	Input signal to noise ratio to a device
$S_m(t)$	General form of PSK waveform
S_o/N_o	Output signal to noise ratio of a device
s	Syndrome vector product
T	Noise temperature of a device
T_a	Antenna noise temperature
$T_b(\theta, \phi)$	Brightness temperature
T_e	Effective noise temperature of the receiver
T_F	TDMA frame duration
T_{ground}	Ground noise temperature
T_{sky}	Sky noise temperature
T_0	Ambient temperature
t	Errors that can be corrected

Chapter 6

$\Lambda(x)$	Set of available channels in cell x
$A_{FESact}(i, t)$	Actual coverage area of FES i at time t
$A_{FESpot}(i, t)$	Potential coverage area of FES i at time t
$A_{FEStar}(i, t)$	Target coverage area of FES t at time t
$A_{sat}(j, t)$	Area covered by spotbeam j at time t
$B(i, j, t)$	Equal to 1 when FES i is connected to spotbeam j at time t, 0 otherwise
$C(i, j, t)$	Equal to 0 when FES i cannot be connected to spotbeam j at time t, 1 otherwise
$C_x(i)$	Channel allocation cost function for cell x
$C_x(k, i)$	Channel allocation cost function for channel i due to interfering channel k
D	Re-use distance

$F_D(x)$	Set of channels allocated to cell x on a fixed channel assignment basis
K	Frequency re-use factor
N	Number of channels permanently allocated to a cell
M	Number of system resources
R	Cell radius

Chapter 8

A	Affordability
a	Market take-up rate parameter defined by logistic model
$\text{Area}_{\text{rural}}$	Rural area of a country
Area	Total area of a country
b	Market take-up rate parameter defined by logistic model
c	Coefficient derived from the penetration curve
d	Coefficient derived from the penetration curve
$M_{\text{briefcase\_port}}(t)$	Total portable briefcase-type terminal market at year t
$M_{\text{lap-top\_port}}(t)$	Total lap-top-type terminal market at year t
$M_{\text{maritime\_ships}}(t)$	Total maritime ship terminal market at year t
$M_{\text{mobile}}(t)$	Total mobile market at year t
$M_{\text{palm-top\_port}}(t)$	Total palm-top-type terminal market at year t
$M_{\text{port}}(t)$	Total portable market at year t
$M_{\text{total}}(t)$	Total S-UMTS terminal market at year t
$M_{\text{vehicular\_cars}}(t)$	Total car terminal market at year t
$M_{\text{vehicular\_trucks}}(t)$	Total trucks and lorries terminal market at year t
$N_{\text{office\_staff}}$	Gross potential market for rural office staff
$N(t)$	Number of subscribers at year t
N_{cars}	Gross potential market for cars
N_{IT}	Gross potential market for international business travellers
$N_{\text{rur\_pop}}$	Gross potential market of rural population
N_{ships}	Gross potential market for ships
N_{trucks}	Gross potential market for trucks
$P(t)$	Penetration at year t
$P_{\text{briefcase\_port}}(t)$	Market penetration of briefcase-type terminals at year t
$P_{\text{lap-top\_port}}(t)$	Market penetration of lap-top-type terminals at year t
$P_{\text{mobile}}(t)$	Market penetration of mobile terminals at year t
$P_{\text{palm-top\_port}}(t)$	Market penetration of palm-top-type terminal market at year t
$\text{Population}_{\text{rural}}$	Rural population of a country
Rur%	Percentage of a country's population living in rural areas
S	Saturation level
t	Number of years after service launch
X_{1i}	GDP per capita at year i obtained from historical data
X_{2j}	GDP per capita at year j
Y_{1i}	Tariff of cellular service at year i obtained from historical data
Y_{2i}	Estimated tariff of new (S-UMTS) service at year j

Index